AF807096

Contemporary South Asian Studies

Editor-in-Chief

Paulo Casaca, South Asia Democratic Forum, Brussels, Belgium

This book series features scientific and scholarly studies focusing on politics, economics and changing societies in South Asia. Utilizing recent theoretical and empirical advances, this series aims at providing a critical and in-depth analysis of contemporary affairs and future developments and challenges in the region. Relevant topics include, but are not limited to, democratization processes, human rights concerns, security issues, terrorism, EU-South Asia relations, regional and economic cooperation and questions related to the use of natural resources.Contemporary South Asian Studies (CSAS) welcomes monographs and edited volumes from a variety of disciplines and approaches, such as political and social sciences, economics and cultural studies, which are accessible to both academics and interested general readers. The series is published on behalf of the South Asian Democratic Forum (Brussels), which is one of the most well-known think tanks in Europe focusing on South Asia.

Selected volumes published in this series are indexed by the Web of Science.

More information about this series at http://www.springer.com/series/15344

Saptarshi Mitra · Sumana Bandyopadhyay ·
Stabak Roy · Tomaz Ponce Dentinho
Editors

Railway Transportation in South Asia

Infrastructure Planning, Regional Development and Economic Impacts

South Asia Democratic Forum

Springer

Editors
Saptarshi Mitra
Department of Geography and Disaster Management
University of Tripura
Agartala, Tripura, India

Stabak Roy
Department of Geography and Disaster Management
University of Tripura
Agartala, India

Sumana Bandyopadhyay
Department of Geography
University of Calcutta
Kolkata, India

Tomaz Ponce Dentinho
Atlantic Applied Economics Studies Centre (CEEAplA)
University of the Azores
Angra do Heroísmo, Portugal

ISSN 2509-4173 ISSN 2509-4181 (electronic)
Contemporary South Asian Studies
ISBN 978-3-030-76877-5 ISBN 978-3-030-76878-2 (eBook)
https://doi.org/10.1007/978-3-030-76878-2

© The Editor(s) (if applicable) and The Author(s), under exclusive license to Springer Nature Switzerland AG 2021, corrected publication 2021
This work is subject to copyright. All rights are solely and exclusively licensed by the Publisher, whether the whole or part of the material is concerned, specifically the rights of translation, reprinting, reuse of illustrations, recitation, broadcasting, reproduction on microfilms or in any other physical way, and transmission or information storage and retrieval, electronic adaptation, computer software, or by similar or dissimilar methodology now known or hereafter developed.
The use of general descriptive names, registered names, trademarks, service marks, etc. in this publication does not imply, even in the absence of a specific statement, that such names are exempt from the relevant protective laws and regulations and therefore free for general use.
The publisher, the authors and the editors are safe to assume that the advice and information in this book are believed to be true and accurate at the date of publication. Neither the publisher nor the authors or the editors give a warranty, expressed or implied, with respect to the material contained herein or for any errors or omissions that may have been made. The publisher remains neutral with regard to jurisdictional claims in published maps and institutional affiliations.

This Springer imprint is published by the registered company Springer Nature Switzerland AG
The registered company address is: Gewerbestrasse 11, 6330 Cham, Switzerland

The original version of this book was revised: The author name "Sumana Bandyopadhyay" and the affiliation have been updated. The correction to this chapter is available at https://doi.org/10.1007/978-3-030-76878-2_19.

Contents

Contributors

Sumana Bandyopadhyay Department of Geography, University of Calcutta, Kolkata, India

Madhu Bharti Faculty of Planning, CEPT University, Ahmedabad, India

Pavithra Velechettiar Bhaskaran Meenakashi College of Engineering, Chennai, India

Jayasmita Bhattacharjee Architect and Urban Planner, North Eastern Space Application Centre (NESAC), Umiam, Meghalaya, India

Pinaki Bhattacharya Begum Rokeya University, Rangpur, Bangladesh

Trinity Borgohain Government Model College, Kaziranga, Golaghat, Assam, India

Tomaz Ponce Dentinho Atlantic Applied Economics Studies Centre (CEEAplA), University of the Azores, Ponta Delgada, Portugal

Laura Eboli University of Calabria, Rende, Italy

Ana Ferreira Northern Regional Development and Coordination Commission, Porto, Portugal;
Centre for the Study of Geography and Spatial Planning, Coimbra, Portugal

Vallary Gupta Research and Knowledge Management, Coalition for Disaster Resilient Infrastructure, New Delhi, India

Muhammad Zamir Khan School of Economics, Quaid-I-Azam University, Islamabad, Pakistan

Paulose N. Kuriakose Department of Urban and Regional Planning, School of Planning and Architecture, Bhopal, India

Gabriella Mazzulla University of Calabria, Rende, Italy

Saptarshi Mitra Department of Geography and Disaster Management, Tripura University, Agartala, Tripura, India

Tanya Mittal Deutsche Gesellschaft Für Internationale Zusammenarbeit (GIZ) GmbH, New Delhi, India

Bhaswati Mondal DODL, University of Kalyani, Kalyani, India

Plamen Patarchanov Geology-Geography Faculty, St. Kliment Ohridski, Sofia, Bulgaria

Emilia Patarchanova Faculty of Mathematics and Natural Sciences, SWU Neofit Rilski, Blagoevgrad, Bulgaria

Hugo Silveira Pereira CIUHCT – Interuniversity Centre for the History of Science and Technology, NOVA School of Science and Technology, Almada, Portugal; Department of History, University of York, York, UK

Sara Rachdan Department of Spatial Planning, Technische Universität Dortmund, Dortmund, Germany

Stabak Roy Department of Geography and Disaster Management, Tripura University, Agartala, Tripura, India

Gopa Samanta Department of Geography, The University of Burdwan, Burdwan, India

Sanghamitra Sarkar Centre for Excellence in Studies On Tribes and Marginalized Communities, Utkal University, Bhubaneswar, India

Shuchita Sharmin Department of Development Studies, University of Dhaka, Dhaka, Bangladesh

Chitresh Shrivastva Department of Political Science, Jain (Deemed to be University), Bengaluru, India

Lyuben Stoyanov Geology-Geography Faculty, St. Kliment Ohridski, Sofia, Bulgaria

Hülya Zeybek Vocational School of Transportation, Eskisehir Technical University, Eskisehir, Turkey

Abbreviations

ACDI	Average Composite Dimension Index
ADB	Asian Development Bank
AGC	Average Generalized Cost
AGFI	Adjusted Goodness of Fit Index
ALCO	American Locomotive Company
ASEAN	Association of Southeast Asian Nations
ASHB	The Assam State Housing Board
BART	Bay Area Rapid Transit
BCG	Boston Consulting Group
BGTA	Bombay Goods Transport Association
BRI	Belt and Road Initiative
BRO	Border Road Organisation
BRTS	Bus Rapid Transport Systems
BTK	Baku-Tbilisi-Kars
CAGR	Compound Annual Growth Rate
CART	Classification and Regression Tree Approach
CBA	Custom Brokers Associations
CBD	Central Business District
CDM	Clean Development Mechanism
CFI	Comparative Fit Index
CFSs	Container Freight Stations
CIDB	Construction Industry Development Board
CMA	Chennai Metropolitan Area
CMDA	Chennai Metropolitan Development Authority
CMRL	Chennai Metro Rail Limited
CONCOR	Container Corporation of India
CSE	Centre for Science and Environment
CSI	Customer Satisfaction Index
CSLA	Container Shipping Line Associations
CSSUTP	Centrally Sponsored Scheme of Urban Transport Planning
CTT	Comprehensive Traffic and Transportation
CY	Container Yard
DB	Deutsche Bahn
DEA	Data Envelopment Analysis

DFA	Distribution-Free Approach
DFC	Dedicated Freight Corridor
DFCCIL	Dedicated Freight Corridor Corporation of India Limited
DI	Dimension Index
DMCI	Delhi-Mumbai Industrial Corridor
DMRC	Delhi Metro Rail Corporation
DMUs	Decision-Making Units
DPD	Direct Port Delivery
DPE	Direct Port Entry
DPR	Detailed Project Report
ECO	Economic Cooperation Organization
EDM	Ex-situ Development Model
ELU	Existing Land Uses
EMUs	Electric Multiple Units
EU	European Union
EWS	Economically Weaker Section
EXIM	Export-Import
FCI	Food Corporation of India
FCL	Full Container Load
FDH	Free-Disposal Hull
FDI	Foreign Direct Investment
FFFAI	Federation of Freight Forwarders Associations of India
FMCG	Fast-Moving Consumer Goods
FSI	Floor Space Index
GFI	Goodness of Fit Index
GHG	Green House Gas
GLS	Generalized Least Squares
GMC	Guwahati Municipal Corporation
GoI	Government of India
GPS	Global Positioning System
GTFS	General Transit Feed Specification
HCV	Heavy Commercial Vehicle
HDIP	Hydrocarbon Development Institute of Pakistan
ICD	Inland Container Depot
ICT	Information and Communication Technologies
IDM	In-situ Development Model
IITTM	Indian Institute of Tourism & Travel Management
IMO	International Maritime Organization
INDCs	Intended Nationally Determined Contributions
INLP	Integer Non-Linear Programming
IR	Indian Railways
IRSDC	Indian Railway Station Development Corporation
IT	Information Technology
ITI	Istanbul-Tehran-Islamabad
ITS	Intelligent Transport System

JICA	Japan International Cooperation Agency
JNNURM	Jawaharlal Nehru National Urban Renewal Mission
JNPT	Jawaharlal Nehru Port
JNR	Japanese National Railways
JR	Japanese Railways
JR-West	West Japan Railway Company
KMA	Kolkata Metropolitan Area
KMC	Kolkata Municipal Corporation
KMDA	Kolkata Metropolitan Development Authority
KMR	Kolkata Metropolitan Region
LAN	Local Asymptotic Normality
LAP	Local Area Plan
LCA	Life Cycle Assessment
LCL	Less than Container Load
LDHV	Low Density High Value
LIG	Low Income Groups
LP	Linear Program
MaaS	Mobility as a Service
MIG	Middle-Income Group
ML	Maximum Likelihood
MLA	Member of the Legislative Assembly
MoCI	Ministry of Commerce and Industry
MoHPA	Ministry of Housing and Urban Poverty Alleviation
MoHUA	Ministry of Housing and Urban Affairs
MoR	Ministry of Railways
MoRTH	Ministry of Road Transport and Highways
MoU	Memorandum of Understanding
MoUD	Ministry of Urban Development
MRTS	Mass Rapid Transport System
MTP	Metropolitan Transport Project
NFR	North-Eastern Frontier Railway
NH	National Highway
NHAI	National Highway Authority of India
NLP	National Logistics Policy
NRIC	National Railway Infrastructure Company
NUTP	National Urban Transport Policy
OHT	Overhead Electric Traction
ONGC	Oil and Natural Gas Corporation
PEST	Political, Economic, Social and Technological
PESTLE	Political, Economic, Social, Technological, Legal and Environmental
PKM	passenger-kilometres
POL	Petroleum, Oil, & Lubricants
PPP	Public-Private Partnership
PR	Pakistan Railways

RDSO	Research, Designs and Standards Organisation
RFQ	Request for Qualifications
RITES	Rail India Technical and Economic Service
RMR	Root Mean Residual
RMSEA	Root Mean Square Error of Approximation
SCI	Supreme Corporation of India
SCM	Supply Chain Management
SCOT	Strengths, Challenges, Opportunities and Threats
SEM	Structural Equation Modelling
SEM-MIMIC	Structural Equation Multiple Cause Multiple Indicator
SFA	Stochastic Frontier Analysis
SJ	Swedish State Railways
SRTC	State Road Transport Corporation
SS	Station Supervisor
SSA	Sub-Saharan Africa
TA	Transporters Associations
TAR	Trans-Asia Rail
TCPO	Town and Country Planning Organization
TEN-T	Trans-European Transport Network
TEUs	Twenty-Foot Equivalent Unit
TFA	Thick Frontier Analysis
TFC	Total Final Consumption
THC	Terminal Handling Charges
TITR	Trans-Caspian International Transport Route
TKM	Ton-kilometres
TN	Tamil Nadu
TOD	Transit-Oriented Development
TRACECA	Transport Corridor Europe-Caucasus-Asia
TTC	Total Transport Cost
UIC	International Union of Railways
ULS	Un-weighted Least Squares
UMTA	Urban Metropolitan Transport Authority
UN	United Nations
UNCRD	United Nations Centre for Regional Development
UNECE	United Nations Economic Commission for Europe
UNESCAP	United Nations Economic and Social Commission for Asia and The Pacific
VFC	Value Capture Financing
WFC	Western Freight Corridor
WIM	Weight in Motion
WIPGR	West of India Portuguese Guaranteed Railway
WIPGRC	West of India Portuguese Guaranteed Railway Company
WLS	Weighted Least Squares

List of Figures

Impact of Metro-Rail Projects on Land Use and Land Value in Indian Cities—The Case of Chennai

Privatization of Indian Railway Services: The Story so Far

List of Tables

Intermodality—Towards Enhancing Rail Freight Transportation Prospects

Assessing Perceptions of Railway Service Quality: A Compendium of Literature Studies

Slums on Railway Land in Guwahati City, Assam: A Sociological Review

Railway Modernisation in India: A South Asian Case Study

What Role for Railways in the Eurasian Supply Chains?

The Role of Railway Transport Systems in Modern Multiscale Spatial Development. Bulgaria in the Orient Express

Privatization of Indian Railway Services: The Story so Far

Introduction: Railway Transportation—Regions, Economy and Development

Saptarshi Mitra, Sumana Bandyopadhyay, Stabak Roy, and Tomaz Ponce Dentinho

Railways constitute the most important developmental instrument of a region, province or country (Wagner, 2012). Railway transportation is not an "essential precondition" but a "fundamental precondition" for regional development in regions without access to water transport (and even when such access is present railway transportation to the interior regions is always needed).

The development of railway transportation not only reduces transport costs in terms of both money and time, but also contributes to the integration of different regions. Railway transport systems cannot be detached from regional economics (Aldagheiri, 2010). The advancement of transport networks is based on the interaction between the development of a transport system and economic development (Norton, 1984). Railways connecting markets improve integration and generate revenues (Chaudhary & Bogart, 2013). Railway transport systems provide an inclusive economic activity in terms of both passengers and freight mobility, which generates employment both directly or formally and indirectly or informally (Uma & Shruthi, 2014).

Regional economics as affected by high-speed railway transport system constitutes an important component in socioeconomic appraisals (Blanquart & Koning, 2017). Although improvements in railway transportation can generate benefits, the regional distribution of those benefits varies. Railway transportation reduces costs, increases business efficiency and promotes a sustainable economy (Yang et al., 2019).

Railway transportation combines market accessibility with labour mobility, which leads to agglomerations effects and economic development. This is why railway

The original version of this chapter was revised: The correct author name "Sumana Bandyopadhyay" and the affiliation "Department of Geography, University of Calcutta, Kolkata, India" have been updated. The correction to this chapter is available at https://doi.org/10.1007/978-3-030-76878-2_19

S. Mitra · S. Bandyopadhyay · S. Roy
Department of Geography, University of Calcutta, Kolkata, India

T. P. Dentinho (✉)
Atlantic Applied Economics Studies Centre (CEEAplA), University of the Azores, Ponta Delgada, Portugal
e-mail: tomaz@sadf.eu

© The Author(s), under exclusive license to Springer Nature Switzerland AG 2021, corrected publication 2021

S. Mitra et al. (eds.), *Railway Transportation in South Asia*, Contemporary South Asian Studies, https://doi.org/10.1007/978-3-030-76878-2_1

transportation is continuously encouraged—for its growing importance and potential contributions to regional growth and development (Chen & Silva, 2013), which imply changes on land use and travel behaviour (Kasraian et al., 2016). The development of rail transport systems is seen as one of the most effective ways to shape sustainable urban growth and mobility (Połom, 2018). Railways also play a very important role in world trade and commerce. In meeting increasing demand and supply, railways become an integral part of regional development in terms of both freight and passenger transportation (Roy & Mitra, Railway Stations of Tripura, India: An Assessment of Infrastructural Conditions, 2020).

Railway services and supporting infrastructure play an important role in the growth and development of regional economies (Gani, 2017). The basic railway infrastructure supports cross-border trade, which influences regional cooperation in terms of both trade and investments, including issues such as tariff and nontariff barriers to trade and foreign direct investment or FDI (Kuroda et al., 2006). In recent times, more and more attention has been given to railway systems as a mass transport mode for regional development. Various factors are at play: railway capacity, station infrastructure, degree of urbanization, global trade and private ownership of vehicles, modal exchange of public transport and other important indicators of logistical support for regional development and economic growth.

The objective of this book is to highlight the varying magnitude of railway transportation as well as the effects and externalities of its mechanisms across geographical spaces. Keeping in mind regional development, economic growth and urban metamorphosis, each chapter centres on a specific regional entity through a scientific analysis of a given railway transport system. Railway transport is related to contemporary issues such as railway infrastructural facilities, service quality, passenger mobility, passenger satisfaction, rail freight transportation, market demand, railway management systems, technological development, urban transportation, railway planning, railway and environmental susceptibility, future trends of development and cross-border trade.

The book is structured in four parts. The First Part involves a few elements of the historical evolution of Railway Transport Systems. Chapter 2, written by Sara Rachdan, focuses on the recent revitalization of railway stations, transformed from transport-complementary facilities into specialized and multifunctional transportation mega-hubs. The chapter focuses on the cases of the Berlin Hauptbahnhof in Germany, the Osaka-Umeda Station in Japan and the Habibganj Railway Station in India. The aim is to reveal the evolution of railway stakeholders and the enlarged involvement of the private sector. Chapter 3, written by Hugo Silveira Pereira, recalls that the history of railways links is strongly associated with nineteenth-century colonialism, which involved the integration and control of foreign lands and their resources.

The second part of the book includes six chapters focusing on the ways that railway transportation affected development and urbanization. Chapter 4, developed by Madhu Bharti and Pavithra Velechettiar Bhaskaran, looks at the impact of metro networks in land uses and values in Chennai. The chapter provides evidence that the increment in land value is greater in the project formulation stage and inversely proportional to the distance from the metro node, which is common because the value of land may decrease along with accessibility to public transportation. Bhaswati Mondal and Gopa Samanta write Chapter 5, which centres on waiting times, distance between stations, and commuters' features in two non-connected sections of the Kolkata metro. Muhammad Zamir Khan proposes in Chapter 6 to analyse energy efficiency of railways in Pakistan, arguing that railway represents a remarkable increase in energy efficiency per passenger. Rail efficiency is also the focus of Chapter 7, this time applied to Eastern India by Stabak Roy and Saptarshi Mitra. These authors demonstrate the reduction of rail efficiency in the region associated to the reduction of timber exports. Chapter 8 approaches again the role of metro rail systems in India—which Paulose N. Kuriakose and Jayasmita Bhattacharjee strongly advocate to promote a better quality of life in Indian cities. The second part finishes with Chapter 9, named Intermodality towards Enhancing Rail Freight Transportation Prospects, wherein the authors Tanya Mittal and Paulose N. Kuriakose defend a strategy focussed on Intermodality that allows modal shift towards green fuels.

The third part of the book analyses passengers' perceptions and mobility. Chapter 10, authored by Laura Eboli and Gabriella Mazzulla, looks into the perceived quality of railway service and ranks the more relevant features involved. The role of railways in rural development is the title of Chapter 11, written by Ana Ferreira, compares the railway connections between rural and urban areas in India and in Bangladesh to conclude that the lack of connectivity between rural and urban areas affects negatively the development of rural areas. Chapter 12, developed by Sanghamitra Sarkar and Tomaz Ponce Dentinho, looks into the design of Kolkata's metro networks and argues that these have a strong impact on the spatial profile of incomes and on the social organization of space in the metropolitan area. The connection between Railways and Tourism is the focus of Pinaki Bhattacharya and Shuchita Sharmin in Chapter 13, which analyses the untapped opportunities involved in the Bangladesh-India Rail Connectivity. In Chapter 14, Trinity Borgohain addresses the problem of slums on railway lands—looking in-depth into the Guwahati in Assam, India.

The last part of the book proposes a prospective analysis of railways in South Asia. Chitresh Shrivastva, in Chapter 15, focuses on Foreign Direct Investment in the modernization of railways in India. Chapter 16, written by Hülya Zeybek, develops the idea of a middle corridor in Eurasian connections. Somehow related to this idea of a middle Eurasian corridor, Plamen Patarchanov, Emilia Patarchanova and Lyuben Stoyanov propose advances in the Bulgarian Section of the Orient Express. Finally, in Chapter 18, Paulose N. Kuriakose and Vallary Gupta evaluate the effects of privatization of railways in India.

This compilation of texts on railways in South Asia is a very interesting exercise, creating a broad image on railways in the South Asian region and able to inform both researchers and stakeholders regarding the knowledge and wisdom accumulated on this topic. This potential effectiveness justified the efforts by authors, editors, publishers and readers.

References

Aldagheiri, M. (2010). The expected role of railways in the economic development of Saudi Arabia. *Urban Transport, 16*, 157–167. https://www.witpress.com/Secure/elibrary/papers/UT10/UT10015FU1.pdf.

Blanquart, C., & Koning, M. (2017). The local economic impacts of high-speed railways: Theories and facts. *European Transport Research Review, 9*(12), 1–14. https://doi.org/10.1007/s12544-017-0233-0

Chaudhary, L., & Bogart, D. (2013). *Railways and Indian economic development*. London: London School of Economics. https://blogs.lse.ac.uk/southasia/2013/04/29/railways-and-indian-economic-devpment/.

Chen, G., & Silva, J. A. (2013). Regional impacts of high-speed rail: A review of methods and models. *Transportation Letters, 5*(3), 131–143. https://doi.org/10.1179/1942786713Z.00000000018

Gani, A. (2017). The Logistics Performance Effect in International Trade. *The Asian Journal of Shipping and Logistics, 33*(4), 279–288. https://doi.org/10.1016/j.ajsl.2017.12.012

Goodbody Economic Consultants. (n.d.). *Transport and regional development*. Goodbody Economic Consultants. http://www.irishspatialstrategy.ie/docs/pdf/Transport%20and%20Regional%20Development.pdf.

Kasraian, D., Maat, K., & Wee, B. V. (2016). Development of rail infrastructure and its impact on urbanization in the Randstad, the Netherlands. *The Journal of Transport and Land Use, 9*(1), 151–170.

Kuroda, H., Kawai, M., & Nangia, R. (2006). *Infrastructure and regional cooperation*. Tokyo: Asian Development Bank. https://ppiaf.org/sites/ppiaf.org/files/documents/toolkits/Cross-Border-Infrastructure-Toolkit/Cross-Border%20Compilation%20ver%2029%20Jan%2007/Resources/Kuroda%20Kawai%20Nangia%20-%20Infrastructure%20and%20Regional%20Cooperation.pdf.

Norton, W. (1984). *Historical Analysis in Geography*. Longman.

Połom, M. (2018). Urban transformation in the context of rail transport development: The case of a newly built railway line in Gdańsk (Poland). *Journal of Advance Transportation, 1–16*, https://doi.org/10.1155/2018/1218041.

Roy, S., & Mitra, S. (2016). Infrastructural Status of Railway Transport System in North East India: A geographical Analysis. *Asian Journal of Spatial Science, 4*(1), 89–100.

Roy, S., & Mitra, S. (2020). Railway Stations of Tripura, India: An Assessment of Infrastructural Conditions. In S. Bandyopadhyay, C. R. Pathak, & T. P. Dentinho (Eds.), *Urbanization and Regional Sustainability in South Asia* (pp. 177–200). Springer.

Uma, H. R., & Shruthi , B. R. (2014). An analysis of Indian railways contribution towards employment generation. *International Journal of Advanced Research inManagement and Social Sciences, 3*(3). https://www.academia.edu/6757990/an_analysis_of_indian_railways_contribution_towards_employment_generation.

Wagner, L. (2012). *Infrastructure lessons for economic growth and business success*. Area Development.

Yang, X., Lin, S., Zhang, J., & He, M. (2019). Does high-speed rail promote enterprises productivity? Evidence from China. *Journal of Advanced Transportation, 1–20*, https://doi.org/10.1155/2019/1279489.

Chronological Evolution of Railway Transport System

Post-industrial Revitalisation of Railway Stations: The Path to Commercialisation

Sara Rachdan

1 Introduction

Since the early nineteenth century, stations have been designed in a grand and iconic manner. However, only in the twentieth century were they recognised as significant components of every urban landscape. Nevertheless, railway stations were still seen as temporary spaces for passengers; improvements in design were rarely appreciated beyond their temporary function as transitional spaces. Only towards the 1980s, post the advents of high-speed rail, did railway stations gain notable significance. They then came to be seen as promising sources of economic gain—and there followed a strong wave of revitalisation.

With revitalisation endeavours came the great need for investment and financing, which in return played a strong role in the privatisation of many railway operators—and railway stations. However, it should be emphasised that the complexities involved in the redevelopment of railway stations can be traced back to several different components that much surpass privation alone (Bertolini & Spit, 1998). Privatisation was not involved in all cases of railway station redevelopment—particularly in Europe, where privatisation was often not deemed necessary. The main driver for the changes was probably the new spatial perception of railway station buildings, now assessed for their potentiality as urban centres.

Railway station redevelopments were often part of grander urban regeneration projects implemented through state funds. However, in the 1990s privatisation did make its way into railway station redevelopment, particularly in Germany and the UK within Europe. Privation also featured in India in the twenty-first century. In the late 1970s, privatisation had already greatly affected countries with a high dependency on rail infrastructure, namely Japan.

S. Rachdan (✉)
Department of Spatial Planning, Technische Universität Dortmund, Dortmund, Germany
e-mail: sara.rachdan@tu-dortmund.de

© The Author(s), under exclusive license to Springer Nature Switzerland AG 2021
S. Mitra et al. (eds.), *Railway Transportation in South Asia*, Contemporary South Asian Studies, https://doi.org/10.1007/978-3-030-76878-2_2

By the end of the twentieth century, a new trend arose linked to high-speed rails and titled "Station Renaissance". With the revival of railway stations, we witnessed the development of new forms of stations. These new stations were to be entirely user-centric, modernised and more attractive to users. The purpose of this trend was to transform stations from their bleak and underwhelming states into charismatic, accommodating spaces. High-speed trains were considered an emblem of modernism; with the coming of the new model of train travel, new private railway operators pressed for an entirely new approach to the design of station buildings (Kido, 2013).

2 Literature Review

Railway redevelopment, whether relative to the development of terminals for high-speed rails, the introduction of light rail systems, or commercial development through the construction of station amenities such as hotels and shopping centres or offices, have become staples in the effort towards urban revitalisation (Bertolini & Spit, 1998; Bruinsma, et al., 2008). Nonetheless, there lacks sufficient studies on railway station area redevelopment, with regard to post-industrial restructuring of urban cores (Peters, 2009).

As regards the shift in station design following the emergence of high-speed rail in the post-industrial era, Kido (2013) and Peters (2009) argue that stations have become more user-friendly and that they now satisfy user needs, besides providing several functions beyond their traditional transportation function.

Moreover, it should be noted that many station redevelopments are privately implemented, causing a shift in power from the public domain to the private domain, as commented by Weizsacker and et al. (2005). Although there are often many disadvantages to privation, said privatisation of station redevelopments may work in favour of the public as identified by Zacharias et al. (2011).

3 Context on the Study Area

Since their creation railway stations have proven to constitute indispensable components of a city's core. In the last decades, railway station operators have made great efforts to develop and commercialise railway stations. Thus, in the interest of analysing the global approach to "station renaissance", this study uses examples of largely commercialised railway stations from both Europe and Asia as case studies. With regard to the European case study, Germany's "Flagship" Berlin Hauptbahnhof was chosen. As for Asia, the highly praised Osaka-Umeda Station in Japan, along with the still-evolving Habibganj Railway Station, following the recent shift towards station redevelopment in India, were chosen.

4 Objectives of the Study and Methodology

This study aims to investigate the evolution of railway stations relative to their commercialisation, following the conception of high-speed rail transport, by reviewing both successful and unsuccessful case studies of railway station revitalisation. Using historical reviews, reports and both empirical and qualitative studies, the selected case studies are analysed sequentially from an initial point of early twentieth-century developments up to late twentieth-century "Station Renaissance"—and, eventually, up to present-day mega-developments. A particular focus was laid on stakeholder evolution and the privatisation of rail operators and railway stations.

This study uses an integrative review method so as to critically review and re-conceptualise the existing literature, reports and theory on the developments of railway stations in regard to commercial development and station revitalisation. The review is undertaken using combining perspectives and insights from different fields and research traditions, such as economic geography and spatial development, history, governance and policy-making and urban planning.

5 Study of Railway Station Developments in Germany, Japan and India

5.1 "Station Renaissance" in Germany, From Deutsche Bundesbahn to Deutsche Bahn AG and the Development of the Berlin Central Station

5.1.1 The Fall of Deutsche Bundesbahn

Founded in 1949, The Deutsche Bundesbahn was a characteristic federal property under the control of the German ministry of transport. During this federal control, a board of directors was responsible for its business and economic management. The Deutsche Bundesbahn had to be compensated for public service obligations and special tariffs for certain user groups. However, the board did not maintain either the obligation to be profitable, cover costs, or the necessary freedom to pursue such goals (Peter, 2008). Due to its several restrictions and heavy market access regulation, the Deutsche Bundesbahn suffered many losses in market share, which led to its substantial financial decline. By the early 1990s, the federal railway operator had reached a point of extreme debt and financial failure (Peter, 2008), which led to the need for privatisation. In such a context we witnessed the rise of the Deutsche Bahn AG.

Fig. 1 South facade of Berlin Central Station (*Source* Janericloebe/Wikimedia. https://commons.wikimedia.org/wiki/File:Berlin_Hauptbahnhof_006.JPG)

5.1.2 The Privatisation of the German National Railways and the Formation of the Deutsche Bahn AG

The German national railways were officially privatised under the name of Deutsche Bahn AG in 1994; however, they remained within state possession. In the first phases of privatisation, the DB AG was subdivided into four categories, each dealing with a specific form of transport—local and regional passenger transport, long-distance passenger transport, freight transport and, finally, infrastructure. By 1999, these divisions led to the subsidisation of DB AG into 5 companies, all falling under a single DB AG holding: the DB Regio AG, the DB Reise und Touristik AG, the DB Cargo AG, the DB Netz AG and the DB Station & Service AG. Thereafter, the Deutsche Bahn AG's subsidiaries were reiterated. In2011 they were settled as Arriva, DB Bahn, DB Dienstleistungen, DB Netze, and DB Schenker.

The redevelopment of German railway stations began in the mid-1990s, under the control of the DB AG subsidiary, DB Netze, which is responsible for infrastructure and operations. The DB Netze is itself subdivided into five organisms: the DB Netze Fahrweg, the DB Netze Energie, the DB Netze Personenbahnhöfe, the DB Projekt Bau and the DB Station & Service (this last specifically responsible for passenger operation and railway stations). One of DB Netze's first and most successful railway station redevelopments is the Leipzig Central Station, renovated in 1996 (Fig. 1).

5.1.3 The "Emergency Program" and the Development of Germany's "Flagship" Railway Station, Case Study: The Berlin Central Station

In 2002, according to an "Emergency Program" developed by the DB AG, "station renaissance" had finally begun in Germany (Kido, 2013). The driver was the need for a renovation and beautification of the stations. However, the programme focussed not only on renovation, but also the construction of new, modern railway buildings capable of representing the modernisation and rebranding of the DB AG holding. With this came the construction of the Berlin Hauptbahnhof. The new station was part of a mega-project in urban redevelopment— the station alone cost a whopping 1.2 billion Euros. The station was advertised by DB AG as "the largest and most modern crossing station in Europe" and considered a "flagship" railway station—meaning that its design was to impress both architecturally and structurally (Peters, 2009).

The overall area of the station covers a total of 175,000m^2, 100,000m^2 of which are contained within the building structure itself. Nearly 25 per cent (23,000m^2) is entirely dedicated to commercial spaces. A notable element of the architecture is the station's transparency, as the structure is primarily constructed from glass and steel—described as "a large "cathedral" of glass and steel" (Baker, 2006). The space is thus drowsed with generous amounts of light.

Accommodated within a glass shell is a multileveled space. The basement level dedicated to the Berlin underground; the upper two levels are entirely set for commercial usage; the highest level is dedicated to the high-speed trains. The Berlin Central Station was developed with the added notion of being an Einkaufsbahnhof or a "shopping station", meaning able to generate revenue. Several units were planned for offices, hotels and shops to be rented to private companies. However, the DB AG now faces unanticipated consequences as the station's bid to become a significant retail hub has been quite unspectacular. Though travellers may appreciate the "mall", Berliners complain it's far too inaccessible in comparison with other shopping centres (Baker, 2006). Consequently, the DB AG failed to find tenants to occupy its commercially set spaces and was forced to take over the space itself (Kurpjuweit, 2010). Since then, the station project has been outlined as too "expensive and unprofitable" by Bertram Subtil, Head of Marketing at Arcadis, in an interview conducted by the Morgenpost in (2018), which centres on Subtil's economic study of the Berlin Central Station comparatively to world's top metropolitan railway stations, Subtil had remarked "With a total of 1.2 billion Euros in construction costs, the station was very expensive", followed by "If the station does not sustain itself, future investments will be difficult or difficult to convey". The station has indeed performed poorly.

Figure 2, taken from the report released by Subtil (2018), illustrates the station's performance in comparison with 27 of the world's top railway stations. The Berlin Central Station ranks at n°15.

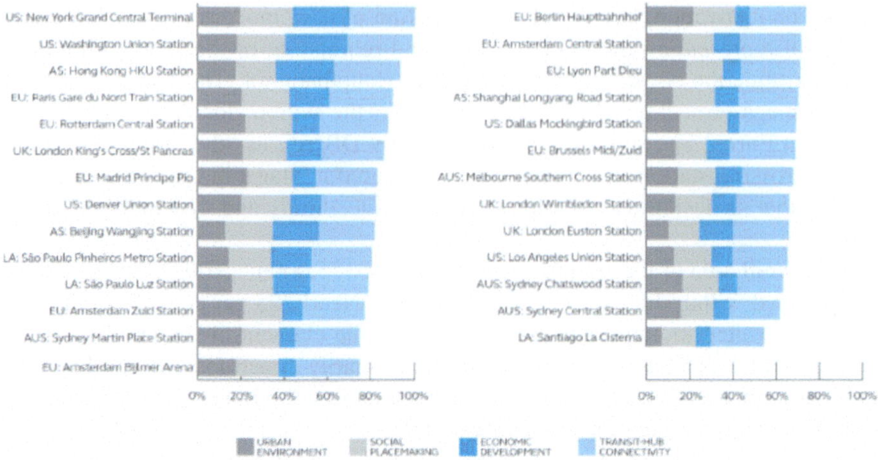

Fig. 2 The Berlin Hauptbahnhof's performance

5.2 The Split of the Japanese National Railways, an Ideal Case of Privatisation and "Station Renaissance". Case Study: The Osaka Central Station

5.2.1 Years of Japanese Railways, from Government to Public to Private

Established in 1920 by the Ministry of Railways, the Japanese Government Railways, often called the "Ministry Lines", was managed entirely by the ministry as the sole operator of the railway until 1949, when the Government Railways were reorganised as a public corporation called Japanese National Railways, as part of the nationalisation of Japan (Imashiro, 1997).

Until 1987 and under the "State-Owned Railway Law", all Japanese railways, including the Shinkansen, were owned by the government. Private railway companies were permitted to utilise the railway, but only within limited regional areas (Ministry of Land, Infrastructure, Transport & Tourism). By the 1960s, financial troubles began when JNR incurred hefty amounts of infrastructural costs as a consequence of a political decision to create a new high-speed railway system. This would eventually lead to JNR's accumulation of ¥37 trillion (~ €217.7 billion) of debt (Japan Times, 2017). Ultimately, the JNR had finally dismantled and was divided into 7 private companies (Table 1).

Table 1 The Seven JR Companies

No	*Name of Railway*	*Type of Operation*	*No*	*Name of Railway*	*Type of Operation*
1	JR Hokkaido	Passenger transport	5	JR Shikoku	Passenger transport
2	JR East	Passenger transport	6	JR Kyushu	Passenger transport
3	JR Central	Passenger transport	7	JR Freight	Freight

Source Terada (2001)

5.2.2 The Rebirth of the JNR as "Japan Railways" and the Phases of its Privatisation

One of the core challenges that JNR faced was the incredibly large scale of the organisation having to be managed and operated by a single public entity. This led to the need for privatisation (Kim & Huang, 2019). The privatisation of JNR was undertaken in two phases. The first phase, initiated in 1987, was mostly symbolic, as most company shares were still owned almost entirely by the state. The second took place in 1993, when part of the more profitable shares was sold on the stock market (Weizsacker et al., 2005). In 1991, Japan's high-speed line, the Shinkansen, the world's first "Bullet" train (which became operational in 1964 when owned by the Shinkansen Holding Corporation—SHC), was bought out by JR Central, JR East and JR West (Weizsacker et al., 2005). As of 2019, JR Central, JR West, JR East, JR Kyushu, and JR Hokkaido all retain ownership over the Shinkansen (Kim & Huang, 2019).

Over time, the operation of the Shinkansen by the JR firms allowed for the introduction of faster services, leading to more flexible management and investments for the expansion of their businesses. At present, many of the mega stations owned by JR generate revenue through non-railway-related operations such as real estate and commercial establishments in or within proximity of the station (Japan Times, 2017).

In Japan, the relationship between retail and railway stations goes back to the early twentieth century, when the first private railway station was established in Hankyu in 1907. The Hankyu Railway Station, now known as the Osaka-Umeda Station or the Osaka Central Station, was the first railway station's commercial complex—a department store was opened at the terminal station in 1929 (Kuchiki et al., 2017).

5.2.3 The Osaka-Umeda Station: "Osaka Station City"

Since the early 1990s, the Osaka-Umeda Station has been noted as one of Japan's most commercially successful railway stations, with the station's northern Umeda terminal area and the southern Abedo terminal area being heavily packed with shopping and entertainment facilities (Edgington, 1990). Completed in 2011, JR West intended to unify the station's two terminal areas by employing a new key project set to serve as a landmark and gateway to the city of Osaka, the "Osaka Station City" (JR-West, 2014)—presently identified as one of Japan's largest and most successful

Fig. 3 The New Osaka Station building (*Source* Beeboys/Shutterstock, https://insideosaka.com/new-osaka-station-building/

station mega-project development. The core initiatives of the "Osaka Station City" followed three key measures set by the JR West. The first was the "sales of good and food services", which entailed the renovation of retail shops in the station and the development of areas within proximity of the JR West service areas. Second, the "Department store"—here the West Wing of the "Osaka Station City" was to be renovated. This implied that department stores were to be specialised and hold a competitive advantage so as to leverage the strengths of the shopping centre and department store in the station. Along with this, new sales targets were set, with a return profit of approximately ¥80 billion (~ €667 million) from the redeveloped West Wing of the station. Lastly, the "Real Estate Business" was contemplated—entailing the advancement of condominiums as well as overall development and renovation of areas surrounding the station (JR-West, 2014) (Fig. 3).

The "Osaka Station City" is designed as a "city". The station contains many streets and eight unique plazas (Kido, 2013). In terms of commercial spaces, the station contains a shopping mall comprising hundreds of shops across 12 floors—as well as a movie theatre, a fitness centre, a hotel and several department stores.

5.3 "Station Renaissance" in India, Indian Railways: Redevelopment in the Twenty-First Century. Case Study: Habibganj Railway Station

5.3.1 Indian Railways, a Process of Reverse Privatisation

Unlike the cases of the Deutsche Bahn AG and Japan Railways, the Indian Railways originated as a private railway network. As India was still under British rule when rail was first introduced in 1853 through a major British investment of approximately £95 mil (~ €4.5 billion according to today's currency) between 1845 and 1875

Table 2 The Zonal Railway Divisions

No	*Zonal Railway*	*Abv*	*No*	*Zonal Railway*	*Abv*
1	Central Railway	CR	10	Western Railway	WR
2	Eastern Railway	ER	11	South Western Railway	SWR
3	Northern Railway	NR	10	West Central Railway	WCR
4	North Central Railway	NCR	13	North Western Railway	NWR
5	Northern Eastern Railway	NER	14	South East Central Railway	SECR
6	Northeast Frontier Railway	NFR	15	East Coast Railway	ECoR
7	South Eastern Railway	SER	16	East Central Railway	ECR
8	South Central Railway	SCR	17	Metro Railway	MTP
9	Southern Railway	SR	18	South Coast Railway	SCoR

Source Standing Committee on Railways (2018)

(Macpherson, 1955), private British railway companies owned the majority of rails in India.[1] It was not until 1869 that Lord Lawrence, the Viceroy of India between 1864 and 1869, had recommended that the state seize ownership of the railways. Following a policy shift, the state proceeded to purchase rail from private railway companies. Eventually, by the late 1920s, the state owned approximately 70% of rail in India (Saritha, 2013). By 1947, India had finally gained independence and the subcontinent had split into two countries, India and Pakistan. As a result, the existing railway was divided between the two countries and the Indian State was finally granted full control of all railways in India".

5.3.2 Post-Independence Re-Organisation of the Indian Railways

Following independence, the Ministry of Railways was established and the existing rail network was divided into six zones. By 2003, the six existing zones were further split into an additional six zones. As of 2019, there exist a total of eighteen zonal railways operated by the Indian Railways (Dutta, 2020; Table 2):

Aside from the Zonal divisions, the Ministry of Railways is split into eleven administrative undertakings: the Dedicated Freight Corridor Corporation, the Indian Railways Catering and Tourism Corporation, the Konkan Railway Corporation, the Indian Railway Finance Corporation, the Mumbai Rail Vikas Corporation, the Railtel Corporation of India—Telecommunication Networks, RITES Ltd.—the Consulting Division of Indian Railways, the IRCON International Ltd.—Construction Division, the Rail Vikas Nigam Ltd, the Container Corporation Ltd, and the Rail Land Development Authority. Accounting for the several production units, there are a total of 68

[1] Value of £1 from 1945–2020: According to the Office for National Statistics composite price index, today's prices in 2020 are 4,244.44% higher than average prices since 1945, meaning £1 in 1945 is equivalent to about £43.44 in 2020. According to today's currency, Britain invested a total of £4.126.800.000 (~ €4.554.295.212) in India's railway network.

Table 3 Seven Corridors Planned for High-Speed Railway Construction

Corridor	*Route*	*Length (km)*
1	Pune–Mumbai–Ahmedabad	Approx. 680
2	Delhi–Chandigarh–Amritsar	Approx. 480
3	Delhi–Agra–Lucknow–Varanasi–Patna	Approx. 1,000
4	Howrah–Haldia	Approx. 140
5	Hyderabad–Dornakal–Vijayawada–Chennai	Approx. 780
6	Chennai–Bengaluru–Ernakulam–Thiruvananthapuram	Approx. 1,020
7	Delhi–Jaipur–Jodhpur	Approx. 530

Source National High-Speed Rail Corporation Ltd. (2015)

operating divisions within the Indian Railways (Standing Committee on Railways, 2018).

5.3.3 'Superfast Trains' to Bullet Trains

In 1969, India witnessed its first high-speed train, the Rajdhani Express, which ran between New Delhi and Howrah at an operating speed of 115kmph. By 1971, the Ministry of Railways decided to increase the Rajdhani's operating speed to 130 kmph. Shortly after, in 1988, the Shatabdi arrived—a superfast train operating at 155 kmph, running between New Delhi and Jhansi (Shrirvastva & Mahmoud, 2019).[2] Moving towards the twenty-first century and India's rapid economic growth, in 2009 the "Indian Railways Vision 2020" for the development of passenger transport had been established. In other words, India had finally begun its transition to high-speed rail transport.

Through several studies carried out by RITES Ltd, seven routes were recognised as candidates for high-speed railway construction at a cost of €1.4 billion, with a projected 14 per cent return on investment (Shrivastva & Mahmoud, 2019; Table 3). The Mumbai-Ahmedabad line was the first to be constructed. In 2013, India and Japan had announced their decision to conduct a joint study on the Mumbai-Ahmedabad high-speed railway, which resulted in the Ministry of Rail signing a MoU (Memorandum of Understanding) with the JICA (Japan International Cooperation Agency) for Technology Transfer of High-Speed Railways (JICA Final Report, 2015). This signed agreement entailed India receiving its very own Shinkansen. However, as a means to execute the HSR developments, PPPs (Public–Private Partnerships) were established as a model for investments (Pillai, 2013), urging private funding to fill the need for the privatisation of Indian Railways.

[2] According to the International Union of Railways (UIC), trains with a Maximum Operating Speed of 160kmph are regarded as superfast trains.

5.3.4 The First Steps Towards the Twenty-First-Century "Station Renaissance" in India

As part of the "Indian Railways Vision 2020", India announced its plans to catch up with world-class station standards and declared its aim to redevelop 400 stations within 100 cities in its first phase of redevelopment—which is to be executed through a partnership with the Ministry of Urban Development, as well as with several PPPs (Sriraman, 2010).

Additionally, the Ministry of Railways appointed the Boston Consulting Group (BCG) as the strategic advisor for India's twenty-first-century "station renaissance". The focus of these mega transformations concerns improving passenger amenities, facilities and overall station infrastructures (BCG, 2017). Following world station standards, as in the cases of the Berlin Hauptbahnhof and the Osaka-Umeda Station, the development of iconic, multifunctional railway station complexes, with particular emphasis on commercial utilisation of station space is underway. The first 48 stations were planned for development/redevelopment in India (Table 4).

5.3.5 India's First Multi-Modal Commercial Railway Station Complex—The Habibganj Railway Station (Bhopal)

Planned to be completed in 2022 with a total area of approx. 400,000 sqm and an expected cost of €40 million for commercial development, the Habibganj Railway Station will be India's first commercial railway station implemented through a traditional PPP model. A private sector player, Bansal Pathways Habibganj Pvt. Ltd. was selected to execute the station redevelopments—under the technical supervision of Germany's gmp Architekten. Additionally, the group was granted a 45-year long encroachment-free land lease for commercial and real estate developments (Garg & Chaudhry, 2017). Nonetheless, the Indian Railways remains the official operator of the railway station area, which totals 130,548sqm, amounting to ~ 32% of the overall space (Fig. 4).

As concerns commercialisation, while it is difficult to compare the developments of Habibganj Railway Station with those of the Osaka-Umeda Station—specifically due to the scale of the projects—the station seems to be inspired by the Japanese "station city" trend with the project titled "Bonsal City Centre"(after the name of its private developer, Bansal Pathways). Additionally, the railway station is planned to contain functions similar to that of the Osaka-Umeda Station. 7% (~47,857 sqm) of the overall station's built area is allocated to residential development; ~ 28% (~ 133,494 sqm) is allocated to commercial space such as trade centres; an anchor store, service apartments, a hospital, budget and luxury hotels, a convention centre, and the Bansal One shopping mall. The redevelopment scheme additionally includes plans for public green space, as well as several recreational areas, and a large percentage of 38% (~153,516 sqm) of the overall area allocated towards private parking (IRSDC, 2014). One redevelopment that is particular to the Habibganj Railway Station redevelopment is the establishment of an underground passenger network leading to

Table 4 List of Railway Stations Set for Development/Redevelopment

No	*Name of Station*	*Zonal Railway*	*Implement Agency*	*Status (as of 2020)*
1	Lokmanya Tilak	CR	By bid (private)	Planned
2	Pune	CR	By bid (private)	Planned
3	Thane	CR	By bid (private)	Planned
4	Howrah	ER	By bid (private)	Planned
5	Visakhapatnam	ECoR	By bid (private)	Planned
6	Kanpur Central	NCR	By bid (private)	Planned
7	Allahabad	NCR	By bid (private)	Planned
8	Kamakhya	NEFR	By bid (private)	Planned
9	Udaipur City	NWR	By bid (private)	Planned
10	Faridabad	NR	By bid (private)	Planned
11	Jammu Tawi	NR	By bid (private)	Planned
12	Secunderabad	SCR	By bid (private)	Planned
13	Vijayawada	SCR	By bid (private)	Planned
14	Ranchi	SER	By bid (private)	Planned
15	Chennai Central	SR	By bid (private)	Planned
16	Kozhikode	SR	By bid (private)	Planned
17	Yeshwantpur	SWR	By bid (private)	Planned
18	Bangalore Cantonment	SWR	By bid (private)	Planned
19	Mumbai Central	WR	By bid (private)	Planned
20	Bandra Terminus	WR	By bid (private)	Planned

(continued)

Table 4 (continued)

21	Borivali	WR	By bid (private)	Planned
22	Indore	WR	By bid (private)	Planned
23	Bhopal	WCR	By bid (private)	Planned
24	Surat	(WR)	IRSDC	Planned
25	Habibganj	WCR	IRSDC	On-going
26	Gandhinagar(Gujrat)	WR	IRSDC	On-going
27	Bijwasan (Delhi)	NR	IRSDC	On-going
28	Anand Vihar (Delhi)	NR	IRSDC	On-going
29	Shivaji Nagar (Pune)	CR	IRSDC	Planned
30	Chandigarh	NR	IRSDC	On-going
31	Amritsar	NR	IRSDC	On-going
32	Gwalior	NCR	IRSDC	Planned
33	Gandhinagar (Jaipur)	NWR	IRSDC	Planned
34	Nagpur	CR	IRSDC	Planned
35	Baiyyappanahalli (Bengaluru)	SWR	IRSDC	Planned
36	Tirupati	SCR	NBCC + RLDA	Planned
37	Delhi Sarai Rohilla	NR	NBCC + RLDA	Planned
38	Nellore	NR/ SCR	NBCC + RLDA	Planned
39	Madgaon	SWR	NBCC + RLDA	Planned
40	Charbagh (Lucknow)	NR	NBCC + RLDA	Planned
41	Gomtinagar (Lucknow)	NER	NBCC + RLDA	Planned
42	Kota	WCR	NBCC + RLDA	Planned
43	Thane New	CR	NBCC + RLDA	Planned
44	Ernakulam	SR	NBCC + RLDA	Planned
45	Puducherry	SR	NBCC + RLDA	Planned
46	Safdarjung Delhi	NR	RLDA + IRCON	Planned
47	Sabarmati (Ahmedabad)	WR	IRSDC	Planned
48	Thakurli (Mumbai)	CR	IRSDC	Planned

Source Indian Railways Institute of Civil Engineering (2017)

Fig. 4 Envisaged 3-D Model of the Habibgang Railway Station (*Source* Boston Consulting Group)

separate arrival and departure halls aiming to reduce passenger congestion (IRSDC, 2014), thus enabling the station to function similarly to an airport.

Furthermore, concerning the design and construction of the railway station, while some have claimed that the station takes a similar design approach to the Heidelberg Hauptbahnhof (India Times, 2018), there seems to be more similarity in design to the Berlin Hauptbahnhof, as the station also contains a massive "glass dome". This is particularly true as concerns the large glass facade and atrium planned between the floors (with several bridges running through each level), which allow natural light to flow through the entire station.

As a final point, in order to steer the Habibganj Railway Station in a more eco-friendly direction, the Indian Railways partnered with the Ministry of Urban Development (MOUD) as part of the India Smart City Mission (2016) of the Government of India. Accordingly, the transportation complex is planned to have solar power and rain-water harvesting provisions.

6 Conclusion

Over the last three decades, the emergence of high-speed trains incited a new perspective on railway stations as urban spaces. The new transportation technology has pushed forth "station renaissance", which has transformed stations from transport facilities to massive, multi-modal transportation hubs. These transformations have urged railway stations to become a core focus in many mega-urban development projects. Accordingly, modern railway stations are now seen as "gates" to their cities, essentially functioning as cities within cities.

Furthermore, as concerns developmental costs, while in some cases state funding may be applicable (thus avoiding the interference of private entities such as in the case of the Berlin Hauptbahnhof), privatisation often allows the creation of specialised, user-centric multifunctional spaces as is the case in Japan. Privatisation is often regarded as unfavourable, as it may create several limitations for the public use of spaces; however, when executed appropriately it may favour the public.

Finally, while it is too soon to see the effects of railway station revitalisation in India and the consequences of involvement by private developers, having discussed the role of privatisation in Germany and Japan, it is plausible to consider a similar future in the case of railway station development. However, this study, along with several others on the topic, does not discuss the social implications attached to these developments and the process of privatisation. In this regard, the societal shift created with the development of railway stations should be considered in further studies.

References

Baker, J. (2006). *Berlin central station. Architecture week.* http://www.architectureweek.com/2006/1108/design_1-1.html.

Bertolini, L., & Spit, T. (1998). *Cities on rails: The redevelopment of railway station areas.* E&FN Spon.

Bruinsma, F., Pels, E., Rietveld, P., Priemus, H., & Van Wee, B. (Eds.) (2008). *Railway development: Impacts on urban dynamics.* Physica-Verlag Heidelberg. DOI:https://doi.org/10.1007/978-3-7908-1972-4.

Edgington, W. (1990). Managing industrial restructuring in the Kansai region of Japan. *Geoforum, 21*(1), 1–22. doi:https://doi.org/10.1016/0016-7185(90)90002-N.

Garg, A., & Chaudhry, R. (2017). *Indian railways station redevelopment: Transforming railways and creating win-win opportunities.* Boston Consulting Group Reports. https://image-src.bcg.com/Images/BCG-FICCI-IndianRailways-Stations-Redevelopment-Jun-2017_tcm21-161940.pdf.

Imashiro, M. (1997). Dawn of Japanese National Railways. *Japan Railway & Transport Review, 9,* 46–49. https://www.ejrcf.or.jp/jrtr/jrtr10/pdf/history.pdf.

India Smart City Mission. (2016). *The smart city challenge: Smart city proposal.* Ministry of Urban Development.

Indian Railway Station Development Corp. Ltd. (2014). *Development agreement for Habibgaj railway station.* IRSDC.

Indian Railway Station Development Corp. Ltd. (2016). *Redevelopment of Habibganj railway station.* IRSDC Report.

India Times. (2018). *Bhopal's Habibganj railway station set to be redeveloped along lines of Germany's Heidelberg.* https://www.indiatimes.com/news/india/bhopal-shabibganj-railway-station-set-to-be-redeveloped-along-lines-of-germany-s-heidelberg-353551.html.

Indian Railways Institute of Civil Engineering. (2017). *Indian railways stations redevelopment project.* IRICEN Report.

Japan International Cooperation Agency & Ministry of Railway. (2015). *Final report: Joint feasibility study for Mumbai-Ahmedabad high-Speed railway corridor.* JICA.

Japan International Cooperation Agency (JICA), Ministry of Railways, Republic of India (MOR). (2015). *Joint feasibility study for Mumbai-Ahmedabad high-Speed railway corridor.* Final Report: Volume 1.

Japan Times. (2017). *Privatization of JNR, 30 years on.* https://www.japantimes.co.jp/opinion/2017/04/04/editorials/privatization-jnr-30-years/.

Kido, E. (2013). *Stations for people—Recent developments in railway station design*. National Land Culture Research Institute-Construction Technology Research Institute, 53–81. http://202.251.0.3/kokubunken/pdf/publication/2013_10.pdf.
Kim, C., & Huang, M. (2019). *The privatization of Japan railways and Japan post: Why, how and now*. ADB Institute.
Kuchiki, A., Mizobe, T., & Gokan, T. (2017). *A multi-industrial linkages approach to cluster building in East Asia: Targeting the agriculture, food and tourism industry*. Springer Nature.
Kurpjuweit, K. (2010). *Hauptbahnhof bald komplett vermietet*. Tagesspiegel. https://www.tagesspiegel.de/berlin/gewerbeflaechen-hauptbahnhof-bald-komplett-vermietet/1914924.html.
Macpherson, W. J. (1955). Investment in Indian Railways, 1845–1875. *Economic History Review*.
National High Speed Rail Corporation Ltd. (2015). *Joint feasibility study for Mumbai-Ahmedabad high speed railway corridor*. Final Report: Volume 1.
Organisational restructuring of Indian Railways. Available at SSRN 3574155. doi:https://doi.org/10.2139/ssrn.3574155.
Peter, B. (2008). *Railway reform in Germany: Restructuring, service contracts, and infrastructure charges*. TU Berlin.
Peters, D. (2009). *The renaissance of inner-city rail station areas: A key element in contemporary urban restructuring dynamics*. The Soka University of America.
Pillai, G. (2013). *Development of high-speed trains in India*. Ministry of Railways.
Saritha, S. R. (2013). *Colonialism and Modernisation*. The University of Kerala.
Schmidt, F. (2018). Weltweites Ranking: Berliner Hauptbahnhof nur im Mittelfeld. *Morgenpost*. https://www.morgenpost.de/berlin/article214855659/Weltweites-Ranking-Berliner-Hauptbahnhof-nur-im-Mittelfeld.html.
Shrivastva, C., & Mahmoud, A. (2019). High-speed rail corridor: The Indian assessment. *Journal of Management and Public Policy, 10*(2), 21-32. doi:https://doi.org/10.5958/0976-0148.2019.00002.7.
Sriraman, S. (2010). Railway budget 2010–11: Towards vision 2020. *Economic and Political Weekly, 45*(15), 34–38. https://www.jstor.org/stable/25664330.
Standing Committee on Railways. (2018). 19th Report: "Demand for Grant". Ministry of Railways.
Subtil, B. (2018). *Worldwide station comparison study: Germany with Berlin HBF in 15th place in the middle*. Arcadis Design & Consultancy for Natural & Built Assets.
Terada, K. (2001). Railways in Japan: Private & public sectors. Japan Railway & Transport Review. *Railways Operators in Japan, 1*(27), 48–55. http://www.ejrcf.or.jp/jrtr/jrtr27/pdf/s48_ter.pdf.
Weizsacker, E., Young, O., & Finger, M. (2005). *Limits to privatisation: How to avoid too much of a good thing*. Earthscan.
West Japan Railway Company (JR-West). (2014). *Annual report: Operating results by business segment*. https://www.westjr.co.jp/global/en/ir/library/annual-report/2014/pdf/c07.pdf.
Zacharias, J., Zhang, T., & Nakajima, N. (2011). Tokyo station city: The railway station as urban place. *Urban Design International 16*, 242–251. doi:https://doi.org/10.1057/udi.2011.15.

Portuguese Colonial Railways: Agents and Subjects of Railway Imperialism (1880–1915)

Hugo Silveira Pereira

1 Introduction

Railway imperialism is usually associated with the colonial agendas pursued by Britain, France or Germany in Africa and Asia—and with how these core nations used their financial, technical, industrial and financial leverage to impose railway construction over foreign territories and peoples.

Seldom do academics analyse the ways in which smaller peripheral nations tried to engage with railway imperialism—or how (un)successful they were. In this chapter, we contribute to the discussion through a case-study regarding Portuguese railway investments in its then empire, with a specific emphasis on the colony of Goa, India. We aim to show the ways in which Portugal used railway infrastructure as a tool of empire in overseas domains—before the country itself became a victim of railway imperialism by stronger powers, particularly Britain.

We use recent literature regarding the implementation of railway infrastructure in Portuguese former colonies of Africa and Asia, as well as statistics regarding railway construction and operation, so as to determine both their source of financing and the actual returns of investment.[1]

This chapter is divided into five sections. Firstly, we describe the general historical context surrounding the construction of railways in Portuguese colonies. Secondly, we overview the debate regarding railway imperialism. Thirdly, we explain how

[1] Besides official compilations (Portugal, 1912, 1917), railway statistics may be found in the Portuguese Overseas Historical Archive (Portugal, 1899), items: 52 1G, 299 1H, 301 1H, 927 1 N, 2526 1B, 2564 1B, 2565 1B, 2673 1B, 2675 1B and 2756 1B.

H. S. Pereira (✉)
CIUHCT – Interuniversity Centre for the History of Science and Technology, NOVA School of Science and Technology, Almada, Portugal
e-mail: hjs.pereira@fct.unl.pt

Department of History, University of York, York, UK

© The Author(s), under exclusive license to Springer Nature Switzerland AG 2021

S. Mitra et al. (eds.), *Railway Transportation in South Asia*, Contemporary South Asian Studies, https://doi.org/10.1007/978-3-030-76878-2_3

this paper contributes to the literature regarding Portuguese colonial railways and railway imperialism. In Sect. 4, we analyse unpublished statistical data regarding Portuguese colonial railways activity and associate this with related literature on the same subject so as to compare both the financial costs and the financial, economic and political/diplomatic results of railway implementation. The fifth section contains some final remarks.

2 The Implementation of Railways in the Portuguese Empire

Beginning in 1850, Portugal undertook an ambitious programme of public works, spearheaded by railways, so as to modernise the country and bring it closer to its European counterparts. This programme was inspired by the Saint-Simonian ideology, which started its appeal on Portuguese engineers in the 1820s during their trainings in French and Belgian schools (Matos, 2009), Saint-Simonianism advocated for the construction of large public works as well as transportation networks intended to promote the circulation of people, goods and capital (Vleuten, 2006). Additionally, large transportation systems such as railways were considered tokens of progress—a gauge used to measure each nations' worth (Adas, 1989).

In the late 1870s, this strategy was extended to Portuguese colonies in Africa and Asia. The goal was not only to modernise transportation networks but also assert Portuguese sovereignty on those territories—which were coveted by other, more powerful European nations arguing that Portugal lacked the resources to effectively occupy and develop them (Diogo & Laak, 2016). Railway construction in the Portuguese empire began in 1881 in Goa, India. In the following years, it was resumed in Angola and Mozambique. By 1915, total mileage ascended to 2,051 km (Portugal, 1917).

Throughout the years, policymaking regarding the implementation of railways in the Portuguese colonies was erratic and driven by the circumstances of the moment (availability of capital and entrepreneurs, diplomatic pressures, military concerns). Therefore, leasing, financing, construction and operation took on diverse forms (Table 1): private initiative supported by state subsidies, private initiative without any public allowance, construction by state engineers; foreign financing; state financing; a varied array of suppliers (German, French, British, Belgian); and assorted technical standards, most notably gauges (Navarro, 2018; Kerr & Pereira, 2012).

Nonetheless, there was one thing in common in all Portuguese imperial lines: the presence of British influence. In some cases, this influence was minimal, limited to the supply of railway infrastructures; in other instances it was overwhelming and felt at different stages of railway implementation—from financing to construction and further to management. This influence paved the way for an unequal relationship of power between Portuguese and British interests and colonial agendas—and led to the establishment of forms of railway imperialism.

Table 1 Lines built in the Portuguese colonies (1886–1915)

Line	Location	Construction	Lessee	Gauge (cm)	Extension (km)
Lourenço Marques	Mozambique	1886–1890	Edward McMurdo*	106.7	89
Beira	Mozambique	1892–1898	Beira Ry. Co. and Beira Junction Ry. Co.	60**	339
Quelimane	Mozambique	1901	Zambezi Co.	60	28
Swaziland	Mozambique	1905–1909	State	106.7	69
Xai-Xai	Mozambique	1909–1912	State	75	52
Inharrime	Mozambique	1910–1914	State	75	92
Polana	Mozambique	1911	State	106.7	6
Xinavane***	Mozambique	1913–1914	Dick Kerr & Co. Ltd.	106.7	89
Ambaca	Angola	1886–1899	Royal Co. of Rys. Across Africa	100	364
Malange	Angola	1902–1909	State	100	139
Benguela	Angola	1903–1913****	Benguela Ry. Co.	106.7	519
Moçâmedes	Angola	1905–1915*****	State	60	183
Mormugão	India	1881–1888	West of India Portuguese Guaranteed Ry. Co.	100	82
Total					2,051

Notes Decauville lines in the cocoa plantations of S. Tomé not included. *Transferred to the Delagoa Bay East and African Ry. Co. Ltd. in 1887 and nationalised in 1889; **Re-gauged to 106.7 cm in 1900; ***Extended to the Limpopo valley in 1935 (90 km); ****Construction resumed until 1928 when the line reached the border with Congo (830 km); *****Extended to Lubango in 1923 and Menongue in 1961 (465 km in total)
Sources Navarro (2018); Portugal (1917)

3 Railway Imperialism: Definitions and Literature Review

Railway imperialism can be defined as the use of railways (and the industry, capital and expertise accompanying these) so as to dominate foreign territories and resources. It is often associated with nineteenth and twentieth-century colonialism and with *civilising missions* by Western nations in Africa and Asia; however, it can also occur between nations outside the imperial context. Usually, railway imperialism hides behind promises of economic development and overall progress brought about by railways—while in fact most often the gains are exclusively enjoyed by the imperial power involved, with little to no benefits for local economies and populations.

The association between imperialism and railway infrastructure has long been established in academic debate. Headrick (1981) classified railways as a tool of empire in the period of what he calls *New Imperialism* (from the mid-nineteenth century to World War II). Lee (1989) demonstrated the ways in which railways served French economic imperialist forces in China, while Kerr (2007) addressed similar issues as concerns British rails in India. More recently, several authors have analysed the importance of railways in 'gentlemanly capitalism'—a concept that stresses the role played by business interests and London City's financial sector, as well as their alliances with the British government at Whitehall, in the expansion of the British empire (Akita, 2002).

Building on some of this discussion, a group of historians (Davis et al., 1991) developed the concept of railway imperialism, which underscores the specific agency of railways in both formal and informal empire (the former refers to the use of railways so as to impose political rule whereas the latter refers to the establishment of economic dependence). Focusing on different case-studies (colonial Africa, imperial China, North and South America), these authors highlight the role of three different agents: local entrepreneurs who needed railways to exploit colonial resources; private financiers who sought to apply their capital in fairly riskless enterprises (considering they usually benefitted from a guarantee of yield from the state); and metropolitan governments who could use those railways in various imperial strategies.

More recently, Divall (2003), while evaluating these previous references regarding railway imperialism, defined a hierarchy in levels of railway imperialism. His structure ranges from a strategic/competitive level, wherein local elites can resist and play foreign powers against each other, to a level of fully formal military, political and economic control of a territory by an alien nation. It includes the 'informal empire exerted by European core nations over peripheral countries from Europe and South America'. Another recent and interesting development on the topic was provided in the special issue of *Comparativ* edited by Castryck (2015), which added the issue of globalisation to the debate.

In Portuguese historiography, research regarding colonial railways has taken decisive steps in recent years. Arguably, the most detailed work is Navarro's (2018) book, which analyses colonial lines as tools of appropriation in Angola and Mozambique (for the case of the Mormugão track, in Goa, see Kerr & Pereira, 2012). Other studies stress the importance of railways as a political and diplomatic tool used to assert Portuguese presence in Africa, or their role as promoters of globalisation (Diogo & Navarro, 2018; Pereira, 2018b). However, not many works debate railway imperialism in the Portuguese colonial scenario. Clarence-Smith (1985) in his study about the Portuguese empire as a case of economic imperialism makes only short references to imperial railways. Bouene and Santos (2006) and Pereira (2021) engage more profoundly within that framework: the former illustrates how railways were used to control the migration of local populations from south Mozambique to the mines of Transvaal, whereas the latter examines a failed attempt to impose a railway travelling from Macao to imperial China.

In this paper, we aim to contribute in filling this gap with a study regarding the Portuguese colonial railway system, focusing on the case of the Mormugão railway in Goa, India. We also aim to further the debate about railway imperialism through the example of a second-rate colonial country from the European periphery who wished to assert itself in the concert of imperial nations. I argue that even though Portugal managed to use railways to exert formal control (and formal empire) over large tracts of land in its overseas territories, it also intensified its position as a British economic satellite (and as a subject of British informal empire), as it lacked the industrial and financial resources to properly build and operate railways.

4 The Portuguese Colonial Railway Sector: Costs and Benefits

As I mentioned before, by 1915 the Portuguese imperial railway network extended for over 2,000 km (Table 1 and Figs. 1, 2 and 8).

There were specific reasons motivating the construction of each railway. For instance, the decision to lay the Ambaca line was taken in the aftermath of the Berlin Conference and intended to promote the effective occupation of territory, whereas the Moçâmedes railway was built so as to facilitate the military occupation of Angola's southern hinterland. In Mozambique, the Lourenço Marques track was

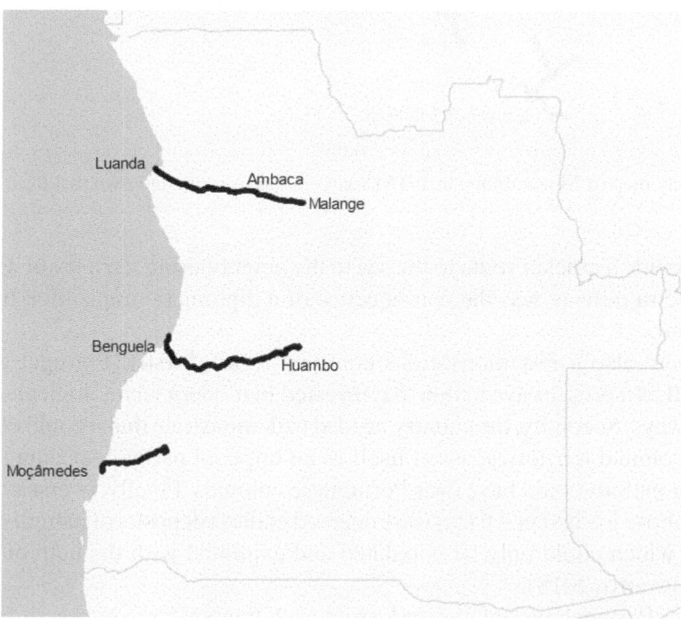

Fig. 1 Railway map of Angola in 1915 (*Source* Sharemap.org and own making)

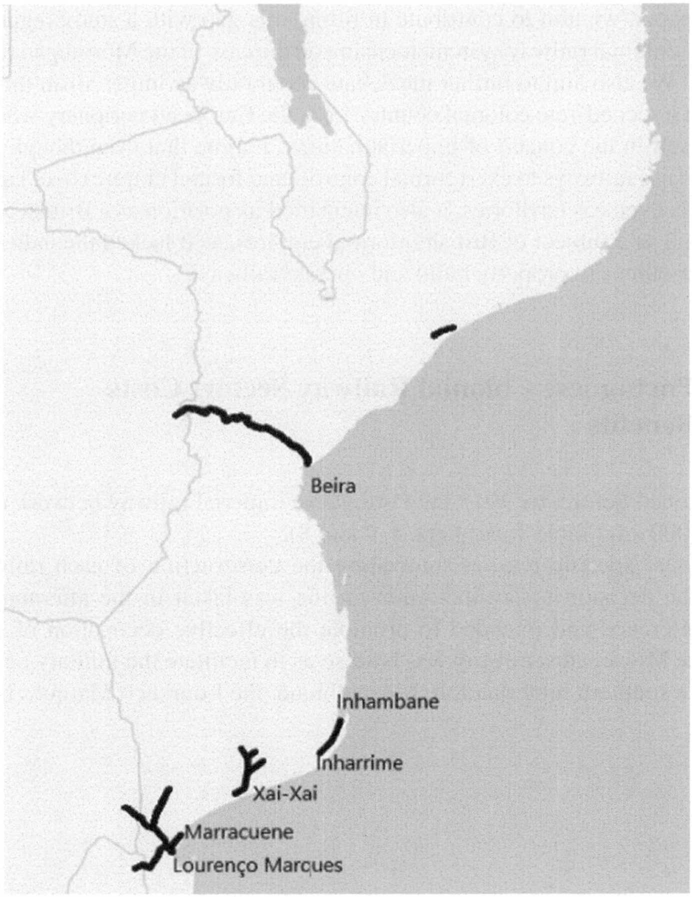

Fig. 2 Railway map of Mozambique in 1915 (*Source* Sharemap.org and own making)

meant to provide a quicker route to the sea to the neighbouring territory of Transvaal, while the Beira railway was the consequence of a diplomatic imposition by Britain over Portugal.

There were also a few motivations common to all. Firstly, Portugal sought to present itself as a progressive nation that invested in modern technological structures such as railways. Secondly, the country needed to demonstrate that it could effectively occupy its colonial territories, assert itself as an imperial nation and shun away any claims other nations could have over Portuguese colonies. Finally, overseas domains (especially those located in Africa) were deemed endless deposits of natural resources and wealth which could only be populated and exploited with the help of railways (Diogo & Navarro, 2018).

However, Portugal lacked the industrial and financial resources to build and operate rail tracks. These had therefore to be found in industrial nations which exported financial capital, some of which were, as mentioned above, also coveting Portuguese colonies. This created an ambiguous situation wherein Portugal used railways to impose imperial rule over colonial populations and territories while at the same time falling under imperial domination from abroad.

Portugal was not a rich country; however, its inclusion in the gold-standard (since 1854) created some stability which, until the partial bankruptcy of 1892, allowed for the establishment of steady flows of foreign direct investment into the country (Duarte & Andrade, 2012; Lopes & Simões, 2017). Yet even after the bankruptcy, Portugal still had to rely on investors from abroad so as to build railways on its own colonies.

The two major investments in the 1880s in African colonies were the railways of Ambaca and Lourenço Marques, both leased to private operators. In the 1890s and early 1900s, the Beira and Benguela lines were also chartered to two private companies. Even though these firms were Portuguese *de jure* (except for the Beira and Beira Junction companies which were seeded in Britain), they were de facto controlled by British agents, as most financing originated in Britain (totalling an amount of €572,000,000—see Table 2): the capital for the Ambaca line was raised with bonds issued in London; the main stockholder of the company which leased the Lourenço Marques line was an American entrepreneur who afterwards sold his stock in Britain; the main financier of the Benguela track was Tanganyika Concessions, a British company with economic interests in Central Africa; finally, the Beira

Table 2 Cost of the lines built in Portuguese colonies (values current in 1915 and 2019)

Line	Total cost (contos, 1915)	Total cost (million €, 2019)
Ambaca	12,459	181
Lourenço Marques	1,608	23
Beira	3,576	51
Benguela	21,827	316
Total	**39,470**	**572**
Lourenço Marques—nationalisation	5,950	104
Quelimane	167	2
Malange	2,666	38
Swaziland	1,862	27
Moçâmedes	2,236	32
Xai-Xai	273	3
Xinavane	801	11
Total	**8,005**	**217**

Notes 1 conto = 1,000,000 réis. In order to convert contos to euros, we used the exchange rate from réis to pounds calculated by Valério (2001). We then used the online tool provided by Officer & Williamson (2020) to convert pounds to 2019 dollars, and the exchange rate from dollars to euros (2019)

Sources Those mentioned in note 1, except figures for the Beira line, which were calculated by Pereira (2019a)

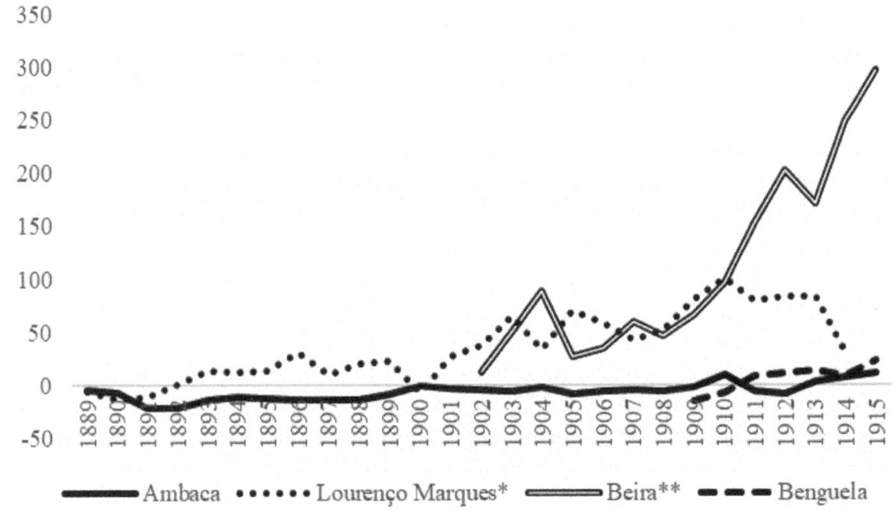

Fig. 3 Operational profit of the imperial lines built and operated by private enterprises (index 100 = profit of the Beira line in 1900) (*Notes* *The Lourenço Marques line was nationalised in 1890; **Includes figures from the line between the border with Rhodesia and Salisbury (*Sources* Those mentioned in note 1. For the Beira line, also those sources mentioned in Pereira [2019a])

companies were part of Cecil Rhodes' giant holding British South African Company. In more or less explicit fashion, these entrepreneurs promoted the imperial agenda designed in Whitehall and thus engaged in forms of the above-mentioned gentlemanly capitalism. In several occasions, this foreign predominance motivated decisions regarding fares, operation or track routes which plainly contradicted Portuguese interests (Navarro, 2018).

The operation of some of these lines generated profit—and served commercial or economic interests of British agents in neighbouring territories (Fig. 3): the landlocked territory of Rhodesia found a shorter connection to the ocean through the Beira line (its operational profit often paid dividends); the Congolese copper mines of Katanga found a shorter way to the Atlantic via the Benguela track. As for the Lourenço Marques line, it supplied the mines of Transvaal with labourers from Mozambique—and exported its ore to the harbour of Lourenço Marques. This line also became a very successful enterprise and its profits were returned to the Portuguese exchequer, as the line had been nationalised in 1890 (Navarro, 2018). The Ambaca line benefitted from a subsidy from the Portuguese state that covered losses and guaranteed a yield to the bondholders which financed its construction (for a financial analysis of the Ambaca deal: Pereira, 2019b).

Moreover, the imperial railway enterprise was an appealing market for European industrial manufacturers (British, German, French and Belgian)—for Portugal lacked a heavy industry able to produce locomotives, rolling stock, rails and other heavy materials. Coal, too, had to be imported. Portuguese manufacturers supplied small utensils for the permanent way (fishplates, treenails, spikes) and offices or household products for members of the colonial administration and settlers (Navarro, 2018; for the case of the metropolitan railway sector see Pinheiro, 1988). As concerns the skilled labour market, British engineers could be usually found in the work yards, especially in those enterprises managed and financed by their own countrymen.

In sum, in this regard, the role of foreign imperial agents helped transform Portuguese colonies into a British economic satellite (to use the terminology of Divall [2003]). Portugal increased its dependence on foreign capital and industrial products from European core nations.

Different Portuguese governments tried to counter foreign influence on the country's imperial tracks (especially in the years that followed bankruptcy and the British Ultimatum of 1890, which demanded the withdrawal of Portuguese forces from the areas between Angola and Mozambique). However, there was a chronic lack of resources, especially at the financial level. The state could only afford to invest in shorter lines—in which a little over 2,000 contos (€113,000,000) were spent. Nationalising tracks was not a feasible solution, considering the costs involved: the nationalisation of the Lourenço Marques line alone mounted to almost 6,000 contos (€104,000,000). It is also important to remember that public subsidies to the Ambaca line ascended in 1913 to over 15,000 contos (€331,000,000)—which were of course diverted from investing in other tracks (Pereira, 2019b). Furthermore, most state lines ran an operational deficit, and those who held operational profits could not cover the financial costs associated with construction. Table 2 illustrates the fact that direct public investment in colonial railways was less than half the values of private sector investments (Fig. 4).

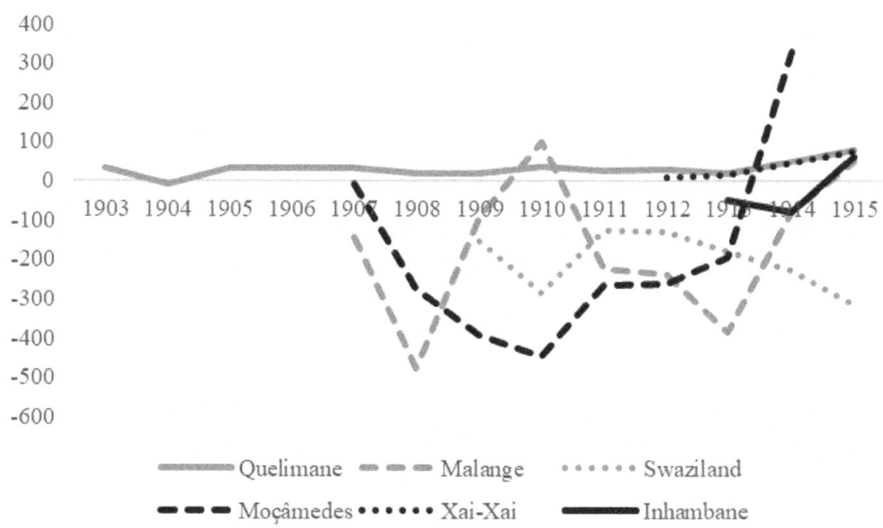

Fig. 4 Operational profits of imperial lines built and operated by the state (index 100 = profit of the Malange line in 1900) (*Sources* Those mentioned in note 1)

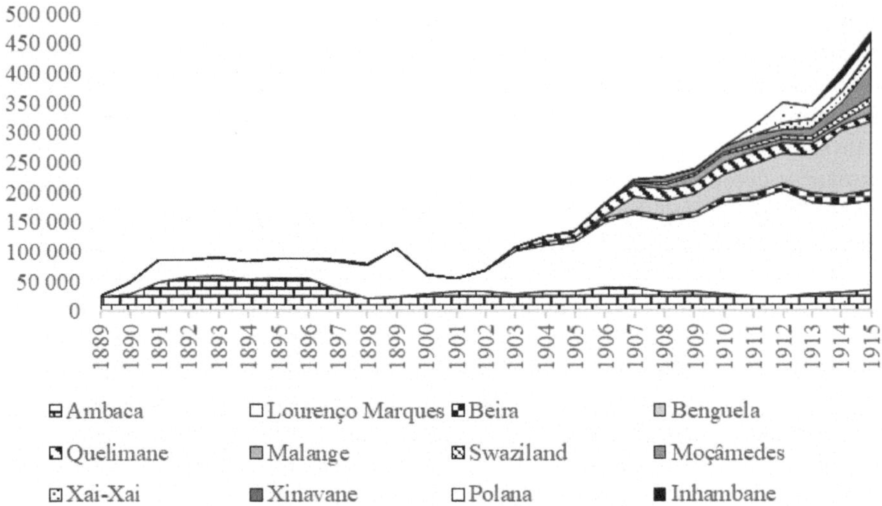

Fig. 5 Passenger transports in railways of the Portuguese colonies of Angola and Mozambique (*Sources* Those mentioned in note 1)

The colonial railway sector also contributed to the Portuguese imperial agenda in Africa; however, its action was limited to the narrow areas of influence of the tracks themselves. Operational statistics denote a repetitive and consistent use of these limited areas, promoting their territorial appropriation (for this concept: Kärrholm, 2012). Between 1889 and 1915, over 15 million passengers used the trains of Portuguese Africa (Fig. 5). In some lines upwards flows (passengers moving towards the hinterland) were higher than downward flows, which hints at the possibility that railways promoted the settlement of European colonists (Pereira, 2018a).

Freight transportation also increased steadily until the eve of World War I (Fig. 6). According to some studies (Navarro, 2018; Pereira, 2018a, 2019b), a substantial part of this freight was composed of railway material; however, we also witness an important share of housing and clothing materials (for settlers) as well as African products (rubber, coffee, grain, cocoa, sugar, etc.).

Although these numbers are far from impressive (and very far from expected returns on the investments involved), they illustrate that railways did promote the occupation of territory and exploitation of its resources, even if only in restricted areas. Additionally, construction and maintenance of these tracks (more or less forcedly) 'employed' thousands of natives (Navarro, 2018), which allowed Portugal to claim it contributed to the *education through work* of local populations and more generally to the European *civilising mission* in Africa (Jerónimo, 2015). Pictures such as that below (Fig. 7) were meant to testify and legitimise these events, in fact hiding the reality of forced labour (cf. Ryan, 1997). All these factors combined promoted railway imperialism, which contributed to the formal control of some territories (cf. Divall, 2003).

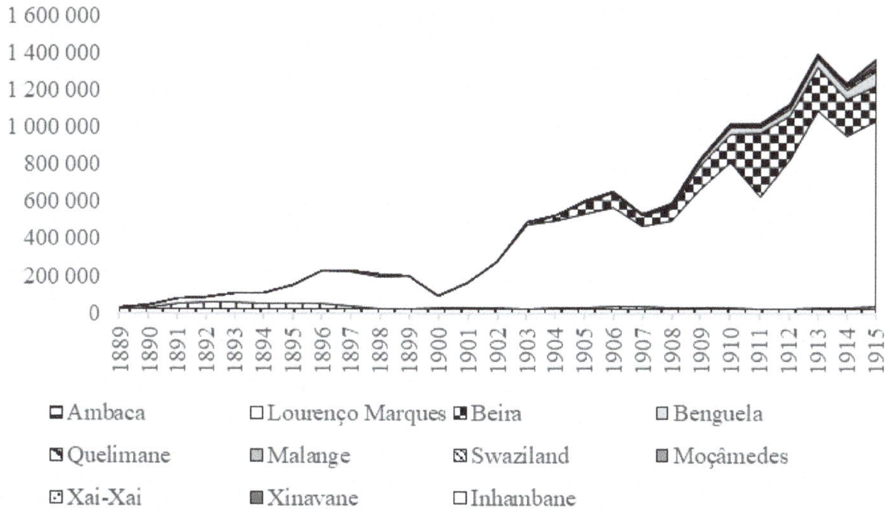

Fig. 6 Freight transported in railways of Portuguese colonies in Angola and Mozambique (in *t*) (*Sources* Those mentioned in note 1)

Fig. 7 Natives employed in the construction of the Lourenço Marques line (*Source* Fowler and McMurdo [1887])

4.1 The Case of the Mormugão Railway

The railway built in Goa, connecting the harbour of Mormugão with the network of British India at the frontier station of Castle Rock, provides another good example of railway imperialism associated with gentlemanly capitalism. Both dynamics mostly favoured Britain (through the Government of India) in the establishment of an informal empire over the Portuguese colony of Goa—with little to no benefit to the territory in question or Portugal's own imperial agenda.

This project (which included harbour improvements in Mormugão) had its inception in a treaty signed by Portugal and Britain in 1878. This treaty solved some grievances (regarding extradition, custom fees, and the production of liquor, salt and opium) between both nations' operations in India. The treaty created a customs union and included a clause (VI) which stated that Britain had to extend a railway built by Portugal in Goa (between the Mormugão harbour and the border) into the territory of British India. In exchange, the Portuguese government prohibited the production of opium and limited the production of liquor and salt in Goa (items smuggled to British India and which competed with local products). The agreement, especially clause VI, was received in Lisbon with delight and considered the first step on the road to regenerate Goa (Kerr & Pereira, 2012).

The opportunities involved were not missed by the Duke of Sutherland, one of the wealthiest men in Britain—with connections to politics, finance and the railway sector. After some conversations with the Portuguese diplomatic attaché in London and the British legate in Lisbon, a contract was signed between the Duke and the Portuguese government to build and operate a harbour at Mormugão and a railway towards the border with British India. Shortly thereafter, the Duke of Sutherland promoted the constitution of the West of India Portuguese Guaranteed Railway Company (hereafter WIPGRC) intended to uphold the contract. In order to support the enterprise, the Portuguese government granted a guarantee of yield on the capital of £800,000 (around €93,000,000)—and 6% on additional capital above that figure (which would be set at £550,000 or around €64,000,000).

Construction of the harbour and the track (initiated in 1881) became an economic opportunity—that is, almost exclusively for British enterprise, industry and expertise. The senior staff hired by the WIPGRC was entirely British (led by chief-engineer Ernest Edward Sawyer). The only Portuguese presence was that of the overseer, engineer Xavier Cordeiro (who went on to have a brilliant career in Portugal in the private sector). Most material for the permanent way came from Britain, whereas the rolling stock was entirely of British origin (Hughes, 1992). Seven years into construction the harbour and railway were fully operational (Fig. 8). Operation in the following years furthered British interests in the WIPGR and offered little advantage to Goa itself.

The treaty that led to the origin of the railway contained other clauses with deleterious effects on the economy of Goa—and advantages to British economic interests. One such clause involved limitations imposed on salt production. Unsurprisingly, there was a drop of 67% in this activity, creating thousands of unemployed. The

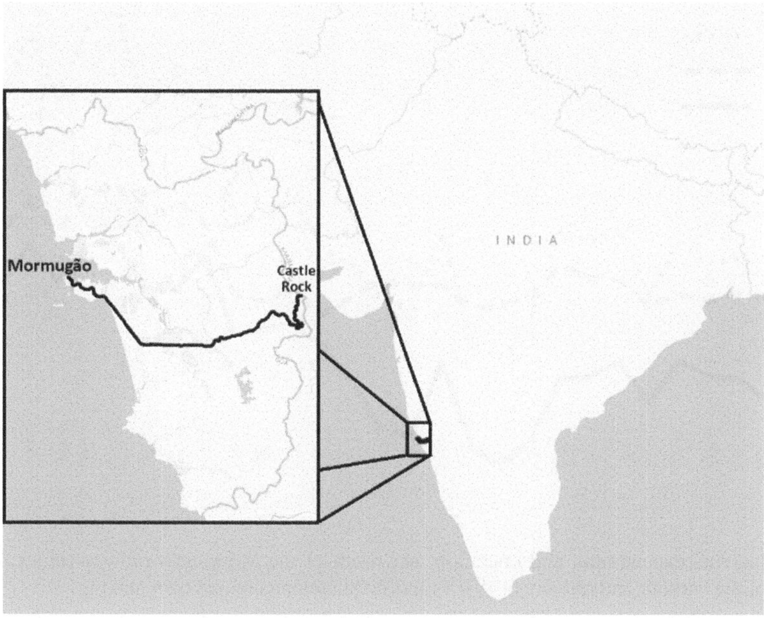

Fig. 8 Location of the Mormugão railway on the Indian subcontinent (*Source* Sharemap.org and own making)

availability of salt in the Goa market decreased as well. When the treaty ceased to have effect in 1890, production was not resumed, for many salterns were then abandoned, while others had been converted to rice cultivation. A similar situation (with similar effects) occurred with the production of liquor from palm trees. Goa, which until then was an exporter of liquor, now had to import it from neighbouring British India so as to cover its own internal market's needs. Production of opium was also terminated; however, this was a residual activity in Goa. In any case, salt, liquor and opium produced in Goa no longer competed with products from British India. Finally, customs revenues dropped, due to the customs union created through the treaty (Albuquerque, 1990).

The operation of the railway was even more harmful to Goa's economy. Thanks to a combination of fares between railway companies in British India, traffic which in normal conditions would flow to Goa was diverted towards Bombay. Hence, the Mormugão line became a *starving railway*, with dire impacts on its profitability. In 1902, the Mormugão railway was leased to a competing company (the Southern Mahratta Railway Company), which slightly improved its economic performance (Fig. 9).

Nonetheless, the guarantee of yield promised by the Portuguese government had to be paid. In some years, the guarantee was paid in full—considering that the operation ran a deficit. By 1904, sixteen years since the beginning of operation, the total amount paid to the WIPGR ascended to over £1,400,000 (€163,000,000)—more than the

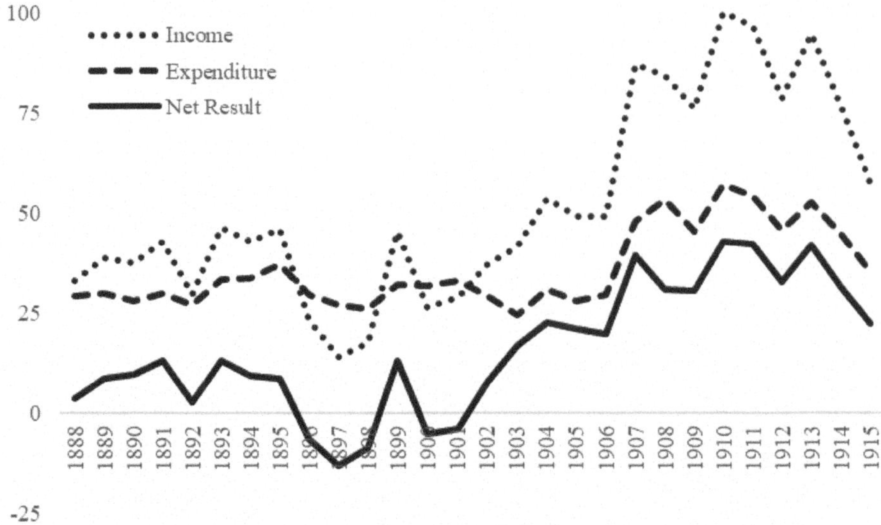

Fig. 9 Income, expenditure, and operation: net result of the Mormugão railway (index 100 = income of the harbour and railway in 1910) (*Source* Sharemap.org and own making)

project's original cost. Payments resumed yearly until 1927, when the net operating profit became higher than the yield guaranteed by the government. Part of this money came from the Goa treasury—however, for the most part, it came from the national exchequer. In any case, this capital could have been invested in the colony instead of being diverted towards the pockets of WIPGR stockholders. Additionally, in order to pay these sums, local taxes and custom fares were raised—which increased the cost of living and harmed overall economic activity. Consequently, many Goa citizens were forced to emigrate. By 1926, around 120 thousand had left the colony for neighbouring territories (Albuquerque, 1990).

The railway's impact in the economy and colonisation of Goa was limited, but not non-existent. It did improve local mobility, which until then was restricted by the absence of proper means of transportation (Deloche, 1980), and it did promote the appropriation of the local territory. Between 1888 and 1915, an average of 277,000 people/year travelled in the railway—for reference, the population of Goa was about 450,000 people (Srivastava, 1990). Freight figures reveal a steady transportation of salt, cotton, grains and assorted seeds.[2] However, most likely, these were commodities originating in British India which, thanks to the railway, found a quicker and easier route to the Goa market, especially after the line was leased to the Southern Mahratta Railway Company. The railway became very useful to support the exploitation of the manganese deposits since their discovery in the late 1900s to the 1940s (Teixeira, 1950). However, this was the only economic benefit Portugal obtained from its attempt at railway imperialism in the colony.

---

[2] Besides the sources mentioned in note 1, see also Arquivo Histórico Ultramarino, Lisbon (Portugal, 1899), items: 743 1I and 2533 1B.

In political and diplomatic terms, however, the railway acted as a token of progress and modernity brought by Portugal to its Indian colony. It was seen as proof of the ability by the Portuguese administration to *civilise* the territory and population (most pick-and-axe labour was done by locals) and effectively invest as well as modernise its overseas domains in the Indian subcontinent. All the same, again, in the long run the endeavour almost exclusively benefitted British interests in the area: those of WIPGR promoters, who regularly received dividends from their investment; and those of the British government, who was able to eliminate competition from Goa's production. British gentlemanly capitalism thus witnessed another success case.

5 Conclusion

Since the mid-nineteenth century, technology was linked with the imperial domination and exploitation of the African and Asian territories and populations—in a process historically known as the *Scramble for Africa*. Within this historical process, railways assumed an important role, so much so that historians often refer to a specific kind of imperialism promoted by railways: railway imperialism.

Portugal, considering its historical traditions and extension of overseas domains, also wished to take a part in the *Scramble*. However, lacking the financial and industrial resources necessary to build railways in its colonies, it was forced to rely on capital and industry of other nations, which were themselves active players in the process.

Consequently, Portugal was placed in a dual position. On the one hand, as an imperial nation with a vast colonial empire, it used railways to strengthen its imperial dominance on overseas domains. It endeavoured this by forcing the natives into labour so as to build those railways (as part of a self-imposed *civilising mission* in Africa) and by using these natives to populate the African hinterland and exploit its natural resources. In this respect, the general goal was to prove Portugal's colonial aptitudes and legitimacy. On the other hand, the country could not but allow some degree of foreign influence from agents (especially British, working closely with the Colonial Office's imperial agenda) which provided financial capital and industrial means; therefore, Portugal was itself subjected to railway imperialism.

The case of the Mormugão railway is particularly revealing of this ambiguous position—and of how the alliance between private initiative and government favoured railway imperialism and informal empire. It can be seen as a token of the Portuguese attempt to modernise its colonies and legitimise its presence in that part of India. Additionally, the railway also favoured the imperial exploitation of manganese deposits in the first half of the twentieth century. Nonetheless, the line proved a much more profitable opportunity for British agents involved in its construction and operation than for anyone else. Investors found an almost riskless undertaking (considering that a yield was guaranteed by the Portuguese government); industrialists and experts found a market for their rails, rolling stock and expertise; British interests in the Government of India managed to restrict the production of salt and liquor which competed in the markets of districts neighbouring Goa.

To conclude, I argue that Portugal was both an agent and a victim of railway imperialism. Enthralled by the sublime associated with railways and by the need to present itself as a modern nation, Portugal placed itself on the hands of British agents almost without scrutiny. In the end, it imposed formal empire in its colonies (even though the impact of colonial railways fell short of the expectations touted by their promoters) but accepted a large degree of informal empire from British colonial interests.

References

Adas, M. (1989). *Machines as the measure of men. Science, technology, and ideologies of western dominance*. Cornell University Press.

Akita, S. (Ed.). (2002). *Gentlemanly capitalism, imperialism and global history*. Palgrave Macmillan.

Albuquerque, T. (1990). The Anglo-Portuguese Treaty of 1878: its impact on the people of Goa. *Indica, 27*(2), 117–124.

Bouene, F., & Santos, M. (2006). O modus vivendi entre Moçambique e o Transvaal (1901–1909). Um caso de "imperialismo ferroviário" [The modus vivendi between Mozambique and Transvaal (1901–1909). A case of "railway imperialism"]. *Africana Studia, 9*, 239–269.

Castryck, G. (2015). Introduction—From railway juncture to portal of globalization: Making globalization work in African and South African railway towns. *Comparativ, 25*(4), 7–16.

Clarence-Smith, G. (1985). *The third Portuguese empire, 1825–1975. A study in economic imperialism*. Manchester University Press.

Davis, C. B., Wilburn, K. E., & Robinson, R. E. (Eds.). (1991). *Railway imperialism*. Greenwood Press.

Deloche, J. (1980). *La circulation en Inde avant la révolution des transports* [Circulation in India before the transport revolution]. École Française d'Extrême Orient.

Diogo, M. P., & Laak, D. v. (2016). *Europeans globalizing. Mapping, exploiting, exchanging*. Palgrave Macmillan.

Diogo, M. P., & Navarro, B. J. (2018). Re-Designing Africa: Railways and globalization in the era of the new imperialism. In D. Pretel & L. Camprubí (Eds.), *Technology and globalisation. Networks of experts in world history* (pp. 105–128). Palgrave Macmillan.

Divall, C. (2003). Railway imperialisms, railway nationalisms. In M. Burri, K. T. Elsasser & D. Gugerli (Eds.), *Die Internationalität der Eisenbahn 1850–1970* [The internationality of railways] (pp. 195–209). Chronos.

Duarte, A. P., & Andrade, J. S. (2012). How the Gold Standard functioned in Portugal: An analysis of some macroeconomic aspects. *Applied Economics, 44*(5), 617–629. https://doi.org/10.1080/00036846.2010.513675

Fowler, C. S., & McMurdo, E. (1887). *Views of Lourenço Marques (Delagoa Bay) and Transvaal Railway*. S. N.

Headrick, D. R. (1981). *The tools of empire. Technology and European imperialism in the nineteenth century*. Oxford University Press.

Hughes, H. (1992). *Indian locomotives: Metre gauge 1872–1940*. The Continental Railway Circle.

Jerónimo, M. B. (2015). *The 'civilising mission' of Portuguese colonialism, 1879–1930*. Palgrave Macmillan.

Kärrholm, M. (2012). *Retailising space. Architecture, retail and the territorialisation of public space*. Ashgate.

Kerr, I. J. (2007). *Engines of change. The railroads that made India*. Praeger.

Kerr, I. J., & Pereira, H. S. (2012). India and Portugal: The Mormugão and the Tua railway compared. In A. McCants, E. Beira, J. M. L. Cordeiro, & P. B. Lourenço (Eds.), *Railroads in historical context: Construction, costs and consequences* (pp. 167–196). MIT Portugal Program, Universidade do Minho & EDP.

Lee, R. (1989). *France and the exploitation of China 1885–1901: A study in economic imperialism*. Oxford University Press.

Lopes, T. S., & Simões, V. C. (2017). Foreign investment in Portugal and knowledge spillovers: From the Methuen Treaty to the 21st century. *Business History, 62*(7), 1079–1106. https://doi.org/10.1080/00076791.2017.1386177

Matos, A. C. (2009). Asserting the Portuguese civil engineering identity: The role played by the École des Ponts et Chaussées, 1825–1866. In A. C. Matos, M. P. Diogo, I. Gouzévitch, & A. Grelon (Eds.), *The quest for a professional identity: Engineers between training and action* (177–208). Colibri.

Navarro, B. J. (2018). *Um império projectado pelo "silvo da locomotiva": O papel da engenharia portuguesa na apropriação do espaço colonial africano. Angola e Moçambique (1869–1930)* [An empire projected by the "whistle of the locomotive": The role of Portuguese engineering in the appropriation of the African colonial space]. Colibri.

Officer, L. H., & Williamson, S. (2020, July 27). *Computing 'real value' Over time with a conversion between U.K. pounds and U.S. dollars, 1791 to present*. http://www.measuringworth.com/calculators/exchange/result_exchange.php

Pereira, H. S. (2018a). O caminho de ferro de Moçâmedes: entre projeto militar, instrumento tecnodiplomático e ferramenta de apropriação colonial (1881–1914) [The Moçâmedes railway: Between militar project, technodiplomatic tool and instrument of colonial appropriation]. *Revista de História da Sociedade e da Cultura, 18*, 157–183. https://doi.org/10.14195/1645-2259_18_8

Pereira, H. S. (2018b). Railways as portals of globalisation: The case of the Portuguese mainland and colonial rail networks (1850–1915). *Comparativ, 28*(5), 121–138.

Pereira H. S. (2019a). O caminho de ferro da Beira em Moçambique (1890–1914): entre antagonismo tecnodiplomático e simbiose económica [The Beira railway in Mozambique (1890–1914): Between technodiplomatic antagonism and economic symbiosis]. *Análise Social, 233*(4), 694–724.

Pereira, H. S. (2019b). The Ambaca railway in Angola: history of a failed public-private partnership (1885–1914 and briefly onwards). *Revista de Historia Industrial, 28*(77), 52–91. https://doi.org/10.1344/rhi.v28i77.28537

Pereira, H. S. (2021). Railway imperialism revisited: The failed line from Macao to Guangzhou. *Technology and Culture, 62*(1), 82–104. https://doi.org/10.1353/tech.2021.0003.

Pinheiro M. (1988). A construção dos caminhos-de-ferro e a encomenda de produtos industriais em Portugal (1855–90) [Railway construction and the orders of industrial products in Portugal (1855–90)]. *Análise Social 24*(101–102), 745–767.

Portugal, Ministério da Marinha e Ultramar. (1899). *Album de estatistica graphica dos caminhos de ferro portuguezes das provincias ultramarinas 1898* [Album of graphical statistics of Portuguese railways in the overseas provinces]. Companhia Nacional.

Portugal, Ministério das Colónias. (1912). *Estatística dos Caminhos de Ferro das Colonias portuguesas de 1888 a 1910. Documentos principais e gráficos* [Statistics of railways of the Portuguese colonies from 1888 to 1910. Main documents and charts]. Imprensa Nacional.

Portugal, Ministério das Colónias. (1917). *Estatística dos Caminhos de Ferro das Colonias portuguesas de 1888 a 1915. Documentos principais e gráficos* [Statistics of railways of the Portuguese colonies from 1888 to 1915. Main documents and charts]. Imprensa Nacional.

Ryan, J. R. (1997). *Picturing Empire. Photography and the visualization of the British Empire*. The University of Chicago Press.

Srivastava, H. C. (1990). Demographic history and human resources. In T. R. d. Souza (Ed.), *Goa through the ages* (pp. 55–77). Concept Publishing Company.

Teixeira, C. (1950). *Notas sobre a Geologia da Índia Portugueza* [Notes regarding the geology of Portuguese India]. Imprensa Moderna.
Valério, N. (2001). *Estatísticas Históricas Portuguesas* [Portuguese historical statistics]. Instituto Nacional de Estatística.
Vleuten, E. v. d. (2006). Understanding network societies. Two decades of large technical system studies. In E. v. d. Vleuten, & A. Kaisjer (Eds.), *Networking Europe. Transnational infrastructures and the shaping of Europe, 1850–2000* (pp. 279–314). Science History Publications.

Railway Transportation: Economy, Urbanization and Develop-ment

Impact of Metro-Rail Projects on Land Use and Land Value in Indian Cities—The Case of Chennai

Madhu Bharti and Pavithra Velechettiar Bhaskaran

1 Introduction

In recent years, urban India has been experiencing tremendous growth due to the mass movement of people in search of jobs and livelihood opportunities (Jaysawal & Sahu, 2014). It is observed that after Indian independence in 1947, the urban population in India increased at a quick pace. It had reached 377 million by 2011 (accounting for 32% of the total population (Census of India, 2011)). As migration to cities to find employment continues, the number of cities and towns keeps growing. As per Census 2011, India has 53 metro cities (the nation counted only four at the time of independence). About 40% of the urban population lives in these 53 metro cities, out of which the largest 10 cities contain about 24% of the urban population. This places a great burden on urban infrastructure and leads to problems linked to haphazard growth such as congestion, environmental degradation, urban sprawl (Doi, 2002). Growth trends in the real estate market correlate with many variables, accessibility to transport systems being one of them (Pojani & Stead, 2015). Home buyers are concerned about the commute distance to work while businessmen want to invest in more accessible locations—thus land close to transport systems is in high demand (CABE, 2006). This increase in demand leads to the transition of land use from residential to commercial and from low intensity to high-intensity use and appreciation of land values (Suzuki et al., 2015).

To address the issues of mass mobility, many cities have encouraged and implemented public transport systems (Montgomery, 2017). Public transport plays a vital role in controlling/directing spatial growth and yielding economic benefits for both citizens and the city (Stjernborg & Mattisson, 2016). Reduction in travel time, cut

M. Bharti (✉)
Faculty of Planning, CEPT University, Ahmedabad, India

P. V. Bhaskaran
Meenakashi College of Engineering, Chennai, India

© The Author(s), under exclusive license to Springer Nature Switzerland AG 2021
S. Mitra et al. (eds.), *Railway Transportation in South Asia*, Contemporary South Asian Studies, https://doi.org/10.1007/978-3-030-76878-2_4

in fuel use and increase in accessibility constitute direct benefits. Indirect benefits include enhancement of land value, land use changes, densification, environmental benefits, etc. (Litman, 2015). These indirect benefits are often uncountable and generally do not reach the government (UNDESA et al., 2012).

In India, there has been a growing interest among policymakers regarding the relevance of rail-based systems. While evaluating different mass transit options for Indian cities, metro-rail systems are often given preference due to the belief that road-based bus systems cannot suitably cater to capacity requirements. In addition to this, metro-rails are perceived to have higher levels of comfort, speed and efficiency than bus systems, making them more attractive to both policymakers and potential users (Goel & Tiwari, 2014). Promoters of metro-rail systems often claim that one of the benefits of the metro-rail is reduced congestion due to the users' shift from road-based motorized modes to metro-rail systems. This mode shift is then claimed to result in reduced air pollution and road accidents. However, the experience of metro-rails in low- and middle-income countries around the world shows a different picture (Goel & Tiwari, 2014).

In the proposal for the twelfth five-year plan, the Planning Commission recommended that all Indian cities with a population of over 2 million start planning rail transit projects, and cities with a population of over 3 million start constructing metro-rails (Planning Commission of India, 2011). The capital cost of construction is 20–30 times that of the Bus Rapid Transit system—depending on whether the metro-rail systems are underground or elevated (ITA Working Group, 2004). According to the Union Minister of State for Housing and Urban Affairs (Ministry of Housing and Urban Affairs, Govt. of India, [MoHUA]) Hardeep S. Puri, over 664 km of metro-rail projects in 15 cities are presently under various stages of implementation, while over 515 km of metro-rail lines are already operational in India. The central government provides financial support for metro-rail projects in the form of a grant to states—up to 10% of the project cost, in the form of 50:50 equity sharing with state governments, or through a viability gap funding to the extent of 20% of the capital cost of public transport projects under the PPP model (Live Mint, 2018).

There is evidence that investment in public transport tends to escalate the rental value of the surrounding properties. Evidence from around the world illustrates how improvement in accessibility has led to an escalation in land values, and how this affects the real estate sector of a city (Zhang et al., 2016). According to studies undertaken in 2007–2008 (Swamy, 2008) with respect to residential and commercial areas, on an average, the land value within 500 m of a metro-rail line increased by 11.3–18.1%. Moreover, land value changes are more consistent and are higher after a metro-rail becomes operational as compared to the case during the construction and planning stages—they increase by 2–4% every year. It was observed that the increase in land value is highly dependent on the income of those occupying the area and whether the area enjoys planned development or not. Another study by Magicbricks.com in 2012 found that some areas in North Delhi witnessed an average appreciation of 30% in capital value after the advent of metro-rail services (Goel & Tiwari, 2014). Similarly, appreciation (25–30%) was also observed in property values in West Delhi. Metro-rail implementation has had a huge impact on real estate prices, both along its corridor and in its influence zone. In the larger picture, it improves the standard of living of a large segment of the urban population and also constitutes a catalyst for sustainable development across large urban areas. This increment in land value and change in land use varies from place to place, at various time frames. However, in India, metro-rails are a relatively recent infrastructure investment; sufficient evidence from cities other than Delhi is not available.

2 Objectives

This chapter aims to assess the impact of metro-rail projects on land use and land value in Chennai. This is achieved by studying the development of metro-rail projects accompanying the growth of the city and by analysing the spatial–temporal changes in land use and land value in the areas around the metro-rail nodes. This study tries to build a relation between the development of metro-rails and proximate land value gains.

From the literature studied, it is evident that the impact of transit services on land values depends on various parameters: the distance of the node from the city's CBD (Deng & Nelson, 2010), the ridership of the transit service in that particular node (Andersson et al., 2010) and the populational density around the node (Cervero & Dai, 2014). These three parameters were considered when selecting the nodes for the study. The nodes on the Chennai metro-rail network Phase 1, Corridor 1 were categorized based on these parameters and the selected nodes were studied to find the spatial–temporal change in land use and land value. From all the nodes along the corridor, four nodes were selected for detailed study.

To analyse the change in land use through time, existing land uses (ELU) before metro-rail projects (2006) were compared with present land uses (based on a survey in 2019). While analysing the change in land value through time, land values during various phases of metro-rail development (before project formulation, during construction and post-construction) were compared. The land use and land value of plots within various buffer zones of the metro-rail (250, 500, 750 and 1000 m) were analysed to establish a relationship between the impact of metro-rails on land use and land value and the distance from the nodes.

3 Literature Review

The integration of land use and transportation is a dynamic process, resulting in spatial and temporal changes (Aljoufie et al., 2011). The development of transportation improves accessibility, which, in turn stimulates changes in land use pattern and escalates the land/property/rental values in surrounding areas (Roukouni et al., 2012). The city's real estate sector responds positively to these improvements, mainly in the form of higher property prices, zoning permissions, densification, etc. (Global

Transport Knowledge Practice, 2018). These are short-term changes; land use change occurs in the longer term. The areas which previously had poor infrastructure will benefit more than those already enjoying good infrastructure (Cervero, 1992).

Many similar studies have been undertaken and it has been found that investment in transport accelerates development and generates economic benefits. At the same time, many empirical studies reveal that, in practice, outcomes are complicated by many external factors other than accessibility (Giuliano, 2004). Transport services have an impact on land value and density in their influence areas and this impact decreases with an increase in distance. The study by Robert Cervero for Santa Clara, California, through the hedonic pricing method, found that 45% of land value premium is observed within a quarter-mile (Cervero, 1992). The commuter rail service had an impact on the provision of affordable housing and an increase in density as well (Boarnet et al., 2017). In Buffalo, New York, residential properties located within half a mile radius of the light rail transit system are valued 2.31 USD higher for every foot closer to the node (Hess & Almeida, 2007). A study of the BRT system in Bogota, Columbia found appreciable land value benefits. For every five minutes of walking closer to a BRTS node, property prices increased by 0.12% (Perdomo, 2011).

In some cases, the socio-economic character of the area around the transit node plays an important role (Pojani & Stead, 2015). For example, a study concerning the city of Atlanta (Nelson, 1992) found that transit proximity increased property value in low-income parts of the city but also decreased values in high-income parts of the city (Lambert, 2009). In Miami, the metro-rail study found that accessibility to transit nodes had no or very little impact on housing values due to low ridership (Gatziaff & Smith, 1993). Andersson et al. (2010) found that high-speed transit systems in Taiwan had little or no impact on property values due to high fares. The monthly commute cost is 70% of the median monthly wage in Taiwan (Andersson et al., 2010).

Although enhanced accessibility is seen to increase property values along transit corridors, the study of the Beijing BRTS suggests that peripheral areas which previously lacked alternative mobility opportunities of a Mass Transit System have experienced greater change due to a larger marginal increase in urban mobility (Deng & Nelson, 2010).

In San Francisco, the Bay Area Rapid Transit (BART) found considerable variation in land price impacts. Downtown commercial properties reaped high gains, while many suburban residential settings experienced no discernible impact due to restrictions and regulations (Cervero & Landis, 1997).

When it comes to land use, non-residential uses are seen near the transit node as was shown in a study of seven metro-rail nodes of the Athens metro-rail line. In a buffer of 250 m, an increase in retail activity was observed at the non-domestic properties; other positive results involved the growth in pedestrian flow and enhancement of employees' mobility, as well as a general upgrade of the area's environment (Roukouni et al., 2012).

Though these literature studies originate from various economic, cultural and political backgrounds and are undertaken through different methods, a few common conclusions can be derived. Accessibility, transport investments and linkages are the key factors influencing increments in land value (Cervero, 1992). The impact is different in different neighbourhoods within the same city. Peripherical areas which previously lacked connectivity witness a greater impact than core city areas. The impact is also dependent on the locality's income group as well as transit fares (Zhong & Li, 2016). Positive impacts are inversely proportional to the distance from the node (Cervero, 1992). Non-residential uses are found nearer to the node but they diminish as the distance from the node increases. Apart from accessibility, other external factors such as policies, regulations, new developments affect the levels of impact due to transit investments (Cervero & Landis, 1997).

4 City Profile

Chennai is the capital city of the state of Tamil Nadu (TN), India and is the country's fourth metropolis (Fig. 1). With a population of 8 million, it is the sixth most populous city in India (Census of India, 2011). It contains the 36th largest urban population in the world. The city and its adjoining region are known as the Chennai Metropolitan Area (CMA). Chennai has a diverse economic base, with leading automobile and assembly industries and a high IT (Information Technology) sector growth after the dotcom boom (National Informatics Centre, 2017). This economic centre attracts both skilled and unskilled migrants from both inside and outside the state (Jeyaranjan, 2017).

Economic growth, employment opportunities and rapid urbanization have bought a steady increase in car ownership, which stood at 29.08 million in 2018 (SESEI, 2019). Chennai suffers from road congestion, air and noise pollution and increased trip duration (IIHS, 2015).

A Comprehensive Traffic and Transportation (CTT) study indicated that the city needed high capacity metro-rail systems (Sekar & Karuppannan, 2012). Metro-rail was held to provide safe, fast, reliable, accessible, convenient, comfortable, efficient and affordable public transport services for all, in a sustainable manner (Chennai Metro Rail Limited [CMRL], 2019). The Delhi Metro Rail Corporation (DMRC) prepared a feasibility study for the project in 2003, to be constructed in two phases. The construction of Phase I started in 2009 and was completed in 2019 with 45 km running across the city. Phase 2 (108 km) of the metro-rail project is expected to become operational from 2029 (CMRL, 2019).

5 Analysis and Discussion

The Phase I-Corridor I originates from the Central Business District (CBD) of Chennai's central railway station which, with a total length of 22 km, is the oldest part of the metro-rail network (CMRL, 2019). The Phase I-Corridor I runs from

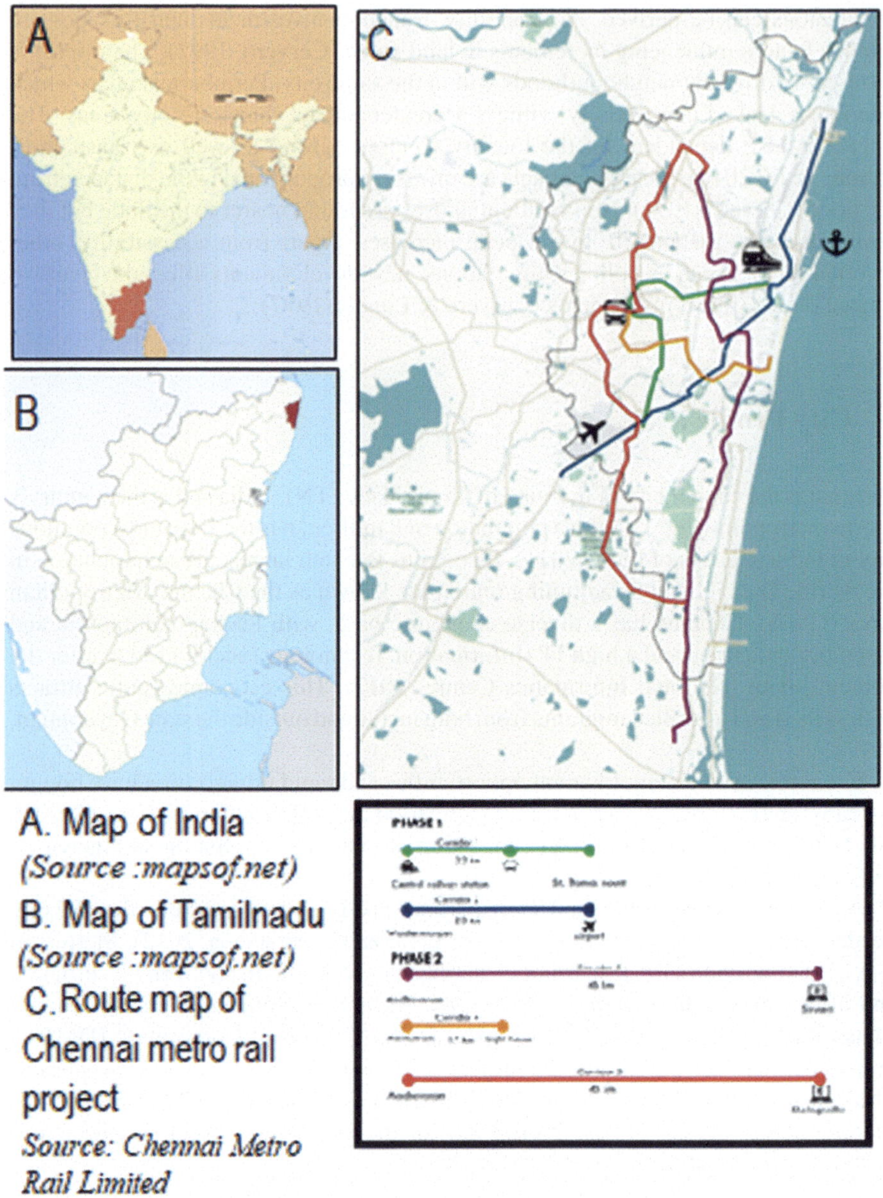

Fig. 1 Location map of the study area (*Source* CMRL [2019])

Table 1 Categorizing nodes based on their distance from the CBD

< 4 km	4 km–8 km	8 km–12 km	> 12 km
Central	Kilpauk	Koyambedu	Ekkaduthangal
Egmore	Pachayappa's College	CMBT	Alandur
Nehru Park	Shenoy Nagar	Arumbakkam	St. Thomas Mount
	Anna Nagar East	Vadapalani	
	Anna Nagar Tower		

(*Source* CMRL (2019))

Chennai Central Railway station to St. Thomas Mount. On this corridor, there are 17 nodes, at an average distance of 1.29 km from each other (CMRL, 2019). Parameters such as the distance from the CBD (Deng & Nelson, 2010), average ridership/day at each node (Andersson et al., 2010) and building density (Cervero & Dai, 2014) were taken into account. From all nodes along the corridor, four nodes were selected for detailed study. The selected nodes were Egmore, Anna Nagar Tower, Vadapalani and Ekkaduthangal.

5.1 Distance from the CBD

See Table 1.

5.2 Ridership (Average Riders/Day)

See Table 2.

Table 2 Categorizing nodes based on ridership

1200	1200–2400	2400–3600	> 3600
Kilpauk	Anna Nagar Tower	Alandur	Thirumangalam
Pachayappa's College	Nehru Park	CMBT	
Arumbakkam	St. Thomas Mount	Koyambedu	
Egmore		Vadapalani	
Shenoy Nagar		Central	
Ekkaduthangal			

(*Source* CMRL (2019))

5.3 Building Density

See Table 3.

In order to obtain an unbiased result, the most repeated node from each column was selected for detailed study. For example, Egmore node repeats itself three times in the first column and hence was selected. Likewise, Anna Nagar Tower, Vadapalani and Ekkaduthangal were selected. The selected nodes are highlighted in red in the route map of Chennai metro-rail (Fig. 2).

Egmore is an institutional area with an intra-state railways node, a government hospital, a museum, a sports complex, etc. This node is very close to the Central Railway node, where a plaza with a multi-modal transit hub is under construction. Anna Nagar is a high-income group residential area with independent villas and new apartment buildings. The land value here is one of the highest in the city. Vadapalani is a residential cum commercial area. The land value of the locality has increased greatly with the opening of a new mall, Forum. Ekkatuthangal has seen a shift from industrial to IT-based development after 2006.

6 Impact on Land Use

The Chennai Metropolitan Development Authority (CMDA) had prepared an existing land use plan in 2008 during the preparation of the master plan of 2026. This land use map is compared with present land use (Primary Survey, 2019) of the plots within 1000 m from the node so as to infer the impact of the metro-rail on land use.

Figure 3 shows the change in land use for Node 1: Egmore. It was observed that there was a shift from residential to commercial use and an annual increase of 7% in commercial use within this node. There was not much change within 500 m of the node due to the institutional use of the land. New commercial developments could be seen along the corridor of the metro-rail node.

Table 3 Categorizing nodes based on building density

< 0.28	0.28–0.34	0.34–0.40	> 0.40
Egmore	Nehru Park	Thirumangalam	Vadapalani
Koyambedu	Pachayappa's college	Arumbakkam	
CMBT	St. Thomas Mount	St. Thomas Mount	
Kilpauk	Anna Nagar Tower	Shenoy Nagar	
Alandur			
Central			

(*Source* CMRL (2019))

Fig. 2 Corridor I with selected nodes (*Source* CMRL (2019))

Figure 4 illustrates the change in land use for Node 2: Anna Nagar Tower. With a 30% increase in commercial use, this node experiences maximum impact. The major increase of commercial land use can be observed along the metro-rail corridor, especially in the west due to the influence zone of the Thirumangalam node. The impact cannot be compared on the basis of distance because the nodes are so close to each other that there is an overlap in the influence zones.

A similar study was undertaken for the other two nodes (Vadapalani and Ekkatuthangal). It was found that there was an increase in commercial activity near the nodes. In Vadapalani, there was only a 3% annual increase in commercial use; however, a certain amount of previously locked land (under-utilized land) was converted into a high-rise residential apartment building (Trellis, AVM Asta) and a mall (Forum Mall). Ekkaduthangal had undergone a huge change along the metro-rail corridor. With the establishment of IT industries, new corporate and IT-based offices started developing in this node. Industrial use was converted into residential use. With all these changes in land use, the character of the neighbourhood was transformed from an industrial area into the city's IT hub (Figs. 5 and 6).

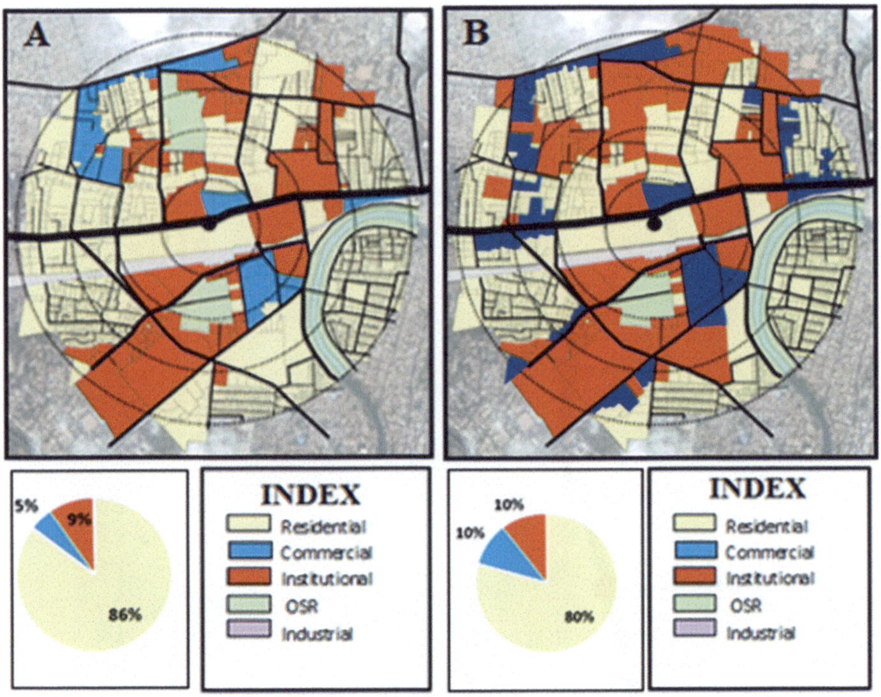

Fig. 3 Land use of Egmore in 2006 (**A**) and 2019 (**B**) (*Source* **a** Existing land use map, 2006, CMDA and **b** ELU primary survey, February 2019)

To summarize the analysis: land use along the metro-rail corridor was found to have witnessed an evident change through time. The connectivity provided by the metro-rail has generated high demand for commercial land use along the corridor and has resulted in land use conversion from residential to commercial.

Figure 7 shows the comparison of land use before and after the metro-rail. An overall annual commercial increase of 7% can be observed, with a 0.4% annual decrease in residential use and 3.5% annual decrease in industrial use.

7 Impact on Land Value

Investment in transport infrastructure within an area leads to improved citizen access. These accessibility benefits can be observed by the increment of land value in monetary terms (Tari et al., 2015). In order to study the impact of metro-rail on land values in Chennai, two approaches were adopted. The first approach consists of analysing the impact of the metro-rail as a function of time; the second consists of analysing its impact as a function of distance. The guideline value from the state's registration

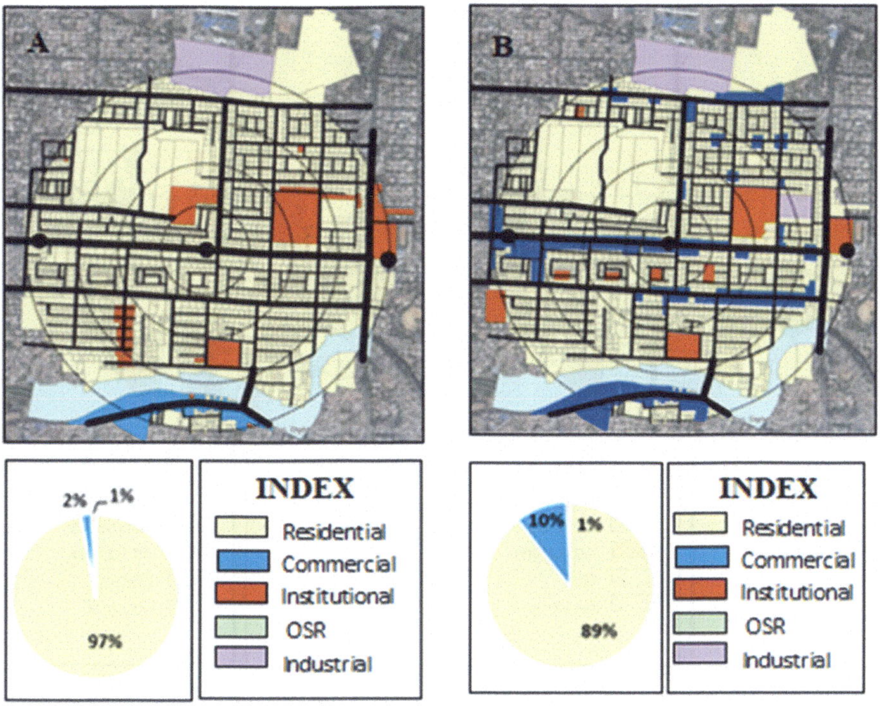

Fig. 4 Land use of Anna Nagar in 2006 (**A**) and 2019 (**B**) (*Source* **a** Existing land use map, 2006, CMDA and **b** ELU primary survey, February 2019)

department portal was verified with the market value and was considered as the land value for the purposes of this study.

7.1 Land Value Increment as a Function of Time

In order to analyse the impact of the Chennai metro-rail as a function of time, guideline values for the period between 2002 and 2019 were collected. The gazettes provide street/survey number wise guideline values. As per the available guideline values, the base year was considered as 2002. The areas within a buffer of 1000 m were chosen as study area following the example of similar studies.

The detailed project report for the Chennai metro-rail was prepared in 2007. Construction started in 2009, followed by operation between 2015 and 2018. In order to study the metro-rail's impact, the land values corresponding to the phase of the metro-rail were selected. The guideline values were separated into 2002–2003 (before metro-rail), 2007–2012 (project formulation), 2012–2017 (construction) and

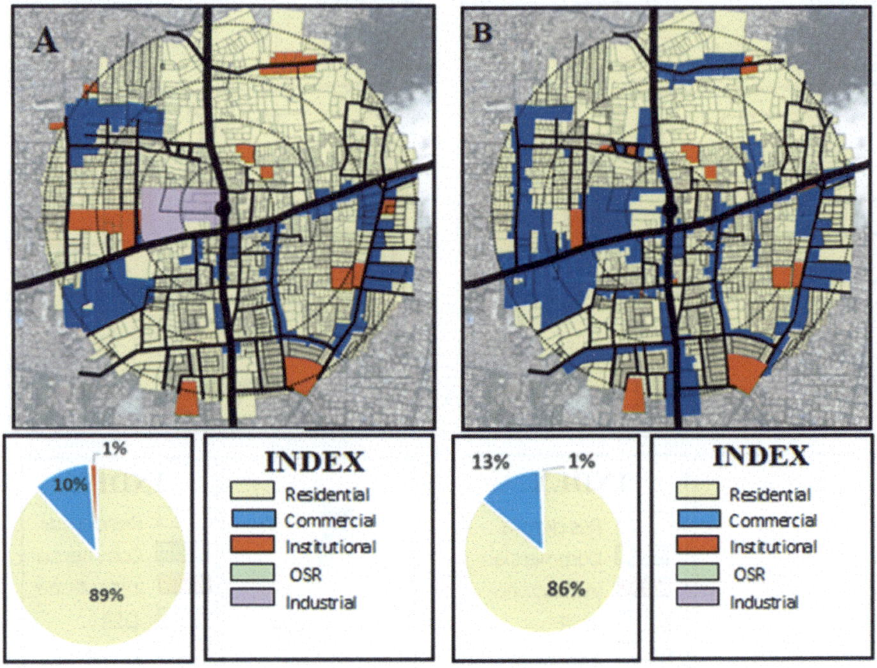

Fig. 5 Land use of Anna Vadapalani in 2006 (**A**) and 2019 (**B**) (*Source* **a** Existing land use map, 2006, CMDA and **b** ELU primary survey, February 2019)

2017 onwards (operation). The change in land values during different time frames was plotted for different study nodes for analysis.

7.1.1 Node 1: Egmore

During the project formulation phase (2003–2007), there was an annual increment of 13% in land values. Growth dropped to 7% during the construction phase (2007–2015). The plots within the 500 m buffer zone witnessed only 3–4% growth. Growth escalated to 11% during the operation phase. Although this node observes much less ridership, the proximity to the main transit hubs of the city (2 km from the node) adds to the accessibility benefits of the plots. These findings are represented in the form map below (Fig. 8):

7.1.2 Node II: Anna Nagar

In the Anna Nagar Tower, an annual increment of 7% can be observed during the project formulation phase (2003–2007). After that, there was much lower change

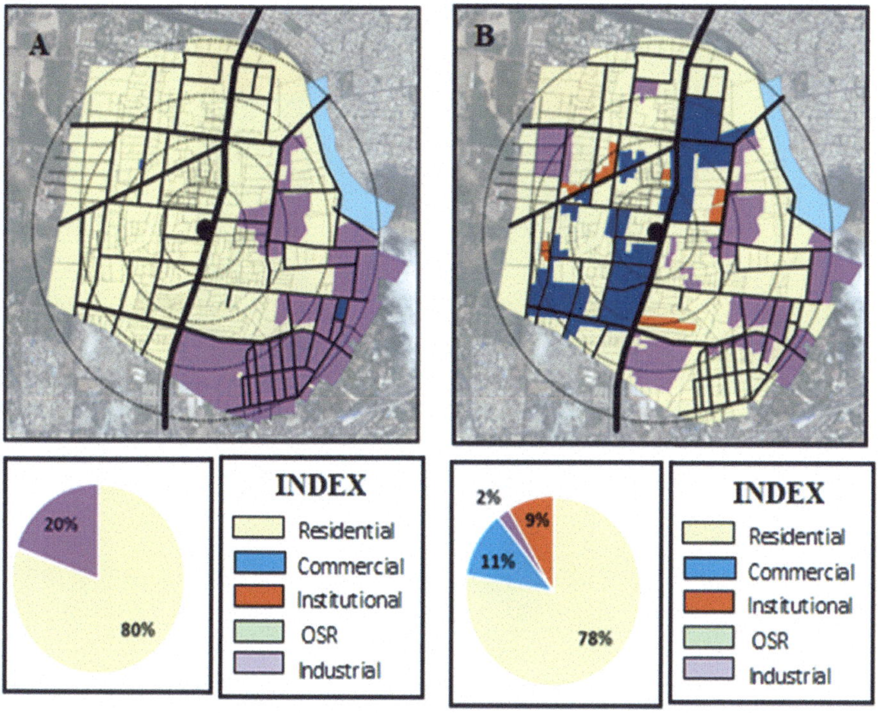

Fig. 6 Land use of Anna Ekkaduthangal in 2006 (**a**) and 2019 (**b**) (*Source* **a** Existing land use map, 2006, CMDA and **b** ELU primary survey, February 2019)

in land values. The construction phase shows an increase of only 0.4%, while after the operation, there was a 4% increase in land value. Although the site has many commercial developments and a fair amount of ridership, the metro-rail did not have much impact on land value. According to the market study, Anna Nagar is a high-income group locality; land values here are one of the highest in the city—leading to saturation of land value in the land market (Fig. 9).

7.1.3 Nodes III & IV: Vadapalani and Ekkaduthangal

Similar studies were carried out for the other two nodes (Vadapalani and Ekkaduthangal) and the results were more or less on the same pattern as the previous nodes. In Node 3: Vadapalani, during the initial project formulation phase (2002–2007), there was an annual increment of 7% in land value. The major increase was observed within the 500 m buffer zone. Growth further increased during the project construction phase (2007–2015) to 18%. This increment was due to the opening of Forum mall located in the 250 m buffer zone. The mall is considered as an external factor for the increase in land value.

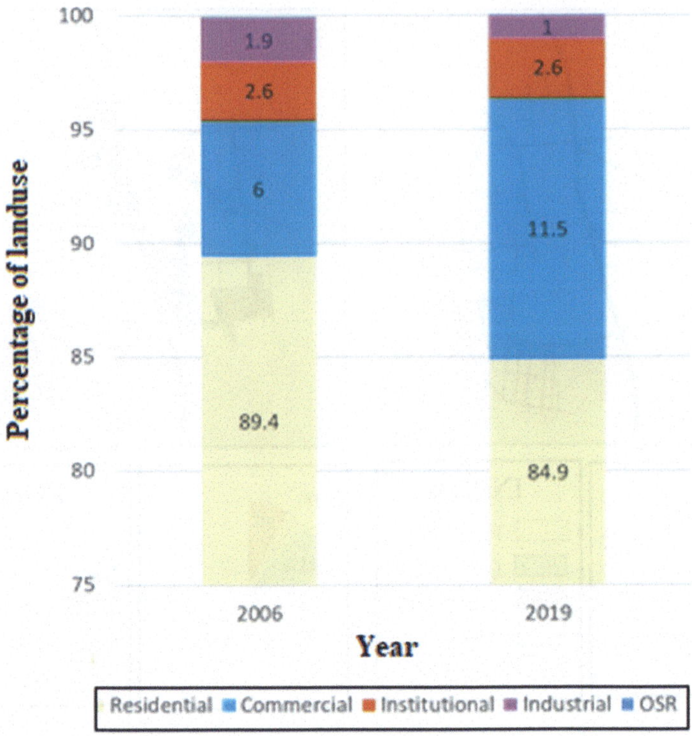

Fig. 7 Comparison of land use in 2006 and 2019 for the selected nodes (Egmore, Anna Nagar Tower, Vadapalani, Ekkaduthangal) (*Source* **a** Existing land use map, 2006, CMDA and **b** LU primary survey, February 2019)

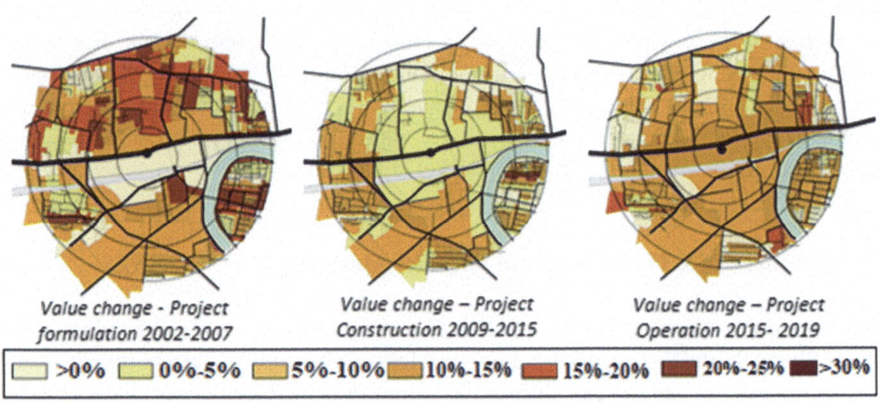

Fig. 8 Land value increment under different stages in Egmore (*Source* Guideline value [Tamil Nadu Registration Department Portal, 2019])

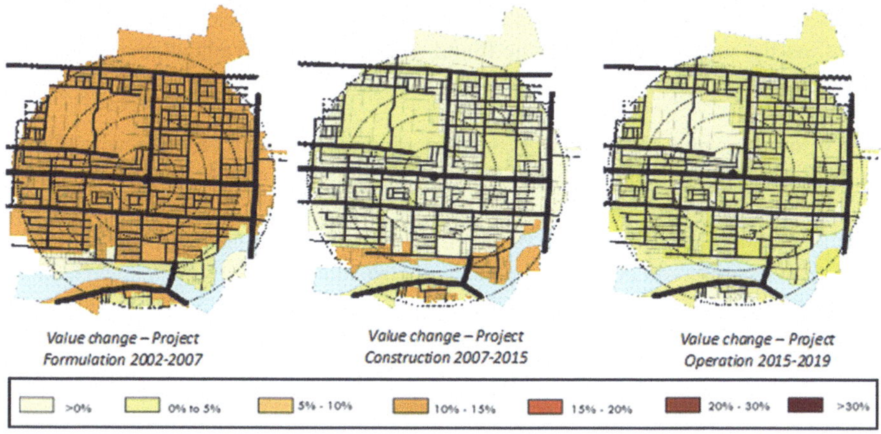

Fig. 9 Land value increment under different stages in Anna Nagar (*Source* Guideline value [Tamil Nadu Registration Department Portal, 2019])

In Node 4: Ekkaduthangal, an annual increment of 12% was observed during the project formulation phase (2002–2007). Growth gradually decreased after that, to 10% during the construction phase (2007–2015). The decrease is evident along the metro-rail corridor. Because the corridor is elevated in this study area, the nuisance caused by the construction can constitute a reason for the decrease in growth rates. The growth rate again increased to 11% after the corridor became operational (Fig. 10).

Accessibility benefits in monetary terms are observed in the form of increased land prices. The four nodes have responded promptly to transportation improvements due to the metro-rail, with an increment in land prices.

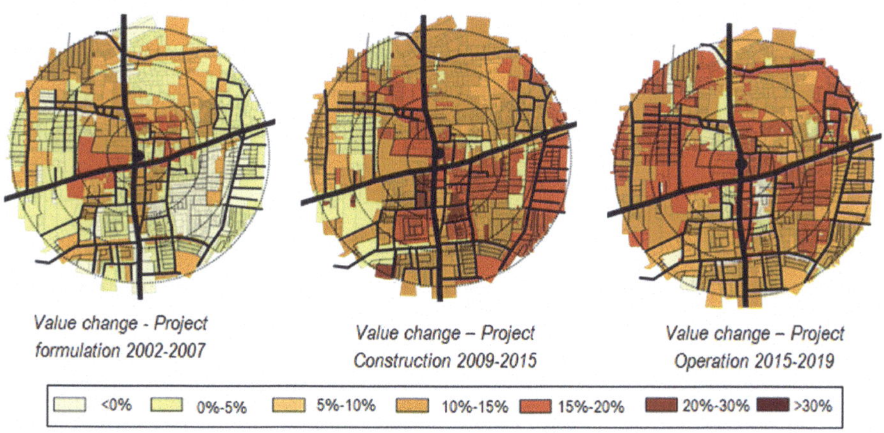

Fig. 10 Land value increment under different stages in Vadapalani (*Source* Guideline value [Tamil Nadu Registration Department Portal, 2019])

This increment is not uniform for all nodes. Anna Nagar Tower reveals a minimal impact, as the area presents saturated land prices, whereas Vadapalani reveals the highest impact, which is probably due to the opening of a new mall near the metro-rail node.

In Egmore, Anna Nagar Tower and Ekkaduthangal, the increment was higher during the project formulation stage (7–13%) and lower during the construction phase (0.4–18%) of the metro-rail. The value further increases in the operation phase (4–11%). The comparative Table 4 explains the changes taking place during the various stages of formation and construction of the metro-rail project.

Table 4 Land value changes in nodes with respect to time

Node/ Project	Formulation stage	Construction stage	Post-construction	Remarks
Egmore	13% increase	7% increase	11% increase	The initial increase in land value leads to the conversion of previously low utilized land to high-rise residential apartments. The proximity to a transport hub (Chennai node) also adds to the value increase
Anna Nagar Tower	7% increase	0.4% increase	4% increase	There is a low impact because Anna Nagar is already a high-income locality where land values are saturated
Vadapalani	7% increase	18% increase	11% increase	The increase during the construction stage is due to an external factor, the opening of Forum mall near the metro-rail node
Ekkaduthangal	12% increase	10% increase	11% increase	The initial increase in land value led to the development of a new IT hub as well as corporate buildings and hotels

(*Source* Guideline value [Tamil Nadu Registration Department Portal, 2019])

7.2 *Land Value as a Function of Distance*

Each node is located in different parts of the city—though all are located on Phase 1 of Corridor 1 of the Chennai metro-rail. Egmore is close to the CBD, whereas the other nodes are farther. The nature of each node is different and, considering that Chennai is a multi-nucleus city, theories of land value and rent are difficult to apply. Land values are different for each node and so is the increment in land values.

Land values are inversely proportional to distance (Deng & Nelson, 2010). This can be observed in the study areas of this research as well. In Egmore, the plots within the 500 m buffer showed greater increments compared to the plots outside. The same can be observed in Vadapalani and Ekkaduthangal as well. There is no such pattern found in Anna Nagar. The increment is more or less similar in all plots. In order to better understand the relationship between land values and distance from the node, land values along the cross-section passing through the metro-rail node were plotted in Fig. 11, showing the east–west section, and in Fig. 12, showing the north–south section.

In the east–west section, it can be seen that land values within the 500 m buffer have increased during the project formulation stage. Land values for the plots outside the 750 m buffer had reduced on the west part of the section. By contrast, the land values of the plots outside the 750 m buffer had increased. This is due to the proximity of those plots to the transport hub (Chennai Central). The section passing north–south reveals that land values within the 500 m buffer were greater. Again, the value reduces as distance increases.

Land values are constant, with a few exceptions. In the east–west section, values dropped near the 750 m buffer due to it constituting a low-influence zone of the metro-rail. Values increased in the plots near the 1000 m buffer line, due to the overlap of the influence zone of adjacent nodes (Anna Nagar East and Thirumangalam).

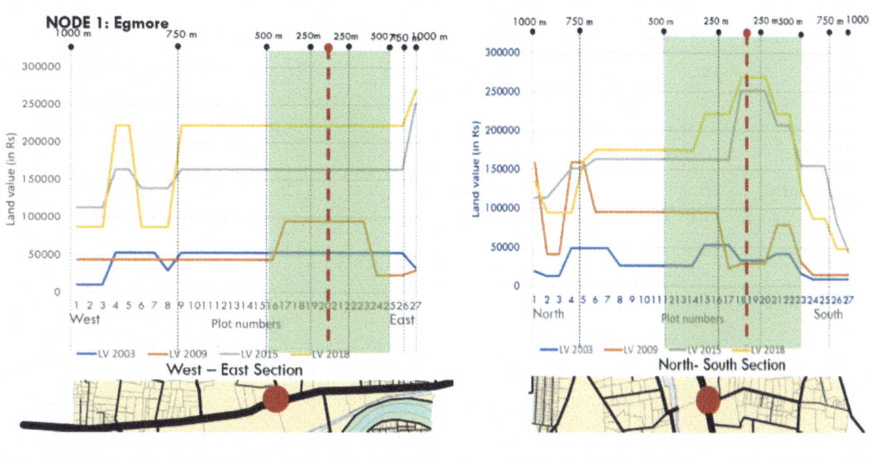

Fig. 11 East–West Section and North–South Section of Egmore Node (*Source* Guideline value [Tamil Nadu Registration Department Portal, 2019])

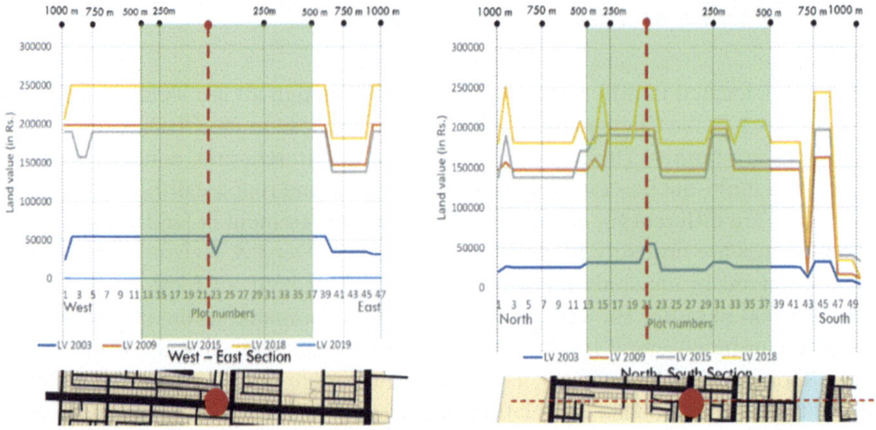

Fig. 12 East–West Section and North–South Section of Anna Nagar Node (*Source* Guideline value [Tamil Nadu Registration Department Portal, 2019])

In the north–south section, land values did not show any pattern concerning distance. The sharp decrease at 750 m in the south is due to the proximity to informal settlements near the Cooum River. Table 5 shows the increase in land value at various distances from the node.

After analysing land value as a function of distance from the metro-rail node, an inverse relationship can be observed in all the three nodes, except Anna Nagar Tower (Fig. 13). At the Anna Nagar Tower, regardless of the distance from the node, a uniform increase in land values is seen from 250 to 1000 m. Egmore has shown a

Table 5 Land value changes respect to distance from the node

Node/Distance from the node	Within 250 m (%)	250 m to 500 m (%)	500 m to 750 m (%)	750 m to 1000 m (%)	Notes
Egmore	15	22	21	10	Impact decreases with an increase in distance
Anna Nagar Tower	16	16	15	15	No change
Vadapalani	26	25	16	11	Impact decreases with an increase in distance
Ekkaduthangal	26	29	17	14	Impact decreases with an increase in distance

(*Source* Guideline value [Tamil Nadu Registration Department Portal, 2019])

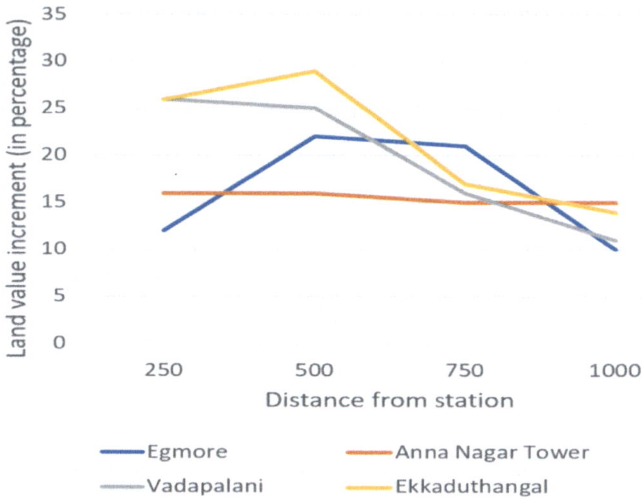

Fig. 13 Relationship between land value change and distance from the node (*Source* Guideline value [Tamil Nadu Registration Department Portal, 2019])

decrease in land values within 250 m due to institutional use (railway) and the value has decreased after a distance of 750 m from the node. Other nodes in this study have shown a decrease as one moves away from the node.

8 Conclusions and Recommendations

8.1 Conclusion

Just like any other transport investment, the Chennai metro-rail has led to benefits in terms of improved accessibility and mobility. These benefits have been directly enjoyed by users in terms of reduced travel times and increase in comfort level. Apart from these direct benefits, the Chennai metro-rail has delivered certain indirect benefits in terms of land development, measured through a change in land use and increment in land value, as summarized below.

Impact on land use

It has been observed that land use changes along the metro-rail corridor are significant. Increased demand for commercial use near the metro-rail node has led to a change from residential to commercial uses. Many previously unused/vacant land parcels have been put to use due to an increase in demand. Though the conversion is inversely proportional to the distance from the node, the said conversion is more evident along the metro-rail corridor. Land use changes within a distance of 500 m are more prominent and the intensity of the impact decreases as we drift away from the node.

Egmore has seen a lesser change because land use is predominately institutional and the land is owned by Indian Railways. Anna Nagar and Ekkaduthangal have seen an over 30% of land use conversion from residential to commercial. There is an overall annual increase of 7% in commercial activity in all the study nodes in the Phase 1-Corridor 1 of Chennai metro-rail.

Impact on land value

Accessibility benefits are reflected in monetary terms and are observed in terms of increased land and property prices (Debrezion et al., 2007). The Chennai real estate market has responded promptly to transportation investment on the metro-rail project with an increase in land value.

Land value increments vary for different nodes at different stages of the project. The impact during the project formulation stage is greater when compared to the construction and operation stage. Among the four study nodes, Vadapalani revealed maximum increase (7–18%), partly due to the upcoming Forum mall within 250 m from the node, while Anna Nagar revealed the least impact due to the saturation of land prices. This increment in land value decreases as we move away from the node. Maximum impacts are seen within 500 m from the node. This is evident in three study nodes—Egmore, Vadapalani and Ekkaduthnagal. However, Anna Nagar has shown stable land values, regardless of its distance from the node, again because it already had high land values. To an extent, the findings from Chennai are in line with the results of similar studies in Bogata, Buffalo, New York and Shanghai: the impact decreases with an increase in distance (Cervero, 1992).

In a similar study for the city of Atlanta by Nelson (1992), it was found that transit proximity increases property values in lower-income areas but decreases the value in higher-income areas. In the case of Chennai, such a pattern is observed in Anna Nagar.

Deng and Nelson (2010) observed that in the case of the Beijing BRT, the impact was greater in the periphery area—which previously lacked connectivity—than in areas that were closer to the CBD. Similarly, in Chennai, Node 4: Ekkaduthangal, the farthest from the CBD, showed the maximum impact. Impacts decrease as the distance from the CBD decreases in all nodes except Node 1: Egmore. The reason for the unusual result is that Egmore is closer to the new multi-modal transit hub in the central node; this external factor thus plays a role.

In the study for San Francisco, it was proved that besides transit and accessibility, there are several external factors such as policies, regulations, new developments and neighbourhood amenities that are jointly responsible for land value escalation (Cervero, 1992). This holds for Chennai too as, unlike other nodes, in Node 3: Vadapalani there has been a huge increment in land value during the construction stage of the metro-rail. The increment is due to the opening of Forum mall very close to the node.

8.2 Recommendations

Implementation of Transit-Oriented Developments

The impact of the metro-rail on land development indicates the need for integration of transport planning in the city's master plans. In many international cities and even Indian cities such as Delhi, Bangalore, Mumbai, Ahmedabad, Kochi and Transit-Oriented Development (TOD) is used as a method for bringing transport and land development together (Agarwal et al., 2014). The National TOD policy for India prepared by the National Urban Development Ministry is in its draft stage (MoUHA, 2011). The policy sets up certain guidelines for sustainable, inclusive, mixed-use and high-density development around the transit nodes.

We recommend the implementation of the National TOD policy's provisions on metro-rail corridors in Chennai. The metro-rail's influence zone should be demarcated and special regulations for high density, mixed-use development should be provided. The influence zone should be marked up to 800 m from the edge of the node (national TOD policy). Local Area Plans (LAPs) for the same would be prepared and executed by the Chennai Metropolitan Development Authority (CMDA).

Provision of additional Floor Space Index (FSI)

In order to meet the increasing demand for land and built space, there is a need to increase the Floor Space Index (FSI) of the plots within the influence zone of the metro-rail node. This will provide the much-needed built space for new commercial and residential use and halt the artificial increase in prices. A higher FSI will also help in unlocking land values near the transit corridor (Suzuki et al., 2015), thus increasing the density around transport nodes and reducing the urban sprawl (Karteek, 2015).

Regulatory changes

According to the National TOD policy (Draft 2014), the minimum FSI for a sustainable Transit-Oriented Development (TOD) must be between 3 and 5 or even higher for bigger cities (MoUHA, 2011). The exact FSI would be determined only after a detailed analysis of demand, infrastructure availability, density, and after land use zoning has been undertaken, which lies beyond the scope of this paper. The additional FSI would be complemented with regulatory changes in minimum plot size, maximum ground coverage and maximum height restriction so as to utilize the additional FSI and encourage redevelopment in the influence zone.

The demarcation and local area plan (LAP) of influence zones would be reflected in the master plan and the development regulations would be amended with a separate section for the TOD zone.

References

Agarwal, N., Rathi, S., Kalra, K., Gupta, M., Pal, S., Lakshmi, A., Bhattacharya, S., & Mishra, A. (2014). Review of urban transport in India. Institute of Urban Transport (India) and Center for Science Technology and Environmental Policy (CSTEP).

Aljoufie, M., Zuidgeest, M., Brussel, M., & van Maarseveen, M. (2011). Urban growth and transport: Understanding the spatial and temporal relationship. *WIT Transactions on the Built Environment, 116*(14), 315–328.
Andersson, D. E., Shyrb, O. F., & Fub, J. (2010). Does high-speed rail accessibility influence residential property price? Hedonic estimates from southern Taiwan. *Journal of Transport Geography, 18*(1), 166–174. https://doi.org/10.1016/j.jtrangeo.2008.10.012.
Boarnet, M. G., Bostic, R., Williams, D., Santiago-Bartolomei, R., Rodnyansky, S., & Eisenlohr, A. (2017). *Affordable housing in transit-oriented developments: Impacts on driving and policy approaches*. A National Center for Sustainable Transportation Research Report. A White Paper from the National Center for Sustainable Transportation.
CABE (Commission for Architecture & the Built Environment). (2006). *What home buyers want: Attitudes and decision making among consumers*. https://www.mae.co.uk/assets/pdfs/151123_Mae_Architects_What_Home_Buyers_Want_Housing_Architecture.pdf.
Census of India. (2011). *Primary census abstract*. New Delhi: Census Commissioner of India and Registrar General of India. https://censusindia.gov.in/2011-common/censusdata2011.html.
Cervero, R. (1992). *Land market impacts of urban rail transit and joint development: An empirical study of rail transit in Washington, D.C. and Atlanta*. University of California Transportation Center, Working Papers qt5890b3xz.
Cervero, R., & Dai, D. (2014). BRT TOD: Leveraging transit-oriented development with bus rapid transit investments. *Transport Policy, 36*, 127–138. https://doi.org/10.1016/j.tranpol.2014.08.001.
Cervero, R., & Landis, J. (1997). Twenty years of the Bay Area rapid transit system: Land use and development impacts. *Transportation Research Part A: Policy and Practice, 31*(4), 309–333. https://doi.org/10.1016/S0965-8564(96)00027-4.
CMRL. (2019). *Chennai metro rail limited*. https://chennaimetrorail.org/schematic-map/.
Debrezion, G., Pels, E., & Rietveld, P. (2007). The impact of railway stations on residential and commercial property value: A meta-analysis. *The Journal of Real Estate Finance and Economics, 35*, 161–180. https://doi.org/10.1007/s11146-007-9032-z.
Deng, T., & Nelson, J. D. (2010). The impact of bus rapid transit on land development: A case study of Beijing, China. *World Academy of Science, Engineering and Technology, 66*, 1196–1206.
Doi, K. (2002). Cities and transportation. *Land Use Policy*.
Gatziaff, D., & Smith, M. (1993). The impact of the Miami metrorail on the value of residences near station locations. *Land Economics, 69*(1), 54–66. https://doi.org/10.2307/3146278.
Giuliano, G. (2004). *Land use impacts of transportation investments. The Geography of urban transportation*. The Guilford Press.
Global Transport Knowledge Practice. (2018). https://www.gtkp.com/index.php.
Goel, R., & Tiwari, G. (2014). *Promoting low carbon transport in India: Case study of metro rails in Indian cities*. UNEP Risø Centre on Energy, Climate and Sustainable Development, Technical University of Denmark. https://wedocs.unep.org/bitstream/handle/20.500.11822/31445/CSMRIC.pdf?sequence=1&isAllowed=y.
Hess, D., & Almeida, T. (2007). *Impact of proximity to light rail rapid transit on station-area property values in Buffalo*. Urban Studies.
IIHS. (2015). *Urban transport in India. Challenges and recommendations*. IIHS RF Paper on Urban Transports.
ITA Working Group. (2004). *Underground or aboveground? Making the choice for urban mass transit systems*. Tunnelling and Underground Space Technology.
Jaysawal, N., & Sahu, S. (2014). Urbanization in India: An impact assessment. *International Journal of Applied Sociology, 4*(2), 60–65. https://doi.org/10.5923/j.ijas.20140402.04.
Jeyaranjan, J. (2017). Chapter 13: The life and times of migrant workers in Chennai. Springer Singapore. In D. N. Reddy, & K. Sarap (Eds.), *Rural labour mobility in times of structural transformation dynamics and perspectives from Asian economies*. https://doi.org/10.1007/978-981-10-5628-4_13.

Karteek, G. (2015). Is FSI dependent on land availability and densities? A comparative review of FSI in Indian cities. *European Journal of Sustainable Development, 4*(2), 27–34. https://doi.org/10.14207/ejsd.2015.v4n2p27

Kilpatrick, J. A., Throupe, R. C., Carruthers, J. I., & Krause, A. (2007). The impact of transit corridors on residential property values. *Journal of Real Estate Research, American Real Estate Society, 29*(3), 303–320.

Lambert, K. D. (2009). *Transit-oriented development and its effect on property values: An Atlanta case study*. A thesis presented to the Academic Faculty, Georgia Institute of Technology. http://hdl.handle.net/1853/31703.

Litman, T. (2015). Evaluating transportation land use impacts. *World Transport Policy & Practice, 1*(4), 9–16.

Live Mint. (2018, October 10). *15 Indian cities where metro train work is going on*. Live Mint.

Montgomery, E. (2017). *Infrastructure in India—A vast land of construction opportunity*. Price Water House Coopers.

MoHUA (Ministry of Housing and Urban Affairs). (2011). Government of India. http://mohua.gov.in/cms/Urban-Transport-Metro-Rail-Projects.php.

National Informatics Centre. (2017). Government of India https://www.nic.in/blogs/.

Nelson, A. (1992). Effects of Elevated Heavy-Rail Transit Stations on House Prices with Respect to Neighborhood Income. *Transportation Research Record*.

Perdomo, J. A. (2011). *A methodological proposal to estimate changes of residential property value: Case study developed in Bogota*. MPRA Paper 37180, University Library of Munich, Germany.

Planning Commission of India. (2011). Government of India. Twelfth five-year plan. https://niti.gov.in/planningcommission.gov.in/docs/plans/planrel/fiveyr/12th/pdf/12fyp_vol1.pdf.

Pojani, D., & Stead, D. (2015). Sustainable urban transport in the developing world: Beyond megacities. *Sustainability, 7*(6), 7784–7805. https://doi.org/10.3390/su7067784.

Roukouni, A., Basbasb, S., & Kokkalisc, A. (2012). Impacts of a metro station to the land use and transport system: The impacts of a metro station to the land use and transport system: The Thessaloniki Metro case. *Procedia—Social and Behavioral Sciences, 48*, 1155–1163. https://doi.org/10.1016/j.sbspro.2012.06.1091.

Sekar, S. P., & Karuppannan, S. (2012). *Contributions of metro rail projects in the urban dynamics of Indian metro cities: Case Study of Chennai and Bangalore*. 48th ISOCARP Congress, Metro rail and Urban Dynamics in Chennai and Bangalore.

SESEI. (2019). *Indian automobile industry*. https://sesei.eu/.

Stjernborg, O. M., & Mattisson, O. (2016). The role of public transport in society—A case study of general policy documents in Sweden. *Sustainability, 8*(11), 1–16. https://doi.org/10.3390/su8111120.

Suzuki, H., Murakami, J., Hong, Y-H., & Tamayose, B. (2015). *Financing transit-oriented development with land values: Adapting land value capture in developing countries*. Urban Development; Washington, DC: World Bank. https://doi.org/10.1596/978-1-4648-0149-5.

Swamy, H. (2008). *Impact of Delhi metro on real estate*. IUT India.

Tamil Nadu Registration Department. (2019). https://tnreginet.gov.in/portal/webHP?requestType=ApplicationRH&actionVal=homePage&screenId=114&UserLocaleID=en&_csrf=ed8ff27b-df22-42be-9814-eebf15adc295.

Tari, N., Jayanti, D. Y., & Munawir, R. (2015). *Land value capture as financing source for spatial development*. Bandung Institute of Technology.

UNDESA, UNDP, & UNESCO. (2012). UN system task team on the post-2015 UN Development Agenda.

Zhang, X., Liu, X., Hang, J., Yao, D., & Shi, G. (2016). Do urban rail transit facilities affect housing prices? Evidence from China. *Sustainability, 8*(4), 380. https://doi.org/10.3390/su8040380.

Zhong, H., & Li, W. (2016). Rail transit investment and property values: An old tale retold. *Transport Policy, 51*, 33–48. https://doi.org/10.1016/j.tranpol.2016.05.007.

Regional Travel and Commuting Patterns: A Study of the Oldest Suburban Railway Line in Eastern India

Bhaswati Mondal and Gopa Samanta

1 Introduction

Geography has long been treated as a purely regional science. The principal objective of geographers has been to analyse space in the context of singular regions, through areal differentiation. However, as Hartshorne has stated, 'in regional geography ... we strike a limited section vertically through all the surfaces to comprehend the totality of their characteristics in a single area' (1939, p. 433). The process of looking beyond regional contexts may be implemented by using several tools and techniques; this is the case of mobility studies.

Mobility is a crucial need for human beings—people have always had to mobilize themselves so as to search for basic commodities. Presently, the most common reason for travel is the basic need for employment. Thus, commuting is one of the popular forms of mobility in both the developed and developing worlds (see, for example, Chandrasekhar, 2011; Cresswell, 2006; Hoai & Ann, 2010; Jensen, 2009; Ommeren et al., 1997; Punpuing, 1993; Stutzer & Frey, 2007). Commuting is distinguished from other sorts of spatial dislocations in terms of duration (Mahbub, 1997). It bears a pendular character in the mobility pattern—through daily back and forth movement. The Indian Railway system is known as the lifeline of the country. It carries millions of passengers every day, connecting every corner of the nation. The 2011 census showed that nearly 70% of non-agricultural and non-household industrial workers in India commuted to work.

In the state of West Bengal, this value is slightly higher than the national average. West Bengal is the second-ranking state in the country in terms of the transportation of commuters beyond a distance of 50 km. Passenger services are available for both suburban and non-suburban segments. Suburban rail, commuter rail and regional

B. Mondal (✉)
DODL, University of Kalyani, Kalyani, India

G. Samanta
Department of Geography, The University of Burdwan, Burdwan, India

© The Author(s), under exclusive license to Springer Nature Switzerland AG 2021

S. Mitra et al. (eds.), *Railway Transportation in South Asia*, Contemporary South Asian Studies, https://doi.org/10.1007/978-3-030-76878-2_5

rail all play a major role in the public transport system in many of India's major cities. Suburban rail is defined as a rail service between a central business district and its suburbs, a conurbation or other locations drawing large numbers of people on a daily basis (Ministry of Railways, 2014). The trains providing such services can be termed 'suburban trains'. The trains of the Mumbai Suburban Railway, the Chennai Suburban Railway and the Kolkata Suburban Railway are often referred to as 'local trains' or 'locals'.

The suburban line connects several smaller cities and villages with the metropolis. There are spatial diversities in terms of nature of the settlements involved, people's economies, administrative hierarchy and a number of other factors within the hinterland of the suburban railway network. In addition to these differences, the nature of commuter flows gives rise to a pattern. This pattern can be visually represented and geographically explained. As Lonsdale has stated, 'commuting or journey-to-work patterns could form the basis for delimiting networks of overlapping regions across the whole expanse of a territory' (1966, p. 115). Thus, areal differentiation can be made within a suburban rail corridor—itself based on an analysis of commuter flows. This chapter attempts to explore regional travel and commuting patterns by looking at the Howrah–Bardhaman Main Line—a suburban railway line.

2 Context of the Study Area

Geographical studies are traditionally known as locational studies since these need a 'locale' or an area to carry out their surveys and analyses. However, this problem becomes intensified when aspects like mobility are researched. In order to meet the objectives, the present research selects the Howrah–Bardhaman Main Railway Line, the oldest railway line in Eastern India and the second oldest in India as a whole. It belongs to the Kolkata suburban railway network, the largest suburban railway network in the country in terms of route length.

With a track length of 108 km, the Howrah–Bardhaman Main Railway Line connects the three districts of South Bengal—Howrah, Hooghly and Purba Bardhaman—with the metropolitan city of Kolkata. The Howrah station serves as the gateway to the city of Kolkata. At present, it is the largest railway station in Asia. The Howrah and the Bardhaman stations are both terminal stations. Including these two, there are 34 stations on this railway line. Among them, six are junction stations: Howrah, Bally, Sheoraphuli, Bandel, Saktigarh and Bardhaman (Fig. 1). The Bandel station lies almost in the middle of this railway line. The Howrah–Bardhaman Main Line comes under the jurisdiction of the Howrah Division of Eastern Railway. However, among the 34 stations of this railway line, there is no official record of Nimo station, located between Memari and Rasulpur. The Eastern Railway considers it as a 'halt station'.[1]

[1] Meaning, a station where passenger trains are usually halted for some time.

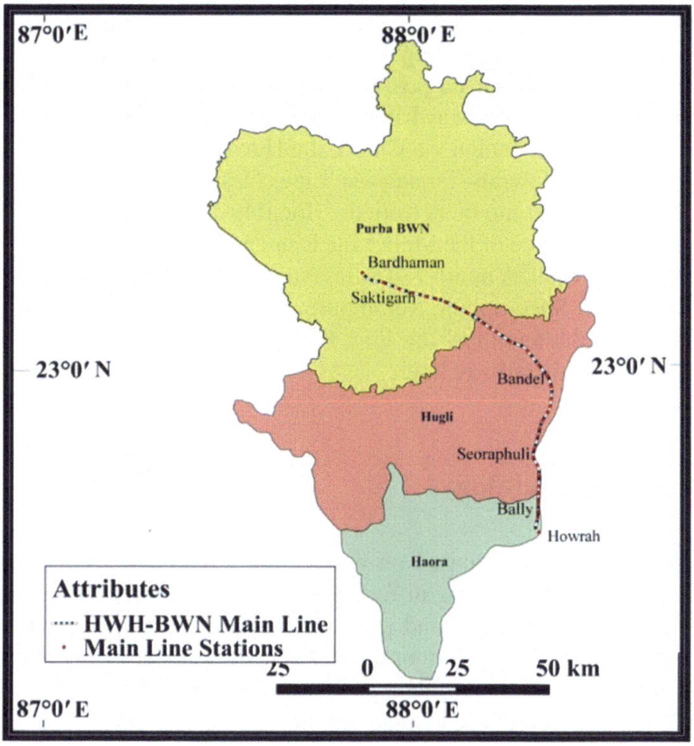

Fig. 1 Howrah–Bardhaman Main Railway Line connecting the three districts with all stations, including the six junction stations (*Source* Prepared by the authors)

The first passenger train on this railway line from Howrah to Hooghly ran on August 15, 1854. The Howrah–Bardhaman Main Line stretches along the levee of the Hooghly River up to the Mogra station—from where it takes a north-west turn and reaches Bardhaman, the seat of the Raj Empire through Pandua, the oldest historical place of Hooghly district; and Memari, a historical service centre (Mitra, 1948; Paterson, 1910). Thus, it links the state's 'rice bowl', Purba Bardhaman, the conurbation of Hooghly and the industrial agglomeration of Howrah, with the state capital—Kolkata. This railway line is classified as a 'B' class line, where trains can run up to a velocity of 130 km/h.

Over five hundred thousand (0.5 million) passengers travel on this railway line daily. Most commuters on the Howrah–Bardhaman Main Railway Line belong to these immediately adjacent three districts. Therefore, those commuters who belong to these three districts and commute for work almost throughout the full duration of the year were considered in the survey conducted. Among the three districts, Purba Bardhaman holds the longest railway track (321 km). However, the density of railway tracks is the highest in the Howrah district.

Among the 34 stations, 21 belong to the Hooghly district, 9 belong to the district of Purba Bardhaman, and the remaining 4 belong to the Howrah district. The Hooghly district holds the most significant position in commuter transportation through the Howrah–Bardhaman Main Railway Line. There are other railway lines in this district, such as the Howrah–Bardhaman via Chord, the Howrah–Katwa Line, the Howrah–Goghat Line and the Howrah–Tarakeswar Line. Nearly three and a half hundred thousand (0.35 million) commuters from the Hooghly district travel through this line every day. The significance of the Main Line is intensified due to the presence of the Hooghly conurbation belt. A number of cities (such as Hooghly-Chinsurah, Chandernagore, Bhadreswar, Sheoraphuli, Serampore, Hindmotor, Uttarpara-Kotrung) are located throughout the length of the line. Every day 182 local trains run through different stations on this railway line.

3 Data and Methodology

The study here conducted is based on both primary data and secondary sources. Primary data were collected using different methods such as questionnaire surveys, formal and informal interviews, and participant and non-participant observations. Data were collected at railway stations as well as in the compartments of the running suburban trains. The collection of data on commuters through the method of observation took place in uncontrolled situations, i.e. in their natural setting. Secondary data were gathered from different official government sources, such as the Transport Guide 2011–2012, originating traffic data from the Divisional Railway Manager's Office of Howrah Division of the Eastern Railway, and the Censuses of India from both 2001 and 2011.

Through the survey, it was found that although the Howrah–Bardhaman Main Railway Line is a continuous suburban rail-route, two different sections within it can be clearly identified based on commuter flows. These are Howrah–Bandel and the Bandel–Bardhaman sections. Thus, the Bandel station serves as the point of division of this railway line. This conclusion has been drawn based on six criteria, as follows:

1. Outflows of commuters;
2. Availability of local trains;
3. Time differences between layovers;
4. Distances between stations;
5. Annual growth in the number of commuters;
6. Commuters' cultural variances.

As mentioned, several quantitative and qualitative methods were used to analyse the data collected. Quantitative methods include several different statistical techniques illustrated in the figures below. This study has also resorted to descriptive statistics. A regression analysis was conducted so as to understand the relation between originating traffic and available train services. It also calculates the average annual growth index of commuters in each of the stations of the Howrah–Bardhaman

Main Railway Line using the following formula:

$$\text{AAGR} = 1/\text{T} * \text{LN}(\text{Yz}/\text{Ya}), \quad (1)$$

where AAGR = Average Annual Growth Rate, T = Times of Increase/ Decrease, LN = Natural Logarithm of a Number, Yz = Daily Average Passenger Volume of the Terminal Year, Ya = Daily Average Passenger Volume of the Initial Year.

To convert the calculated average annual growth rate into percentage form, it is to be multiplied by 100.

In addition to these quantitative methods, qualitative data was utilized to complement conclusions. Namely, narratives and accounts collected from commuters were used. This research has used Microsoft Excel-2010 for all calculations and diagrammatic representations. It has also resorted to Google Earth Pro, QGIS-3.0.2 and ArcMap 10.4 to prepare the maps presented.

4 Outflows of Commuters

The daily originating traffic data recorded at each railway station is a good measure for understanding the outflow rates of commuters. The number of commuters is found to be much higher in the Bandel–Howrah section than in the Bardhaman–Bandel section (Fig. 2). The Bandel–Howrah section of the line carries over four hundred thousand (0.4 million) passengers every day. Each of the stations is frequented by at least 10,000 passengers daily (Table 1). The larger stations such as Chandernagore, Bhadreswar, Konnagar and Serampore host even more commuters. The Serampore Station hosts around 40,000 passengers per day. The largest volume of outgoing commuters in this section is recorded at the Howrah station—nearly 70,000. Thus, the daily average originating traffic in this section is over 25,000 people.

On the other hand, the Bardhaman–Bandel section sees a little over one hundred thousand (0.1 million) passengers daily. The Bardhaman station is the largest in this section, from where around 40,000 commuters travel to different stations of the main line daily (Table 1). Besides Bardhaman, there are two other stations—the Memari and Pandua stations—which record around 10,000 commuters each day. The remaining stations are frequented by a much lower number of day-travellers. Thus, the daily average number of commuters in the Bandel–Bardhaman section is around just 6000 people.

All the stations in the Bandel–Howrah section are under the jurisdiction of the Kolkata Metropolitan Area. It is a continuous urban belt. Serampore city, located in this section of the line, developed before Kolkata. Other cities, like Hooghly-Chinsurah, Chandernagore, Bhadreswar and Konnagar flourished in the pre-colonial period as compact, prosperous settlements—trade being the primary sector in their economies before Kolkata developed into the metropolis that it is today. Therefore, although some accounts state that these areas result from an expansion of Kolkata city into the suburbs, this is not true. At one time, Serampore was the centre of an affluent

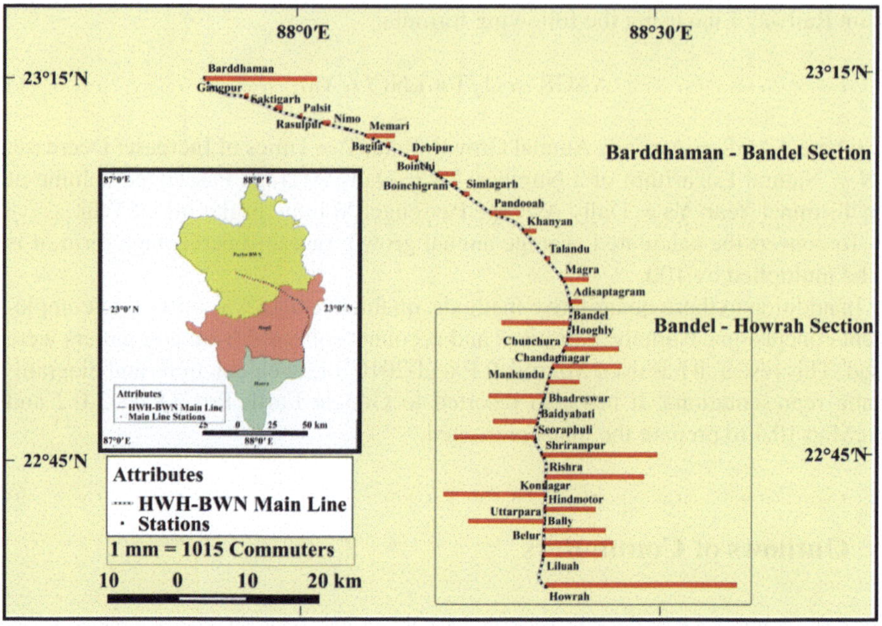

Fig. 2 Daily average originating traffic of Howrah–Bardhaman Main Railway Line (*Note* Data regarding the Nimo Halt station is not available. *Source* Mondal and Samanta, 2021)

Table 1 Differences in commuters' outflows between the Howrah–Bandel and the Bandel–Bardhaman sections

Sections	Total originating traffic/day	Avg. originating traffic/day	Stations with maximum originating traffic/day	Stations with minimum originating traffic/day
Bandel–Howrah	4,33,595	25,506	Howrah (65,971)	Hooghly (9623)
Bardhaman–Bandel	1,00,524	5913	Bardhaman (38,658)	Palsit (956)

Source Calculated based on the average originating traffic, 2005–2006 to 2014–2015, DRM Office, Howrah Division

economy which was later shifted downstream due to the decreasing navigability of the Hooghly River (Chatterjee, 1992). There is a close linkage between the cities of this region and the metropolitan city of Kolkata. People have become dependent on Kolkata for their daily livelihoods.

On the other hand, the Bardhaman–Bandel section is a vast rural belt with few urban pockets. This region possesses fertile soils and potentiality for prosperous agricultural production. Paddy and potato are the two most important agricultural outputs. There are also several other crops coming from the region. Most people of this section are engaged with this sector of the economy—thus being part of a village-based economy. The only two urban centres, Memari and Pandua, also play the role of rural service centres. Therefore, the need to commute for people of the Bardhaman–Bandel section is much lower compared to what is the case in the Howrah–Bandel section.

5 Availability of Suburban Trains

The number of suburban trains, also termed as 'local trains' or simply 'locals', on a railway line depends on the demand in terms of passenger numbers.

There are 49 Bardhaman–Howrah local trains running in between the Bardhaman and the Howrah stations daily. Besides the Bardhaman–Howrah local trains, there are 10 Memari stations Howrah 'locals' running between the Memari and Howrah stations, and 3 Pandua–Howrah local trains running between the Pandua and Howrah daily. Having a significant volume in terms of outflow of commuters from the Memari and Pandua stations, the Pandua locals and Memari 'locals' are run (Fig. 3).

In order to serve the significant volume of traffic in the Bandel–Howrah section, Eastern Railway runs several shuttle trains. There are 71 Bandel 'locals' which run

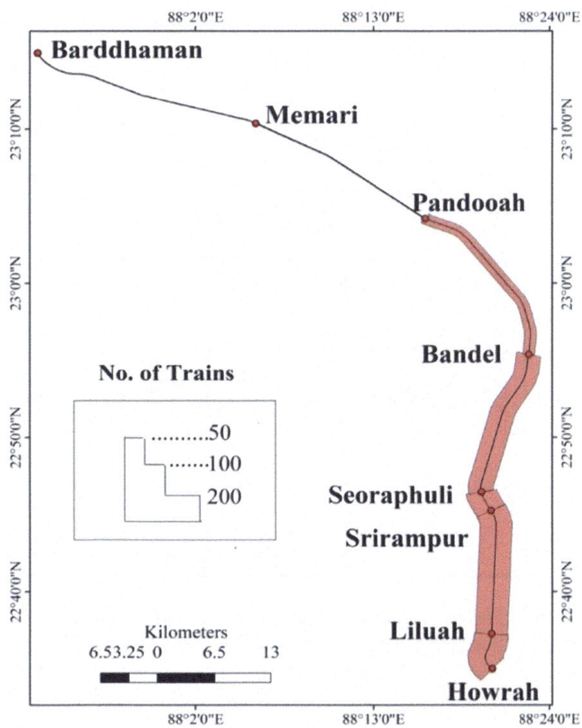

Fig. 3 Flows of local trains in the Howrah–Bardhaman Main Railway Line (*Source* Computed based on Transport Guide, 2011–2012)

in between the Bandel and Howrah stations serving all the intermediate stations. Shuttle trains also run from the Sheoraphuli and Serampore railway stations. The Belur Math 'locals', which connect the Ramakrishna Mission in Belur with Howrah, serve the Liluah station (Fig. 3). Consequently, the flow of local trains along the Howrah–Bardhaman Main Line gradually widens towards Howrah. The two stations of Liluah and Howrah are served by 182 Main Line 'locals', whereas the Bandel and Bardhaman stations are served by 133 and 49 trains, respectively (Fig. 3). This simply means that train frequency in the Bandel–Howrah section is much higher than in the Bandel–Bardhaman section.

In order to provide a notion of the relation between daily average originating traffic and available local trains at each station, a regression analysis was performed (Fig. 4). The correlation coefficient value of +0.66 indicates that the relation is moderate to strongly positive. It explains 44% of the relation. The scatter plots can be categorized into four groups (Table 2). All the stations from Bandel to Liluah lay in an advantageous position. There are a larger number of trains than expected. Bardhaman is the most disadvantaged station in the sense that the number of trains serving this station is much lower than what could be imagined as required.

A simple comparison can explain the situation better: both Bardhaman and Serampore have the same records in terms of daily originating passengers, but while Serampore is served by 171 trains, Bardhaman is served by 49 local trains only. There are

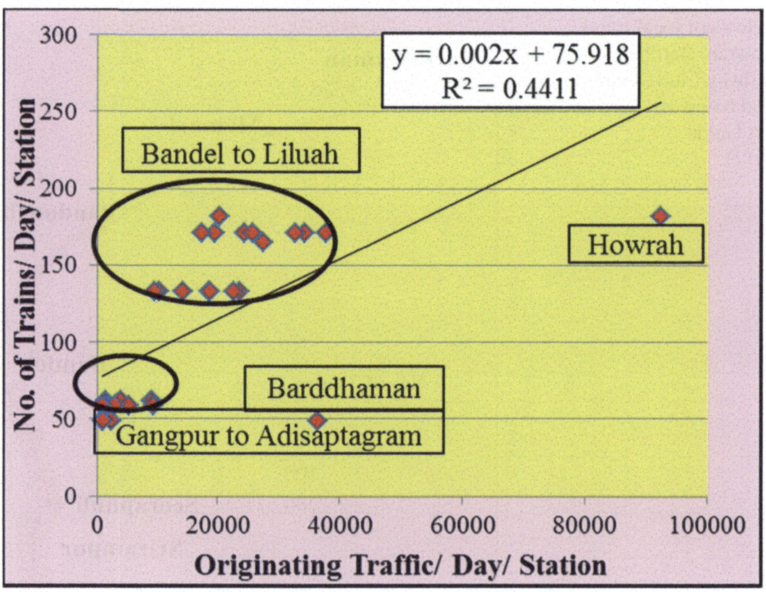

Fig. 4 Relation between daily originating traffic and availability of trains for each station in the Howrah–Bardhaman Main Railway Line (*Source* DRM Office, Howrah and Transport Guide, 2011–2012)

Table 2 Situation of railway stations in relation to traffic-train balance

Situations	Stations
Advantageous	Bandel to Liluah
Requires a few more trains	Gangpur to Adisaptagram
Requires many more trains	Howrah
Most disadvantaged	Bardhaman

Source Computed by the authors

thus an insufficient number of trains in the Bardhaman–Bandel section to cater to local demand.

6 Differences Between Layover Times

The waiting time between two consecutive trains (for a connection) at a particular station is a significant indicator of an efficient transport system. If there is less waiting time, commuters are of course benefited by the frequent train service. It makes the process of mobility smooth, effective and hassle-free. On the other hand, a longer wait for a connection makes the transport system unattractive to commuters. They have to wait for the next train for a longer period and often reach their workplaces late (Cantwell et al., 2009; The Statesman, 2016; *The Anandabazar Patrika*, 2015, 2018).

Within the Howrah–Bardhaman Main Railway Line, the volume of commuters is much higher than in the Bandel–Howrah section. To serve this huge number of commuters, more trains run in this section. On the contrary, in the Bardhaman–Bandel section where the commuters' volume is less, there are fewer trains. Therefore, the difference in waiting time for a connection in the Bandel–Howrah section is much lower than it is in the Bardhaman–Bandel section (Fig. 5).

Generally, at the Bardhaman station, the amount of waiting time for a connection is the longest—nearly one hour and ten minutes. As there are few Memari–Howrah 'locals', the time interval at the Memari station decreases to one hour and five minutes. At the Pandua station, it becomes 45 min. From the Bandel station, this interval falls sharply to 15 min (Fig. 5). In this way, the decreasing tendency continues, and at the Serampore station, there are Howrah-bound main line 'locals' every five minutes.

7 Distances Between Stations

The number of railway stations in an area is a good indicator of the level of citizen mobility, which is in itself an indicator of development. It increases the accessibility and mobility choices of users (Levinson, 1998). In the Bandel–Howrah section, as the demand for accessible transport to Kolkata city has been significant from the very

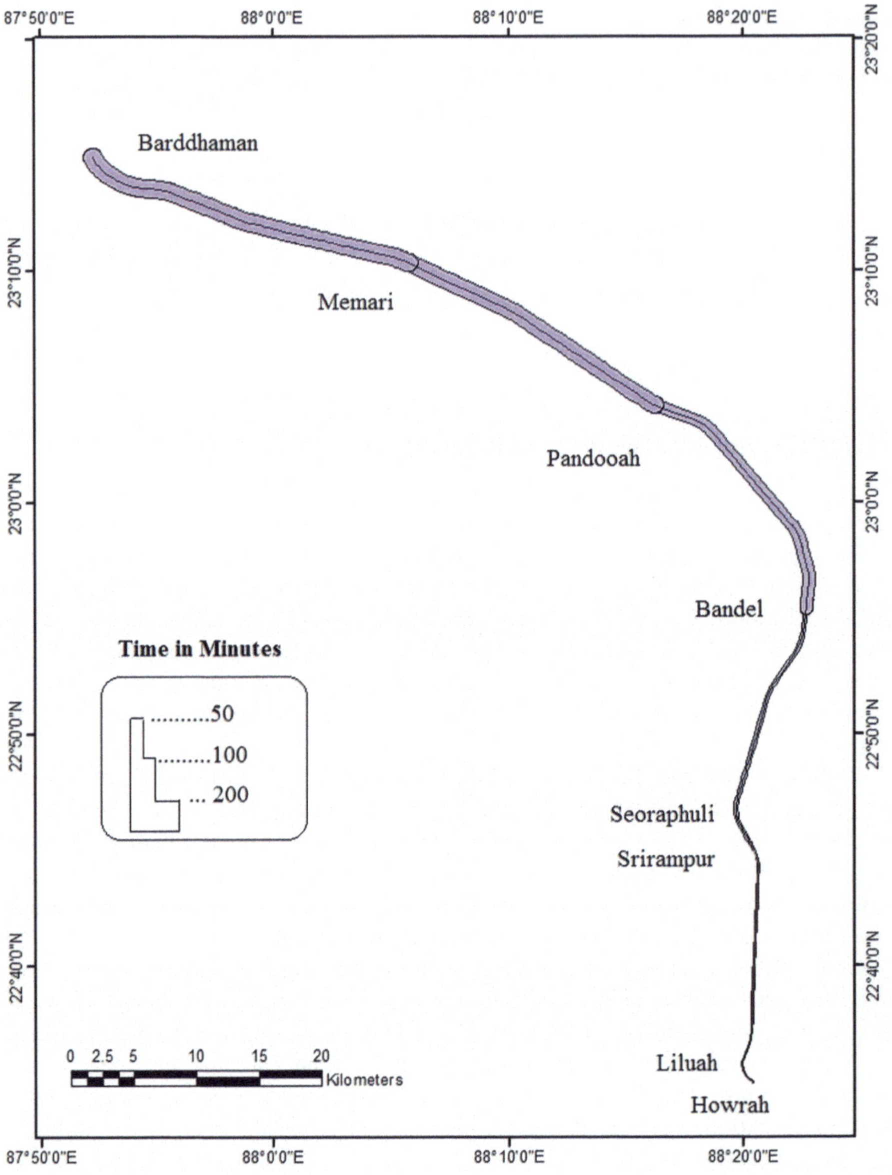

Fig. 5 Inter-train time difference in Howrah–Bardhaman Main Railway Line (*Source* Calculated and computed from Transport Guide, 2011–2012)

Table 3 Differences in inter-station distances in the Main Railway Line (from the Bandel station)

Sections	Route Kilometre	Number of stations	Number of stations in every 10 km	Avg. distance between two stations (km)
Howrah–Bandel	48.3	17	4	2.84
Bandel–Bardhaman	65.8	17	3	3.87

Source Field Survey, 2014–2017

beginning, several suburban stations have been constructed. In this section, there are four suburban stations for every 10-km stretch (Table 3). In the Bardhaman–Bandel section, this number comes down to three. Thus, while in the Bandel–Howrah section two railway stations are situated about three kilometres apart, in the Bardhaman–Bandel section this distance increases to about four kilometres.

8 Average Annual Growth Rates of Commuters

There is a difference in the average annual growth rates of commuters between the Howrah–Bandel and the Bandel–Bardhaman sections (Fig. 6). The average annual

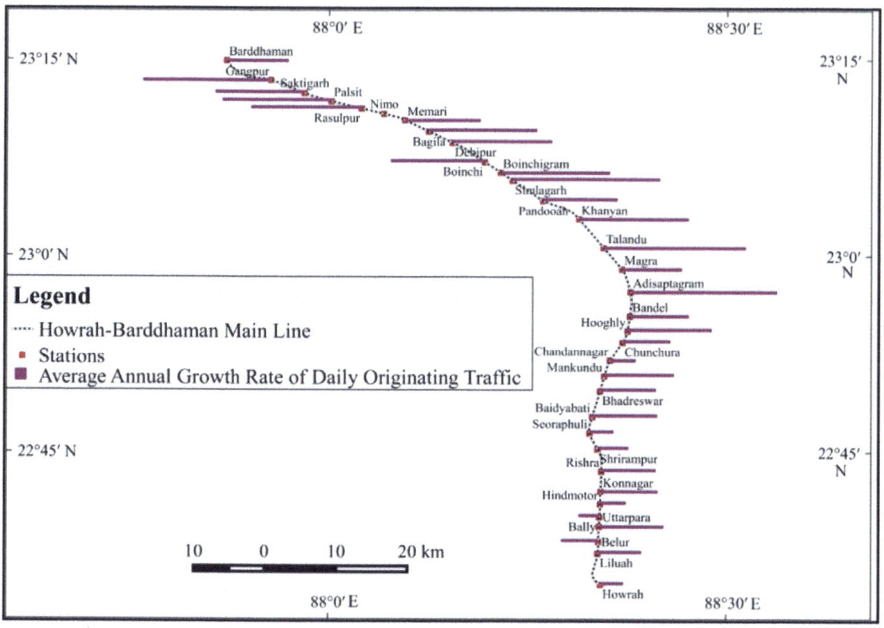

Fig. 6 Average annual growth of daily originating traffic in between 2005–2006 and 2014–2015 (*Note* Data for the Nimo Halt station is not available. *Source* Mondal and Samanta, 2021)

Table 4 *Izzat* monthly and the differential annual growth in the number of commuters

Zones	2009–2010	2010–2011	2013–2014	2014–2015
Howrah–Bandel	4.03	7.14	−0.60	−0.09
Bandel–Bardhaman	16.21	29.00	−5.49	−9.05

Source DRM Office—Howrah, Eastern Railway

growth for the outflow of commuters in the Howrah–Bandel section is three per cent per annum; in the Bandel–Bardhaman section, it is seven per cent per annum. The stations nearer to the metropolitan city experience a slower growth in the number of commuters compared to the more distant stations (Fig. 6).

The stations distant from the metropolis are located in an agriculture-rich region. Most citizens are engaged in farm-based economic activities. However, in recent years, the flow of farm workers outside the agricultural sector has led to greater diversification. In West Bengal, the high man-land ratio makes per capita earnings low. The increased operational costs of agriculture have made it unprofitable, especially for small and marginal farmers (Khasnabis, 2008; Sarkar, 2006). Low returns from agriculture affected the sector's viability (Samanta & Pal, 2018). Therefore, a consistent flow from agriculture to other sectors of the economy has been taking place over the last two decades (Chakraborty & Roy, 2016). Mehrotra et al. (2014) have shown that this shift in the employment structure is taking place towards the relatively unorganized manufacturing and service sectors.

Commuter railway stations connect the villages with cities. Thus, people residing in the villages have the opportunity to work in urban informal jobs. Although these jobs are not always highly remunerated, they do provide an escape to the vicious cycle of poverty and local unemployment (Deshingkar, 2010; Mondal, 2015). Rural educated people also commute outside the area for work. This can be better demonstrated with the help of commuters' growth curve in relation to the 'izzat monthly' (Table 4).

The 'izzat monthly' was a seasonal ticket introduced by the Indian Railways in 2009–2010 to facilitate the ability to commute for workers whose monthly incomes were below INR 2000.[2] It would cost INR 25 per month.[3] With the help of an izzat monthly, one commuter was able to commute up to 100 km daily. With the introduction of the izzat monthly, of course, the number of commuters greatly increased.

The geographical distribution of these commuters was not the same in the Howrah–Bandel and the Bandel–Bardhaman sections of the railway line. The Bandel–Bardhaman section, which serves a rural belt, witnessed a sharp growth of 16% in 2009–2010. In the next year, it witnessed a growth of almost 30% in the number of commuters. However, the izzat monthly did do as well in the Bandel–Howrah section, which is a purely urban belt. In this section, the growth in the number

[2] 2000 Indian Rupees equals to around 23.33 Euro (as on July 16, 2020).

[3] 25 Indian Rupees equals to 0.29 Euro (as on July 16, 2020).

Table 5 Differences in the number of agricultural workers in selected station-centred villages of the Bandel–Bardhaman section, 2001–2011

Mouzas	Main cultivators	Main agricultural labourers
Rasulpur	4	−108
Debipur	−9	−597
Simlagarh	−3	54
Talandu	−9	−192
Bainchi	30	1304

Source Calculated from Census of India, 2001 and 2011

of commuters was four per cent and seven per cent in 2009–2010 and 2010–2011, respectively.

Following some cases of misuse of the izzat monthly, this facility was withdrawn in 2013–2014. This withdrawal had an impact on the number of commuters in almost all stations. The most severe impact was found on the volume of commuters in railway stations lying between Bandel and Bardhaman. In the year the izzat monthly was withdrawn, the number of commuters in this section fell at a rate of five per cent—in the next year, at a rate of nine per cent. Some of the stations experienced a huge growth in commuter numbers in 2010–2011 and then a radical fall in 2014–2015. These were Simlagarh (55.45%, −17.93% respectively), Talandu (36.31%, −15.05% respectively), Debipur (41.55%, −14.85% respectively), Bainchi (37.53%, −15.34% respectively) and Rasulpur (32.56%, −8.08% respectively). In these villages, the number of farm-based workers decreased, and the number of non-farm workers increased between 2001 and 2011 (Table 5). The stations between Howrah and Bandel, on the contrary, recorded a fall in commuter numbers of more than one per cent in both intervals (Table 5).

9 Cultural Variances of Commuters

The word 'culture' is derived from the Latin word 'cult' or 'cultus' meaning cultivating or tilling. According to Hofstede (2001, p. 5), 'Culture is the collective programming of the mind which distinguishes the members of one group or category of people from another'. In a nutshell, the way we live is known as culture. Thus, the way we talk and the way we think are all different components of culture. As the Howrah–Bardhaman Main Railway Line stretches over 108 km, there are cultural differences among the commuters of its two sections. The observational method was used to acquire data on this parameter. The differences have been explained qualitatively. This kind of difference is reflected through several elements, such as language, clothing, ornaments, accessories and perceptions of time.

It was found that commuters in the Bandel–Howrah section have a more polished, articulated and modern urban-based culture. They often use 'mixed languages' in conversations. 'Mixed language' refers to the use of several languages besides vernacular while talking. Generally, Bengali, English and Hindi are used. Commuters in the Bardhaman–Bandel section, on the other hand, generally speak only in Bengali amongst themselves. As stated earlier, the Bandel–Howrah section serves an entirely urban agglomeration.

There are many public and private English-medium schools in these cities. English is the language of teaching in most of these schools. Even students coming from typically Bengali families often choose Hindi as their second language (after English) so as to compete at different national-level examinations. Therefore, through time, they develop fluency in both English and Hindi. Many commuters in this section carry this background. On the other hand, the Bardhaman–Bandel section serves mostly a rural zone. Except for a few schools in the city of Bardhaman, as well as in Memari and Pandua, the education system over the region is mostly run in Bengali. Thus, Bengali is the common language used during conversations.

Commuters in the Bandel–Howrah section appear to have a greater sense of fashion. Since women are taken to be more prone to pursue fashion trends than men (Koca & Koc, 2016), only female commuters were observed so as to identify the differences between the two sections. It is very usual to find female commuters in this section wearing clothing of different trendy styles with matching jewellery. They also appear to use more cosmetics and carry accessories such as purses, bags and sunglasses in the latest fashion. However, in the Bardhaman–Bandel section, women wear traditional dresses, use less jewellery and appear to wear less make-up. This is presumed to happen because, within the urbanized belt of the Howrah–Bandel section, people's purchasing capacity is higher. There are different kinds of jewellery shops, ranging from 'fake' jewellery, eclectic ethnic styles, to pearl, silver and gold jewellery. There are also many fashionable stores of readymade garments and tailoring houses in this zone. However, in the rural belt served by the Bardhaman–Bandel section, less of these options are available.

The perception of time also differs between these two zones. For example, when commuters in the Howrah–Bandel section were asked when they return home, they replied that it was around eight o'clock in the evening. When asked the same question, commuters in the Bardhaman–Bandel section replied that it was around eight o'clock at night. This perception of time differs geographically: eight o'clock may be represented as night or as evening. This difference is rooted in the different cultures of the rural and urban hinterlands. Villagers generally wake up early in the morning and go to bed early as well. Accordingly, they sense time differently from urban commuters who generally wake up late and go to bed late at night. Another reason is that after seven o'clock in the evening, the stations and bus stands in the Bardhaman–Bandel section start to be sparsely populated. It brings the feeling of night-time to commuters.

Thus, combining all these factors, two distinct sections can be identified within the Howrah–Bardhaman Main Railway Line—Howrah–Bandel and Bandel–Bardhaman (Table 6).

Table 6 Regionalization of the Howrah–Bardhaman Main Railway Line

Bases of differences	Howrah–Bandel	Bandel–Bardhaman	Data sources
Daily Avg. Originating Traffic, 2005–2006 to 2014–2015	25,506	5,913	DRM Office, Howrah
Number of local trains per day	Howrah (182), Bandel (133)	Bardhaman (49)	Transport Guide, 2011–12
Differences in layover times (average)	9 min	60 min	Transport Guide, 2011–2012
Number of Stations over every 10 km	4	3	Field Survey
Average annual growth of Commuters (in percentage per annum)	3	7	DRM Office, Howrah
Cultural aspects of commuters	Urban-based culture	Rural-based culture	Observation method

10 Conclusion and Final Remarks

This chapter has reviewed the process of regionalization of a continuous suburban rail corridor based on mobility characteristics. Without this discussion, the concept of mobility remains incomplete (Adey, 2006). The places, stations and railway tracks are immobile; however, the objects, trains and commuters are always mobile. These two factors are intimately attached and complementary. Local identities change with greater mobility characteristics. Within a 108 km length in the Howrah–Bardhaman Main Line, two sections with different characteristics can be identified. Because of its closeness to Kolkata, the Bandel–Howrah section differs significantly from the Bardhaman–Bandel section in terms of the high number of commuters. To serve this huge number of commuters, trains run more frequently. The distances between stations are also lesser in this section compared to the Bardhaman–Bandel section. As the Bandel–Howrah section belongs to a continuous urban area, commuters are found to embody a modern, urban-based culture. However, due to the changing nature of the rural economy, rural-to-urban commuting is increasing in the Bandel–Bardhaman section.

This study has implications on transport policy and regional development. It helps planners identify the stations with fewer trains serving a large volume of out-commuters, and accordingly helps them adopt the necessary measures for these shortcomings to be corrected. It also helps urban planners promote policies for allocating resources and design their plans accordingly.

This kind of research also bears significance in rural development programmes. It builds to an understanding of the shifting nature of rural economies. Suburban local trains connect rural and urban economies. Commuting creates employment opportunities for citizens, even those residing in villages distant from the metropolis. Transport policies have important roles in rural development. This is especially evident within the dynamicity in the number of commuters in the rural belt in relation to fare relaxation in the form of the izzat monthly. A suitable transport policy may be part of regional development policies. These sorts of development programmes are very much needed in developing countries such as India, where policies are still found to be of a localized and sedentary nature.

Acknowledgements This research was carried out with the financial contribution of the University Grants Commission (UGC) in the form of a Junior and Senior Research Fellowship. We are most grateful.

References

Adey, P. (2006). If mobility is everything then it is nothing: Towards a relational politics of (in)mobilities. *Mobilities, 1*, 75–94.
The Anandabazar Patrika. (2015, July 7). Pher train oborodh Haripal e [Train blockade once again at Haripal].
The Anandabazar Patrika. (2018, August 13). Chole mote ekti train, katwai kshubdho yatrira [Only one train plies here, passengers at Katwa agitated].
Cantwell, M., Caulfield, B., & O'Mahony, M. (2009). Examining the factors that impact public transport commuting satisfaction. *Journal of Public Transportation, 12*(2), 1–21.
Chakraborty, S., & Roy, U. (2016). The dynamics of rural non-farm economy in West Bengal during 1991–2001: A statistical analysis. *Geographical Review of India, 78*(3), 229–244.
Chandrasekhar, S. (2011). Workers commuting between the rural and urban: Estimates from NSSO data. *Economic and Political Weekly, 46*(46), 22–25.
Chatterjee, M. (1992, August). *Evolution and growth of municipal towns in Calcutta metropolitan area* (Centre for Urban Economic Studies, Discussion Paper No. 4). University of Calcutta.
Cresswell, T. (2006). *On the move: Mobility in the modern western world*. Routledge.
Deshingkar, P. (2010). *Migration, remote rural areas and chronic poverty in India* (Working Paper 323). Overseas Development Institute. www.chronicpoverty.org/uploads/...files/WP163%20Deshingkar.pdf
Hartshorne, R. (1939). *The nature of geography*. Lancaster: Association of American Geographers.
Hoai, A. T., & Ann, S. (2010). Gender and class in urban transport: The cases of Xian and Hanoi. *Environment and Urbanization, 22*(1), 139–155.
Hofstede, G. (2001). *Culture's consequences: Comparing values, behaviors, institutions, and organizations across nations* (2nd ed.). London: Sage.
Jensen, O. B. (2009). Flows of meaning, cultures of movements–Urban mobility as meaningful everyday life practice. *Mobilities, 4*(1), 139–158. https://doi.org/10.1080/17450100802658002
Khasnabis, R. (2008). The economy of West Bengal. *Economic and Political Weekly, 43*(52), 103–115.
Koca, E., & Koc, F. (2016). A study of clothing purchasing behavior by gender with respect to fashion and brand awareness. *European Scientific Journal, 12*(7), 234–248. https://doi.org/10.19044/esj.2016.v12n7p234
Levinson, D. M. (1998). Accessibility and the journey to work. *Journal of Transport Geography, 6*(1), 11–21.
Lonsdale, R. E. (1966). Two North Carolina commute patterns. *Economic Geography, 42*(2), 114–138.
Mahbub, A. Q. M. (1997). *Mobility behaviour of working people in Bangladesh-rural-rural and rural-urban circulation*. University of Dhaka: Urban Studies programme. ISBN: 984-510-012-0

Mehrotra, S., Parida, J., Sinha, S., & Gandhi, A. (2014). Explaining employment trends in the Indian economy:1993–94 to 2011–12. *Economic and Political Weekly, 49*(32), 49–57.

Ministry of Railways. (2014). *Suburban train services of Indian Railways, with particular emphasis on security of women passengers.* Standing Committee on Railways, 23rd Report presented to the Lok Sabha on 6th February. http://164.100.47.134/lsscommittee/Railways/15_Railways_23.pdf

Mitra, S. K. (1948). *Hugli jelar itihas o Banga samaj* [History of Hugli district and the society of Bengal]. Dey's Publishing.

Mondal, B. (2015). Commuting patterns of workers in a village of Bardhaman district, West Bengal. *Space and Culture, India, 3*(1), 48–66. http://dx.doi.org/10.20896/saci.v3i1.140

Mondal, B. & Samanta, G. (2021). Mobilities in India: The Experience of Suburban Train Commuting. *Springer-Nature*, ISBN: 978-3-030-78349-5 (Forthcoming).

Ommeren, V., Rietveld, P., & Nijkamp, P. (1997). Commuting in search of jobs and residences. *Journal of Urban Economics, 42*, 402–421.

Paterson, J. C. K. (1910). *Bengal district gazetteer, Burdwan district.* Logos Press.

Punpuing, S. (1993). Correlates of commuting patterns: A case-study of Bangkok Thailand. *Urban Studies, 30*(3), 527–545.

Samanta, G., & Pal, B. (2018, April 5). *Mumbai er krishak bikshabh ki Paschim Bange sambhab* [Is a Mumbai-like farmers agitation ever possible in Bengal?]. The Anandabazar Patrika.

Sarkar, A. (2006). The political economy of West Bengal: A puzzle and a hypothesis. *Economic and Political Weekly, 49*(28), 341–348.

The Statesman. (2016, September 10). The small and insignificant railway station of Noli on the outskirts of Delhi sees a mad scramble during the morning rush hour as commuters wait for the single train that wends its way there.

Stutzer, A., and Frey, B. S. (2007). *Commuting and life satisfaction in Germany, Heft 2nd March.* Information zur Rumenwicklung www.bsfrey.ch/articles/456_07.pdf

Analysing Energy Efficiency of Rail and Road Transport in Pakistan Through Data Envelopment Analysis

Muhammad Zamir Khan

1 Introduction

The transport sector's energy consumption in Pakistan is one of the most important sources of growth in total energy consumption. The transport sector (all modes) in Pakistan consumed about 16 million tons of oil equivalent (toe) energy in 2016, which accounts for 19% of total final energy consumption in the country. In comparison with the world average in general and to that of developed regions of the world, the ratio of transport's energy consumption to Total Final Consumption (TFC) in Pakistan is in most cases low. In 2016, the ratios of the nation's transport energy consumption to TFC levels in the world at large (average), Europe, OECD countries, Japan, the UK and the USA were 25.60%, 25.44%, 33.73%, 24%, 32% and 41%, respectively. However, Pakistan's ratio is much higher when compared to other Asian countries such as China (15%), Bangladesh (12%) and India (16%; IEA, 2018b). It is expected that future economic development in Pakistan, along with urbanization and industrialization, may not only accelerate the growth of transport sector but also lead to greater energy consumption.

The growing energy demand by transport may also create serious challenges for environmental sustainability. The transport fuel combustion is the 2nd largest emitter of global carbon emissions, producing 24% of global CO_2 emissions in 2016. The transport sector in Pakistan accounts for about 46.2 million tons of CO_2 emissions as of 2016, which is around 30% of total (national) emissions (IEA, 2018a). Since transport and industry sectors are growing in Pakistan, it is projected that CO_2 emissions associated with these sectors will also continue to grow (Abas et al., 2017). Therefore, it is a major challenge for policymakers and planners in Pakistan to reduce emissions by these sectors, on a priority basis, and particularly those associated with transport sector. Solutions such as incorporating additional renewable energy into the country's energy system (structure), decreasing transport demand or a modal split in

M. Z. Khan (✉)
School of Economics, Quaid-I-Azam University, Islamabad, Pakistan

© The Author(s), under exclusive license to Springer Nature Switzerland AG 2021
S. Mitra et al. (eds.), *Railway Transportation in South Asia*, Contemporary South Asian Studies, https://doi.org/10.1007/978-3-030-76878-2_6

favour of more efficient modes and reducing energy-use in transport through energy-efficient policies may all be considered feasible options for reducing transport's environmental emissions and making transport systems more sustainable.

In the current analysis, the efficiency of a specific transport mode (road, rail, etc.) may be defined as how effective it is in transforming its inputs into outputs. In other words, a transport mode is relatively more efficient than others in a test group if it produces maximum level of outputs for given input levels—or minimizes the input usage for a given output level. Two main approaches are used to estimate the efficiency of transport sectors—parametric and non-parametric approaches (see, for example, Baran & Górecka, 2019; De Borger et al., 2002; Førsund, 1992; Graham, 2008; Markovits-Somogyi, 2011b). Both approaches have merits and disadvantages. The parametric approach is further classified into stochastic frontier analysis (SFA), distribution-free approach (DFA) and thick frontier analysis (TFA). It requires a particular functional form of production frontier for estimations. However, the non-parametric approach does not assume any such prior functional form specification of frontier for estimating efficiencies. Data envelopment analysis (DEA) and free disposal hull (FDH) are the two most frequently used non-parametric techniques or approaches to efficiency measurements.

The efficiency of transport modes has been extensively studied through data envelopment analysis. Some empirical studies have also investigated the transport sector's efficiency in Pakistan using the DEA approach. However, energy efficiency analysis across different transport modes (road and rail, etc.) has not been previously conducted for Pakistan. For instance, Tahir (2013) estimated product efficiency, earning efficiency and financial efficiency for Pakistan Railways (PR), Indian Railways and Chinese Railways with the help of data envelopment analysis and concluded that product efficiency leads to both earning and financial efficiencies. Alam (2017) used data on five inputs (locomotives, coaching vehicles, freight wagons, track-length, labour) and two outputs (number of passengers, freight (tons) of Pakistan's railway over 1950–2013 so as to conduct efficiency assessments. In addition to the CCR-DEA model for measuring efficiency, the study also adopted the super-efficiency model to rank the most efficient years. More recently, Ennen and Batool (2018) used the DEA model to study the technical efficiencies of 12 major airports in Pakistan during 2012. The results indicated that several airports are cost inefficient, mainly due to overinvestment in capacity and overstaffing.

There are three basic transport modes in Pakistan: road, rail and air. However, road and rail are the most dominant modes of transport in the country. For example, road transport carries 90% of passenger and 96% of freight movements (Government of Pakistan, 2012). Each transport mode can be used for two equally important types of transport, e.g. passenger and freight transport—measured through standard indicators of passenger-kilometres (PKM) and ton-kilometres (TKM), respectively. The output of each transport mode (PKM and TKM) requires the coordinated efforts of several inputs such as labour, infrastructure and energy. In the present study, we consider energy as an input to transport; in fact, it is one of the most important inputs to transport modes for facilitating both passenger and freight transport. In many cases, transport energy consumption data is not available exclusively for passenger or freight because the same vehicles or locomotives can carry both. Although it may be possible to estimate the energy efficiencies of both passenger and freight by using some standard norms regarding the allocation of energy consumption, the accuracy of efficiency estimates largely depends on those norms (Ramanathan, 2005). Therefore, we employ a different approach, known as the data envelopment analysis (DEA), which incorporates both passenger and freight transport. The DEA is based on a linear-programming approach and provides efficiency scores for homogenous decision-making units (transport modes) by using a ratio of their outputs to their inputs.

The objective of the present study is to estimate the energy efficiency of two transport modes (road and rail) in Pakistan through the DEA—and to trace the pattern of efficiencies over the period between 1980 and 2018. Energy efficiency of each transport mode is obtained or estimated by comparing its two outputs (PKM and TKM) with one input (energy). We have left out air transport mode from this analysis for two reasons. First, data on domestic air transport outputs such as PKM and TKM is not available. Second, domestic air transport accounts for only a small share of both passenger and freight traffic as compared to road and rail.

The rest of the chapter is organized as follows. Section 2 provides a brief survey of literature on efficiency measurements in the transport sector, particularly as regards energy efficiency. Section 3 offers the DEA's methodology as well as data sources. Section 4 discusses the results of the DEA, while Sect. 5 concludes the study.

2 Literature Review

Charnes et al. (1978) first provided the concept of DEA as an effective tool of performance evaluation. It measures the relative efficiency of homogenous decision-making units (DMUs) by comparing the multiple inputs and outputs. Since the publication by Charnes et al. (1978), the DEA has been extensively used in many applied research fields such as agriculture, banks, economy, education, health and government department (Buleca & Mura, 2014; Deliktas & Günal, 2016; Henriques et al., 2018; Johnes, 2006; Stefko et al., 2018; Toma et al., 2015). Moreover, the DEA method has also been applied in various countries and regions so as to evaluate energy and environmental efficiencies (see, for example, Grigoroudis & Petridis, 2019; Lacko & Hajduová, 2018; Mardani et al., 2018; Rabar, 2017; Wang et al., 2019; Yan et al., 2018).

Empirical studies have widely used the DEA approach to study the efficiency of transport sector. DEA literature on the transport sector is quite diverse and may be further classified as based on transport mode (road, rail, air, seaport, etc.), selection of outputs and inputs for analysis and the type of DEA model used for efficiency measurements. Markovits-Somogyi (2011a) reviewed 69 transport sector studies on all transport modes—road, rail, air, ports and public transport—through DEA applications. The study mainly focused on the selection of inputs and outputs for analysis. Similarly, Cavaignac and Petiot (2017) conducted a more comprehensive review of previous DEA studies in the transport sector by including 461 studies over the period 1989–2016; it also used multiple correspondence analysis so as to describe the main research trends in the transport sector. In one of the earliest studies, Odeck and Hjalmarsson (1996) used the DEA model to measure the relative efficiency of Norwegian public trucks from 1983 to 1985. They did this through four inputs—wages, fuel consumption, costs related to rubber accessories and truck's maintenance costs—and one output, travelled-distance (kilometres). Garcia Sanchez (2009) analysed both technical and scale efficiencies of public bus transport systems with variable returns to the scale-DEA model (VRS or BCC-DEA) in twenty-four Spanish towns and further explored the determinants of technical efficiency via the Tobit regression. Fancello et al. (2014) used both constant and variable returns to scale (also known as CCR and BCC) data envelopment models to measure the relative (technical) efficiency of urban road networks in eight Italian cities. Singh and Jha (2017) employed the DEA method to investigate the efficiency performance of 15 State Transport Undertakings in India during 2003–2014. Similarly, Fitzová et al. (2018) used a sample of 19 Urban Public Transport Systems in Czech Republic from 2010 to 2015 to find the determinants of public transport efficiency. First, the DEA is used to measure the efficiency scores of each system with three inputs—energy, rolling-stock and labour—and one output, passengers. Estimated efficiency scores from the DEA are used further in a Tobit regression analysis to determine the efficiency determinants of a public transport system.

Many DEA efficiency studies are also available for other transport modes or sectors such as air, rail and ports. Therefore, it is impossible to review all of them. The studies measuring efficiency in air transport can be further classified into two ways. First, some studies use commercial activities at airports so as to measure and compare the relative efficiency of selected airports with a DEA analysis (Fung et al., 2008; Lai et al., 2015; lo Storto, 2018; Stichhauerova & Pelloneova, 2019; Wanke et al., 2016). Other studies incorporate the airlines' (input, output) data to assess the relative efficiency of various companies (Hu et al., 2017; Kottas & Madas, 2018; Lozano & Gutiérrez, 2014; Rai, 2013). A similar classification of DEA efficiency studies is also available for rail transport, in which studies either assess the performance of private or public rail transport companies (see, for example, Cantos et al., 1999; Kutlar et al., 2013; Li & Hilmola, 2019; Oum & Yu, 1994; Sharma et al., 2016; Tsai et al., 2015; Wanke & Azad, 2018) or measure the efficiency of railway stations (Duan et al., 2020; Jannah et al., 2020; Kim & Oh, 2009; Román-De la Sancha et al., 2016; Sameni & Preston, 2016; Zhong et al., 2019). Many empirical studies have also explored the technical efficiency of ports using the DEA approach (Cheon, 2008; González & Trujillo, 2009; Kutin et al., 2017; Roll & Hayuth, 1993; Zahran et al., 2017).

However, only a limited number of studies have investigated energy efficiency across different transport modes. For example, Ramanathan (2000, 2005) estimated and compared the relative energy efficiencies of Indian road and rail transport using the DEA method for the period from 1980–81 to 1993–94. Efficiency trends indicated that relative efficiency of rail transport has increased over time, such that rail is considered the most energy efficient mode in a most recent year (1993–94)—while the road's relative energy efficiency in 1993–94 was only about 63%. Moreover, the extended DEA results, based on scenario analysis, showed that huge reductions in energy consumption and related CO_2 emissions could be possible if further modal splits were promoted in favour of the most energy-efficient mode, rail. In a similar study, Lin et al. (2015) incorporated one input (energy consumption) and two outputs (PKM and TKM) in the DEA model to investigate the relative efficiencies of four transport modes—road, rail, aviation and water—in China from 1971 to 2011. The estimated efficiency scores of both rail and water transport were maximum (equal to one) in 2011, meaning that both modes are relatively the most energy efficient during 2011. However, others (road and aviation) were considered as energy inefficient in comparison with rail and water in 2011. Instead of measuring energy efficiency of transportation, Cui and Li (2015) investigated the carbon efficiency of transportation for 15 countries over the period of 2003–2010 through a virtual frontier DEA approach, a model which is superior to both traditional and super-DEA models in terms of differentiating the most efficient decision-making units. Zhang and Du (2017) incorporated two additional inputs (capital, labour) along with energy to analyse the technical efficiency of four transport modes in China via the DEA over the period 1985–2015. They found that efficiency of rail transport is higher than that of competitors, as efficiency scores of rail transport are maximum for most of the periods and have remained relatively stable over time.

Since the transport sector makes significant contributions to environmental emissions, accurate and true efficiency measurements require involving these negative externalities in the analysis. Therefore, some studies have also incorporated the environmental emissions from transport as undesirable outputs in their analysis of energy efficiency evaluation of the transport sector via extended DEA models (Bi et al., 2014; Feng & Wang, 2018; Park et al., 2018; Song et al., 2016; Zhou et al., 2014). Bi et al. (2014) adopted a non-radial DEA approach, a model which differentiates between desirable (transport value-added) and undesirable (transport CO_2 emissions) outputs, so as to evaluate the energy and environmental efficiency of 30 provinces in China from 2006 to 2010. Using a similar non-radial-DEA approach, Song et al. (2016) investigated the environmental efficiency of rail transport in 30 Chinese regions from 2006 to 2011, with two undesirable outputs (CO_2 and SO_2 emissions), and further analysed the impact of rail transport on its environmental efficiency through a beta panel regression analysis. Park et al. (2018) utilized the slack-based DEA method to assess the environmental efficiency of the transport sector in 50 states in the USA over 2004–2012, and they observed that the US transport sector was environmentally inefficient during the investigation period, with an average efficiency score by US states of less than 0.64. Later, Feng and Wang (2018) analysed the performance and determinants of energy efficiency in China's transport sector at the provincial level during 2006–2014. They used global meta-frontier Malmquist and meta-frontier DEA models to account for heterogeneity differences among provinces in their analysis of efficiency assessments. The study found that transport energy efficiencies first decreased between 2004 and 2010 due to poor management and growing regional technological disparities (gap) and then increased thereafter as the technology gap and management efficiency stabilized (Table 1).

Table 1 Summary table of some reviewed DEA transport studies

				Outputs	
Study	Unit	Model	Inputs	Desirable	Undesirable
Odeck and Hjalmarsson (1996)	Trucks	CRS & VRS-DEA	Wages, fuel, maintenance cost, cost of rubber accessories	Distance-travelled (kilometres)	-
Ramanathan (2000)	Road & rail	CRS-DEA	Energy	Passenger-kilometres, Ton-kilometres	-
Garcia Sanchez (2009)	Public buses	CRS & VRS-DEA	Staff, energy, no. of operating buses	Vehicle-kilometres, total seats, fleet-age, service-hours, average no. of stops per route, safety level	-
Bi et al. (2014)	Provincial transport sector	Non-radial DEA & multidirectional efficiency analysis	Capital, energy, employment	Transport value-added	CO_2 emissions
Fancello et al. (2014)	Bus transport	CRS & VRS-DEA	No. of registered vehicles, major attractors, no. of public buses, road safety	Level of service, average time require to reach town-hall, no. of fatal accidents, passengers-carried	-
Cui and Li (2015)	Overall transport sector	CRS, super-DEA, virtual frontier DEA	Transport CO_2 emissions, employment, transport services import value	Ton-kilometres Passenger-kilometres	-
Lin et al. (2015)	Road, rail, air & water transport	CRS-DEA	Energy consumption	Passenger-kilometres Ton-kilometres	-

(continued)

Table 1 (continued)

				Outputs	
Ennen and Batool (2018)	airports	CRS-DEA with weight restrictions	No. of runways, taxiways, terminal size, employees	No. of aircraft movements, domestic, international & total passengers, cargo volume, commercial aircraft movements	-
Park et al. (2018)	State level transport sector	SBM-DEA	Capital expenses, employment, energy	Transport value-added	CO_2 emissions
Song et al. (2016)	Railway	Non-radial DEA	Employment, capital stock, energy	Real GDP	CO_2 emissions SO_2 emissions
Alam (2017)	Railway	CRS & super-DEA	Total locomotives, freight wagons, coaching vehicles, track-kilometres, labour	No. of passengers-carriage, freight-carriage (in tons)	-
Singh and Jha (2017)	Public buses	VRS-DEA	Staff, energy, no. of buses	Passenger-kilometres, Bus-kilometres	-
Zhang and Du (2017)	Road, rail, air & water transport	CRS & VRS-DEA	Employment, transit-mileage (road, rail, water), fixed asset investments	Freight (in tons), passengers (numbers)	-
Feng and Wang (2018)	Regional transport sector	Global meta-frontier DEA	Employment capital, energy	Transport's gross output	CO_2 emissions

3 Methodological Framework and Data

3.1 Data Envelopment Analysis

The DEA model is widely used to measure the relative efficiency scores of homogenous productive units, also known as DMUs. The DEA model is more flexible in the sense that it incorporates multiple inputs and multiple outputs to measure the efficiency of firms. For an efficient DMU, the maximum efficiency score produced by the DEA model is 1, which means that a particular DMU under investigation is 100% efficient.

The basic DEA model, also known as CCR model, is proposed by Charnes et al. (1978) and can be explained as follows:

Suppose that there is a homogenous DMU that consumes inputs and produces outputs. In particular, let assume that a specific DMU under consideration, say k-th DMU, uses $x_{ik}(i = 1, 2, \ldots .r, k = 1, 2, \ldots .n)$ inputs to produce $y_{kp}(p = 1, 2, \ldots .s)$ outputs. The information regarding inputs and outputs can be written in the matrix form as follows:

$\boldsymbol{x_k} = (x_{1k}, x_{2k}, \ldots, x_{rk})^{\mathrm{T}},\ k = 1, 2 \ldots, n$

$$\boldsymbol{y_k} = (y_{1k}, y_{2k}, \ldots, y_{sk})^{\mathrm{T}},\ k = 1, 2, \ldots, n \tag{1}$$

$\vartheta = (\vartheta_1, \vartheta_2, \ldots, \vartheta_r)^{\mathrm{T}}$

$\boldsymbol{\mu} = (\mu_1, \mu_2, \ldots, \mu_s)^{\mathrm{T}}$

where μ and ϑ are the vectors representing the weights assigned to output and input, respectively.

The efficiency of a particular ko-th $(1 \le k_o \le n)$ DMU in the CCR model can be obtained by maximizing the ratio of the weighted sum of its outputs to the weighted sum of its inputs—subject to the constraint that the efficiency score for each DMU is restricted between 0 and 1. Therefore, the output-oriented CCR-DEA model can be written in a standard matrix form as follows:

Max $\frac{\mu^{\mathrm{T}} \boldsymbol{y}_{k_o}}{\vartheta^{\mathrm{T}} \boldsymbol{x}_{k_o}}$

Subject to

$$\frac{\mu^{\mathrm{T}} \boldsymbol{y}_k}{\vartheta^{\mathrm{T}} \boldsymbol{x}_k} \le 1,\ k = 1, 2, \ldots\ldots, n \tag{2}$$

$\mu \ge 0, \vartheta \ge 0$

The DEA model in Eq. (2) is a linear fractional program with multiple solutions. Therefore, it can be converted into a linear program (LP) through CC (Charnes-Cooper) transformation. The linear program can be described as follows:

Max $u^{\mathrm{T}} y_{k_0}$

Subject to

$$v^{\mathrm{T}} x_k - u^{\mathrm{T}} y_k \ge 0,\ k = 1, 2, \ldots, n \tag{3}$$

$v^T x_{k_0} = 1$

$u \geq 0, v \geq 0$

where u and v ($u = t\mu$, $v = t\vartheta$) are new vectors associated with output and input weights, respectively.

We consider n DMUs in a general model. So, to calculate the efficiency of each DMU through the DEA, the linear program in Eq. (3) can be solved n times.

3.2 Data Sources

This study estimates and compares the energy efficiency of road and rail transport in Pakistan using the DEA model, with two outputs ('PKM, TKM') and one input ('energy'). For rail transport, the data on PKM, TKM and energy consumption was obtained from Pakistan Railways (PR yearbook, various issues). However, the energy consumption data is not separately available from that regarding other specific transport modes, e.g. road and air. However, energy consumption in the overall transport sector is available from the Hydrocarbon Development Institute of Pakistan (HDIP; Energy yearbook) by fuel type e.g. aviation fuel, motor spirit, HSD, natural gas, etc. So, we derived the approximate figures of the road sector's energy consumption by subtracting the rail's energy consumption and others (aviation fuel, kerosene oil) from the total energy consumption of the transport sector. Therefore, the road's energy consumption is taken from the HDIP (Energy yearbook, various issues). The remaining road's data of PKM and TKM is obtained from various government sources (Economic Survey of Pakistan, several issues; Pakistan Statistical yearbook, several issues; Oil companies' Advisory Committee, several years). For further information regarding road passenger-kilometres and ton-kilometres, see also Karandaaz Pakistan (2018). The data for efficiency evaluation of rail and road transport ranges from 1980 to 2018.

4 Results and Discussion

4.1 Efficiency Results

The energy efficiency of rail (1980–2018) and road (1980–2016) transport in Pakistan is estimated in such a way that each period is used as an independent decision-making unit (DMU) for efficiency measurements. Time series' data regarding outputs (PKM, TKM) and input (energy) of both rail and road is provided below in Table 2. For rail transport, the data regarding energy consumption reflects a clear declining trend, wherein it was reduced from 0.43 million tons of oil equivalent (mtoe) in 1980 to 0.15 million tons of oil equivalent (mtoe) in 2018. On the other hand, rail's TKM figures have initially decreased over time and thereafter experienced an upward movement.

Table 2 Data description of rail and road transport in Pakistan over 1980–2018

	Rail transport			Road transport		
	Outputs		Inputs	Outputs		Inputs
Years	PKM (billion)	TKM (billion)	Energy (toe) (million)	PKM (billion)	TKM (billion)	Energy (toe) (million)
1980–81	16.4	7.91	0.43	66	18.2	2.20
1985–86	16.7	7.28	0.37	86	26.4	3.11
1990–91	20	5.70	0.24	128	35.2	4.61
1995–96	18.90	5.07	0.18	154.6	79.9	7.07
2000–01	19.6	4.52	0.14	208.4	107	8.05
2005–06	25.62	4.97	0.15	238	124.5	8.76
2010–11	20.61	1.75	0.10	263.8	152.2	11.30
2015–16	21.20	4.77	0.14	282.5	167	14.66
2017–18	24.90	8.08	0.15	N.A	N.A	-

Source Pakistan railway yearbook (various issues), economic survey of Pakistan (various issues), Pakistan statistical yearbook (various issues), Oil companies' Advisory Committee (various issues)

The PKMs also have experienced a mixed pattern of ups and downs but have generally increased over time. So, a positive growth in outputs of rail with less use of inputs (energy) or resources implies energy efficiency in its operations.

Data on the road's input (energy) and outputs (PKM, TKM) depicts a strong positive growth throughout the period from 1980 to 2016. For instance, PKMs (TKMs) have increased from 66 billion (18.2 billion) to 282.5 billion (167 billion) over the period 1980–2016, with an annual average growth rate of 4.12% (6.35%), while energy consumption of the road sector has increased by an average growth rate of 5.40% during the same period. Since outputs, particularly the PKM, have grown faster than the growth in energy consumption, some growth in efficiency is also observed in road transport. Although some guess about patterns of energy efficiency in road and rail transport can be roughly observed by comparing the growth of inputs relative to outputs, it is not a substitute for a more formal mathematical programming approach such as the DEA, which is used next for efficiency measurements.

Next, the energy efficiency of both rail and road transport is estimated over time from 1980 to 2018 and results are reported in Table 3. The DEA was conducted with the help of the DEAP 2.1 software, written by Coelli (1996). The efficiency scores show that rail transport, in general, has experienced an increasing trend in energy efficiency. Two years, 2011–12 and 2017–18, were identified as the most energy-efficient years for rail—as compared to the remaining years, in which the efficiency values have reached the maximum of 100%. For example, rail was only 46.2% energy efficient in 1980–81, as compared to its performance in 2011–12 and 2017–18. This implies that in these two years rail has used its input resources, particularly energy, in a relatively more efficient manner so as to produce or facilitate outputs (PKM, TKM) than in remaining years. A major reason for improvements in energy efficiency of rail

Table 3 Energy efficiency of rail and road transport during 1980–2018

Years	Rail efficiency (%)	Road efficiency (%)
1980–81	36	100
1981–82	32	96
1982–83	36	97
1983–84	37	94
1984–85	36	94
1985–86	43	92
1986–87	50	82
1987–88	57	93
1988–89	62	94
1989–90	54	91
1990–91	51	92
1991–92	56	81
1992–93	61	79
1993–94	59	80
1994–95	69	81
1995–96	63	80
1996–97	68	84
1997–98	52	87
1998–99	76	88
1999–2000	73	89
2000–01	80	94
2001–02	89	95
2002–03	97	96
2003–04	90	94
2004–05	92	90
2005–06	95	100
2006–07	97	100
2007–08	94	87
2008–09	98	94
2009–10	97	93
2010–11	96	93
2011–12	100	90
2012–13	94	92
2013–14	94	93
2014–15	88	88
2015–16	86	79
2016–17	88	N.A
2017–18	100	N.A

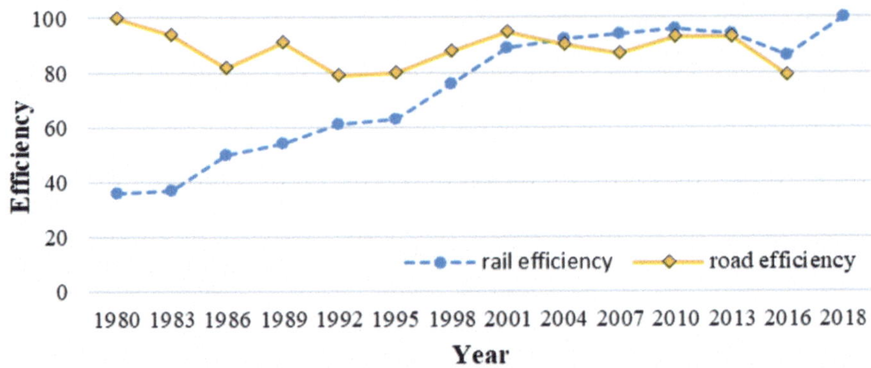

Fig. 1 Energy efficiency patterns of rail and road transport over 1980–2018 (*Source* Author's calculations)

transport in Pakistan over time is that it has gradually shifted from the less efficient coal traction to a more efficient diesel traction.

The estimated energy efficiency values or scores of road transport in Pakistan are also provided in Table 3, next to rail transport. These efficiency values indicate that the road sector is also considered the most energy efficient sector in 1980–81, 2005–06 and 2006–07, which implies that for these three years it has relatively better utilized its input of energy to facilitate the PKM and TKM on roads. The relative performance of road transport in remaining years was relatively less efficient, although efficiency scores of inefficienct years are not very different to those of efficient years. The trend of road energy efficiency shows a mixed pattern of ups and downs. Initially, it witnessed a declining trend for about a decade after 1980–81, and then experienced an upward trend till 2006–07, after which the observed efficiency has decreased.

The same efficiency patterns are shown in Fig. 1. The energy efficiency of rail transport has gradually increased from 36% in 1980 to 100% in 2018. However, the road sector's energy efficiency has decreased from 100% in 1980 to 79% in 2016. This shows that energy efficiency of rail transport in Pakistan has increased over time, while that of road's efficiency has generally decreased over time.

4.2 Sensitivity Results

In addition to energy efficiency estimation, further sensitivity results were also assessed. These show how the performance of rail and road transport during less efficient years could have become efficient by reducing the energy consumption or increasing the passenger-kilometres and ton-kilometres. For inefficient years, the energy (input) and PKM and TKM (outputs) slacks represent the overuse of energy and under-production of PKM and TKM, respectively. The slacks of energy as well as PKM and TKM are given in Table 4. In 1980–81, the energy consumption slack

Table 4 Sensitivity analysis of road and rail transport during 1980–2018

	Rail transport slacks (%)			Road transport slacks (%)		
years	Energy	PKM	TKM	Energy	PKM	TKM
1980–81	64	49	0.00	0.00	0.00	0.00
1981–82	68	32	0.00	3.10	0.00	0.00
1982–83	64	25	0.00	2.40	0.00	0.00
1983–84	64	24	0.00	6.00	0.00	0.00
1984–85	64	25	0.00	5.70	0.00	0.00
1985–86	57	53	0.00	7.20	0.00	0.00
1986–87	50	42	0.00	17.40	0.00	0.00
1987–88	43	47	0.00	6.70	0.00	0.00
1988–89	38	31	0.00	5.90	0.00	0.00
1989–90	46	9	0.00	8.90	0.00	0.00
1990–91	49	0.00	0.00	8.00	0.00	0.30
1991–92	44	1	0.00	18.50	0.00	0.00
1992–93	39	12	0.00	20.80	0.00	0.00
1993–94	41	12	0.00	19.30	0.00	0.00
1994–95	31	18	0.00	19.10	0.00	0.00
1995–96	37	0.00	0.00	19.80	0.00	0.00
1996–97	32	0.00	0.00	15.80	0.00	0.00
1997–98	48	0.00	0.00	12.10	0.00	0.00
1998–99	24	0.00	0.00	11.80	0.00	0.00
1999–2000	27	0.00	0.00	11.30	0.00	0.00
2000–01	20	0.00	0.00	5.10	0.00	0.00
2001–02	11	0.00	0.00	4.40	0.00	0.00
2002–03	4	0.00	0.00	3.40	0.00	0.00
2003–04	10	0.00	0.00	5.50	0.00	0.00
2004–05	9	0.00	0.00	9.10	0.00	0.00
2005–06	6	0.00	0.00	0.00	0.00	0.00
2006–07	3	0.00	0.00	0.00	0.00	0.00
2007–08	6	0.00	0.00	12.30	0.00	0.00
2008–09	2	0.00	0.00	5.70	0.00	0.00
2009–10	3	0.00	0.00	6.40	12.12	0.00
2010–11	4	0.00	0.00	6.20	7.95	0.00
2011–12	0.00	0.00	0.00	9.30	28.15	0.00
2012–13	6	3.74	3.74	8.00	24.51	0.00
2013–14	7	0.00	0.00	6.90	23.01	0.00
2014–15	12	0.00	0.00	11.40	22.85	0.00

(continued)

Table 4 (continued)

	Rail transport slacks (%)			Road transport slacks (%)		
years	Energy	PKM	TKM	Energy	PKM	TKM
2015–16	14	0.00	0.00	20.70	22.74	0.00
2016–17	12	0.00	0.00	N.A	N.A	N.A
2017–18	0.00	0.00	0.00	N.A	N.A	N.A

Source Author' calculations

of rail was 64%, which implies that rail transport in 1980–81 could have been as efficient as its performance in most efficient years (2011–12 and 2017–18) if it had consumed 64% less energy with the same outputs of PKM and TKM. Similarly, the slack of PKM in 1980–81 was 49%. This means that to achieve the same energy efficiency performance of rail sector in 1980–81 as it is in the two most efficient years, the rail sector should have increased the performance of PKM by 49% so as to match the performance of the other factors (TKM and energy consumption).

The slacks in road's inputs (energy) and outputs (PKM, TKM) are also provided in Table 4. Since the energy efficiency of road reached maximum levels in 1980–81, 2005–06 and 2006–07, the slacks in both energy and PKM and TKM in these years are zero—while at least one of these slacks (energy, TKM, PKM) are zero for the remaining, inefficient years. For example, the energy efficiency of road transport in 2015–16 is 79.30%—with slacks of energy and PKM are 20.70 and 22.74%, respectively. This indicates that if road transport in 2015–16 was to be as efficient as it was during 1980–81, 2005–06 and 2006–07, it should either have consumed 21% less energy with same levels of PKM and TKM or it should have realized 23% more PKM with the same performance of energy consumption and TKM.

5 Conclusion

Over the last few decades or so, energy consumption of the transport sector in Pakistan has increased rapidly due to increasing transport sector activities (passenger and freight transport), economic growth and urbanization. It is expected to grow further in future. The transport sector consumes energy mostly in the form of petroleum products, which has direct implications for environmental emissions. Therefore, it has raised major concerns among policymakers and planners in Pakistan regarding environmental sustainability. Moreover, energy is used as one of the most important inputs by various transport modes (e.g. road, rail, air, etc.) for smooth facilitation of passenger and freight transport movements. Estimating and evaluating the energy efficiency of these modes could offer important insights regarding environmental sustainability and sustainability of the transport system. The present study has estimated the energy efficiency of the two most important transport modes (road and rail) in Pakistan; it has further traced the patterns of efficiency over time from 1980 to 2018. In order to calculate estimations, we adopted a non-parametric approach of DEA, which is a linear-programming approach widely used to estimate the efficiencies of homogenous decision-making units by comparing their outputs to inputs. In our case, we used two outputs (passenger-kilometres and ton-kilometres) and one input (energy) for both road and rail as concerns efficiency assessments.

The results, based on efficiency scores, have suggested that the energy efficiency of rail transport has increased over time as Pakistan's railways have moved from less efficient coal traction to the relatively more efficient diesel traction. By contrast, energy efficiency of road transport has generally decreased over time. The estimated relative energy efficiency score of rail transport in a recent year (2018) is relatively maximum (100%), while the road's efficiency score in a recent year (2016) is only 79%. This implies that since rail is relatively more energy efficient than road in terms of facilitating both PKM and TKM, shifting some future passenger and freight traffic from road to rail mode could bring potential energy savings and related lowered CO_2 emissions. This policy implication is in line with the current strategy by the Government of Pakistan, which has decided to increase the share of rail transport from 4 to 20% over the period 2015–2025 (Planning Commission, 2014).

There are also some limitations to the present study. First, since transport modes also generate various types of negative externalities as by-products (such as CO_2 emissions, accidents, air pollution and congestion/road transport), these needed to be incorporated into the analysis for better efficiency measurements. Second, the present study used only energy consumption of both road and rail as one of the most important inputs for carrying passenger and freight transport. However, other inputs such as labour, capital stock and infrastructure also play equally important role in facilitating passenger and freight transport.

References

Abas, N., Kalair, A., Khan, N., & Kalair, A. R. (2017). Review of GHG emissions in Pakistan compared to SAARC countries. *Renewable and Sustainable Energy Reviews, 80*, 990–1016.

Alam, K. M. (2017). Efficiency evaluation of Pakistan railways using data envelopment analysis. *Civil and Environmental Research, 9*(4), 1–6.

Baran, J., & Górecka, A. K. (2019). Economic and environmental aspects of inland transport in EU countries. *Economic Research-Ekonomska Istraživanja, 32*(1), 1037–1059.

Bi, G., Wang, P., Yang, F., & Liang, L. (2014). Energy and environmental efficiency of China's transportation sector: a multidirectional analysis approach. *Mathematical Problems in Engineering*, 2014. https://doi.org/10.1155/2014/539596

Buleca, J., & Mura, L. (2014). Quantification of the efficiency of public administration by data envelopment analysis. *Procedia Economics and Finance, 15*, 162–168.

Cantos, P., Pastor, J. M., & Serrano, L. (1999). Productivity, efficiency and technical change in the European railways: A non-parametric approach. *Transportation, 26*(4), 337–357.

Cavaignac, L., & Petiot, R. (2017). A quarter century of Data Envelopment Analysis applied to the transport sector: A bibliometric analysis. *Socio-Economic Planning Sciences, 57*, 84–96.

Charnes, A., Cooper, W. W., & Rhodes, E. (1978). Measuring the efficiency of decision-making units. *European Journal of Operational Research, 2*(6), 429–444.

Cheon, S. (2008). Productive efficiency of world container ports: A global perspective. *Transportation Research Record, 2062*(1), 10–18.
Coelli, T. (1996). *A guide to DEAP version 2.1: A data envelopment analysis (computer) program.* Centre for Efficiency and Productivity Analysis, University of New England, Australia.
Cui, Q., & Li, Y. (2015). An empirical study on the influencing factors of transportation carbon efficiency: Evidences from fifteen countries. *Applied Energy, 141*, 209–217.
De Borger, B., Kerstens, K., & Costa, A. (2002). Public transit performance: What does one learn from frontier studies? *Transport Reviews, 22*(1), 1–38.
Deliktas, E., & Günal, G. G. (2016). Economic growth and input use efficiency in low, Upper-Middle and high income countries (1991–2011): A data envelopment analysis. *Procedia Economics and Finance, 38*, 308–317.
Duan, Y.-Q., Fan, X.-Y., Liu, J.-C., & Hou, Q.-H. (2020). Operating efficiency-based data mining on intensive land use in smart city. *IEEE Access, 8*, 17253–17262.
Ennen, D., & Batool, I. (2018). Airport efficiency in Pakistan: A data envelopment analysis with weight restrictions. *Journal of Air Transport Management, 69*, 205–212.
Fancello, G., Uccheddu, B., & Fadda, P. (2014). Data Envelopment Analysis (DEA) for urban road system performance assessment. *Procedia-Social and Behavioral Sciences, 111*, 780–789.
Feng, C., & Wang, M. (2018). Analysis of energy efficiency in China's transportation sector. *Renewable and Sustainable Energy Reviews, 94*, 565–575.
Fitzová, H., Matulová, M., & Tomeš, Z. (2018). Determinants of urban public transport efficiency: Case study of the Czech Republic. *European Transport Research Review, 10*(2), 42.
Fung, M. K. Y., Chow, C. K. W., Hui, Y. V., & Law, J. S. (2008). Measuring the efficiency of airports in China with the DEA and endogenous-weight TFP methods. *International Journal of Transport Economics, 35*(1), 45–73. https://www.jstor.org/stable/42747842
Førsund, F. R. (1992). A comparison of parametric and non-parametric efficiency measures: The case of Norwegian ferries. *Journal of Productivity Analysis, 3*(1–2), 25–43.
González, M. M., & Trujillo, L. (2009). Efficiency measurement in the port industry: A survey of the empirical evidence. *Journal of Transport Economics and Policy (JTEP), 43*(2), 157–192.
Government of Pakistan. (2012). *Economic Survey of Pakistan.* http://www.finance.gov.pk/survey/chapters_13/13-Transportfinal.pdf
Garcia Sanchez, I. M. (2009). Technical and scale efficiency in Spanish urban transport: Estimating with data envelopment analysis. *Advances in Operations Research, 2009.*
Graham, D. J. (2008). Productivity and efficiency in urban railways: Parametric and non-parametric estimates. *Transportation Research Part E: Logistics and Transportation Review, 44*(1), 84–99.
Grigoroudis, E., & Petridis, K. (2019). Evaluation of national environmental efficiency under uncertainty using data envelopment analysis. In H. Doukas, A. Flamos, & J. Lieu (Eds.), *Understanding risks and uncertainties in energy and climate policy*. Springer.
Henriques, I. C., Sobreiro, V. A., Kimura, H., & Mariano, E. B. (2018). Efficiency in the Brazilian banking system using data envelopment analysis. *Future Business Journal, 4*(2), 157–178.
Hu, J.-L., Li, Y., & Tung, H.-J. (2017). Operational efficiency of ASEAN airlines: Based on DEA and bootstrapping approaches. *Management Decision, 55*(5), 957–986.
International Energy Agency. (2018a). CO_2 emissions from fuel combustion highlights: 2018 edition.
International Energy Agency. (2018b). *World energy balances.*
Jannah, A. N., Muthohar, I., & Parikesit, D. (2020). Efficiency analysis of commuter line stations (Case study: Tebet Station to Cilebut Station, Indonesia). In *2nd International Symposium on Transportation Studies in Developing Countries (ISTSDC 2019)*, 131–135.
Johnes, J. (2006). Data envelopment analysis and its application to the measurement of efficiency in higher education. *Economics of Education Review, 25*(3), 273–288.
Karandaaz Pakistan. (2018). Bankability of the Transport Sector.
Kim, S.-H., & Oh, S.-M. (2009). Evaluating the performance of freight railway stations using data envelopment analysis. *Journal of the Korean Society for Railway, 12*(2), 315–320.

Kottas, A. T., & Madas, M. A. (2018). Comparative efficiency analysis of major international airlines using Data Envelopment Analysis: Exploring effects of alliance membership and other operational efficiency determinants. *Journal of Air Transport Management, 70*, 1–17.
Kutin, N., Nguyen, T. T., & Vallée, T. (2017). Relative efficiencies of ASEAN container ports based on data envelopment analysis. *The Asian Journal of Shipping and Logistics, 33*(2), 67–77.
Kutlar, A., Kabasakal, A., & Sarikaya, M. (2013). Determination of the efficiency of the world railway companies by method of DEA and comparison of their efficiency by Tobit analysis. *Quality & Quantity, 47*(6), 3575–3602.
Lacko, R., & Hajduová, Z. (2018). Determinants of environmental efficiency of the EU countries using two-step DEA approach. *Sustainability, 10*(10), 3525.
Lai, P., Potter, A., Beynon, M., & Beresford, A. (2015). Evaluating the efficiency performance of airports using an integrated AHP/DEA-AR technique. *Transport Policy, 42*, 75–85.
Li, W., & Hilmola, O.-P. (2019). Belt and road initiative and railway sector efficiency—Application of networked benchmarking analysis. *Sustainability, 11*(7), 2070.
Lin, W., Chen, B., Xie, L., & Pan, H. (2015). Estimating energy consumption of transport modes in China using DEA. *Sustainability, 7*(4), 4225–4239.
lo Storto, C. (2018). The analysis of the cost-revenue production cycle efficiency of the Italian airports: A NSBM DEA approach. *Journal of Air Transport Management, 72*, 77–85.
Lozano, S., & Gutiérrez, E. (2014). A slacks-based network DEA efficiency analysis of European airlines. *Transportation Planning and Technology, 37*(7), 623–637.
Mardani, A., Streimikiene, D., Balezentis, T., Saman, M. Z. M., Nor, K. M., & Khoshnava, S. M. (2018). Data envelopment analysis in energy and environmental economics: An overview of the state-of-the-art and recent development trends. *Energies, 11*(8), 2002.
Markovits-Somogyi, R. (2011a). Measuring efficiency in transport: The state of the art of applying data envelopment analysis. *Transport, 26*(1), 11–19.
Markovits-Somogyi, R. (2011b). Review and systematization of efficiency measurement methods used in the transport sector. *Promet-Traffic & Transportation, 23*(1), 39–47.
Odeck, J., & Hjalmarsson, L. (1996). The performance of trucks—An evaluation using data envelopment analysis. *Transportation Planning and Technology, 20*(1), 49–66.
Oum, T. H., & Yu, C. (1994). Economic efficiency of railways and implications for public policy: A comparative study of the OECD countries' railways. *Journal of Transport Economics and Policy*, 121–138
Park, Y. S., Lim, S. H., Egilmez, G., & Szmerekovsky, J. (2018). Environmental efficiency assessment of US transport sector: A slack-based data envelopment analysis approach. *Transportation Research Part D: Transport and Environment, 61*, 152–164.
Planning Commission. (2014). *Pakistan vision 2025*. Ministry of Planning.
Rabar, D. (2017). An overview of data envelopment analysis application in studies on the socio-economic performance of OECD countries. *Economic Research-Ekonomska Istraživanja, 30*(1), 1770–1784.
Rai, A. (2013). Measurement of efficiency in the airline industry using data envelopment analysis. *Investment Management and Financial Innovations, 10*(1), 38–45.
Ramanathan, R. (2000). A holistic approach to compare energy efficiencies of different transport modes. *Energy Policy, 28*(11), 743–747.
Ramanathan, R. (2005). Estimating energy consumption of transport modes in India using DEA and application to energy and environmental policy. *Journal of the Operational Research Society, 56*(6), 732–737.
Roll, Y., & Hayuth, Y. (1993). Port performance comparison applying data envelopment analysis (DEA). *Maritime Policy and Management, 20*(2), 153–161.
Román-De la Sancha, A., Mayoral, J. M., & Román, L. I. (2016). Modeling urban transfer stations efficiency. *Procedia Computer Science, 83*, 18–25.
Sameni, M. K., & Preston, J. (2016). Evaluating efficiency of passenger railway stations: A DEA approach. *Research in Transportation Business & Management, 20*, 33–38.

Sharma, M. G., Debnath, R. M., Oloruntoba, R., & Sharma, S. M. (2016). Benchmarking of rail transport service performance through DEA for Indian railways. *The International Journal of Logistics Management, 27*(3), 629–649. https://doi.org/10.1108/IJLM-08-2014-0122
Singh, S. K., & Jha, A. P. (2017). Efficiency and effectiveness of state transport undertakings in India: A DEA approach. *Theoretical Economics Letters, 7*(6), 1646.
Song, M., Zhang, G., Zeng, W., Liu, J., & Fang, K. (2016). Railway transportation and environmental efficiency in China. *Transportation Research Part D: Transport and Environment, 48*, 488–498.
Stefko, R., Gavurova, B., & Kocisova, K. (2018). Healthcare efficiency assessment using DEA analysis in the Slovak Republic. *Health Economics Review, 8*(1), 6.
Stichhauerova, E., & Pelloneova, N. (2019). An efficiency assessment of selected German airports using the DEA Model. *Journal of Competitiveness, 11*(1), 135–151.
Tahir, N. (2013). Efficiency analysis of Pakistan Railway in comparison with China and India. *International Journal of Transport Economics/Rivista Internazionale Di Economia Dei Trasporti*, 71–98.
Toma, E., Dobre, C., Dona, I., & Cofas, E. (2015). DEA applicability in assessment of agriculture efficiency on areas with similar geographically patterns. *Agriculture and Agricultural Science Procedia, 6*, 704–711.
Tsai, C. H. P., Mulley, C., & Merkert, R. (2015). Measuring the cost efficiency of urban rail systems an international comparison using DEA and Tobit models. *Journal of Transport Economics and Policy (JTEP), 49*(1), 17–34.
Wang, L.-W., Le, K.-D., & Nguyen, T.-D. (2019). Assessment of the energy efficiency improvement of twenty-five countries: A DEA approach. *Energies, 12*(8), 1535.
Wanke, P., Barros, C. P., & Nwaogbe, O. R. (2016). Assessing productive efficiency in Nigerian airports using Fuzzy-DEA. *Transport Policy, 49*, 9–19.
Wanke, P., & Azad, M. A. K. (2018). Efficiency in Asian railways: A comparison between data envelopment analysis approaches. *Transportation Planning and Technology, 41*(6), 573–599.
Yan, Q., Wang, X., Baležentis, T., & Streimikiene, D. (2018). Energy–economy–environmental (3E) performance of Chinese regions based on the data envelopment analysis model with mixed assumptions on disposability. *Energy & Environment, 29*(5), 664–684.
Zahran, S. Z., Alam, J. B., Al-Zahrani, A. H., Smirlis, Y., Papadimitriou, S., & Tsioumas, V. (2017). Analysis of port authority efficiency using data envelopment analysis. *Maritime Economics & Logistics, 19*(3), 518–537.
Zhang, Z., & Du, L. (2017). Research on evaluation of resource allocation efficiency of transportation system based on DEA. *IOP Conference Series: Materials Science and Engineering, 207*(1), 12056.
Zhong, Y., Chen, J., & Liu, X. (2019). Evaluation model of transfer efficiency in rail transit station. In *2019 12th International Conference on Intelligent Computation Technology and Automation (ICICTA)*, 492–495.
Zhou, G., Chung, W., & Zhang, Y. (2014). Measuring energy efficiency performance of China's transport sector: A data envelopment analysis approach. *Expert Systems with Applications, 41*(2), 709–722.

Rail Freight Transport System in Tripura: An Analysis of Performances and Prospects

Stabak Roy and Saptarshi Mitra

1 Introduction

Rail freight transportation is one of the most important economic driving forces for regional development (Ginevicius & Sinkevicius, 2020). Rail freight transportation is a primary component of all supply chain and logistics systems (Yi, 2018). Generally, rail freight transport is used so as to carry a high volume of goods at a low cost (Zunder & Islam, 2018). The capacity to securely, rapidly transport goods from one place to another, as well as the cost of transportation, constitutes foundations of a viable framework for rail freight transport (Marinov et al., 2013). Rail freight transportation boosts the regional economy by providing access to both international and domestic markets. In recent years, demand for rail freight transport has significantly increased (Islam et al., 2015). In 2017, the value of the Global Rail Freight Transport market was about $294.08 billion and expected to reach $414.96 billion by 2026, thus growing at a Compound Annual Growth Rate (CAGR) of 3.9% (Business Wire, 2019). The continuous growth of rail freight transport is a result of global trade flows, strong regional cooperation and stable domestic demand. The demand for rail freight transport is embryonic, both in its relations with nature, as concerns characteristics of the goods involved and as concerns requirements by stakeholders (Tseng et al., 2005). Freight transportation is highly dependent on demand-driven trade flows and market dynamics (Rodrigue et al., 2013). The world's largest rail freight transportation system lies in the USA—140,000 miles (225,308.16 km) of railway tracks are operated by freight carriers (Research and Innovative Technology Administration, 2010). India holds the fourth position—after the USA, Russia and China—with about 7,000 freight trains, over 65,000 kilometres of routes across 7,146 railway stations (Roy & Mitra, 2016). The expected worth of the rail freight transport market in India is around $307.70 billion—with an expected CAGR of 13.35% by 2020. India has the largest, growing sectors of manufacturing, retail,

S. Roy · S. Mitra (✉)
Department of Geography and Disaster Management, Tripura University, Suryamaninagar, India

© The Author(s), under exclusive license to Springer Nature Switzerland AG 2021

S. Mitra et al. (eds.), *Railway Transportation in South Asia*, Contemporary South Asian Studies, https://doi.org/10.1007/978-3-030-76878-2_7

Fast-Moving Consumer Goods (FMCG) and e-commerce (Newswire, 2015). In the 2018–2019 financial year, Indian Railways hauled 1,222 million tons—mainly, coal (552 million tons), iron ores (117 million tons), cement (105 million tons), mineral oils (43 million tons), food grains (46 million tons), fertilisers (52 million tons) and iron and steel (45 million tons; Mozumder, 2020) With increasing demand and subsequent growth in comprehensive trade, rail freight infrastructure takes on a crucial role in India's economic development (Chandra & Jain, 2007). Rail freight infrastructure and demand are not homogeneously distributed across the country. The demand for rail freight services is higher in industrially developed areas and significantly lower in industrially backward areas (Sharma, 2016). Tamil Nadu, Maharashtra, Gujarat, Uttar Pradesh and Andhra Pradesh are industrially developed states of India; 53% of India's total factories are located in these states (Business Today, 2016). The Government of India delineated 16 states as industrially backward regions—namely, Arunachal Pradesh, Assam, Goa, Himachal Pradesh, Jammu and Kashmir, Manipur, Meghalaya, Mizoram, Nagaland, Sikkim, Tripura, the Andaman and Nicobar Islands, Dadra and Nagar Haveli, Daman and Diu, Lakshadweep and Pondicherry (Planning Commission, 1981). All eight north-eastern states—including Sikkim—fall under this category. Naturally, rail freight demand in this region is not dependent on industrial activities, but on domestic demand. However, this region is also nature's storehouse—containing mineral resources, forest resources, water resources and plant resources; this captivated Britishers (Dikshit & Dikshit, 2014; Pandey, 2008). In accordance with the imperialist interests of the day, the Assam-Bengal Railway was developed in 1892 so as to export tea, oil and other natural resources from this region (Guha, 1968; Hilaly, 2016; Singh et al., 2011). After independence, a re-indemnification of transport infrastructure has taken place. At present, Northeast India has 3789 km railway and 344 railway stations. In recent years, the greatest railway infrastructural development has been taken place in Tripura. The state has huge potentialities in rail freight transportation due to its strategic location.

2 Context of the Study Area

The research has been conducted in Tripura, a north-eastern state of India covering 10,486 sq. km. A landlocked state, it is encircled by Bangladesh to north, south and west (856 km) and by the Indian states of Assam (53 km) and Mizoram (109 km) to north-east and east, respectively (Roy and Mitra, 2020). Tripura is extravagantly a hilly state, with a valley and ridge (*tilla* and *lunga*) topography (Sen et al., 2015). About 60% of its land is hilly, while the remaining 40% is flat land. Even the flat land is broken by many low hills and tillas of 30–60 metres in elevation, covered with trees and shrubs (Saha, 2014). The state has six anticlinal hill ranges—the Baramura, Atharamura, Longtharai, Shakhan, Jampui and Devtamura. Railway tracks cross the major hills of Longtharai (515 m), Atharamura (481 m) and Baramura (249 m) from north to south (Roy & Mitra, 2015). Many rivers, narrow streams, gullies and ravines originate in those hills. The state has 11 major rivers: from north to south, the Longai

(98 km), Juri and Deo (98 km), Manu (167 km), Dhalai (117 km), Khowai (70 km), Haora (53 km), Bijoy (26), Gomati and Muhuri (64 km) and the Feni (Tripura State Control Board, 2013). Road networks in Tripura are very poor, and landslides during the monsoon are a most common phenomenon in the hilly terrains (Sen et al., 2013). In these circumstances, railways have become the prime mode of transportation in Tripura (Roy & Mitra, 2016). According to the Census of India (2011), the total population of the state is about 3,673,917—with a population density of 350.36. The net agricultural area of the state is only 2,077.2 sq. km. (Department of Agriculture, Cooperation & Farmers Welfare Ministry of Agriculture & Farmers Welfare, 2018). Due to the high man-land ratio, huge pressure affects agricultural lands. As a result, Tripura became a food deficit state obliged to import food and other necessary materials from elsewhere in India (De, 2004). The costs of transport by road are much higher than transport through railways. Railways are always a more dependable mode of transportation, especially in a hilly region like Tripura, wherein during the monsoon road deteriorates due to landslides. Railways have played a significant role in freight transport in Tripura since 2016 (after gauge conversion; Singh, 2016). At present, the state contains a 264 km-long railway track with 27 railway stations. The railway line extends from the Churaibari Railway Station (CBZ; 24°26′N and 92°14′E) in the north to the Sabroom railway station (SBRM; 23°00'N and 91°41'E) in the south. There are twenty-five (25) intermediate stations, namely Nadiapur (NPU; 24°23′N and 92°12′E), Dharmanagar (DMR; 24°22′N and 92°10′E), Panisagar (PASG; 24°16′N and 92°09′E), Pecharthal (PEC; 24°11′N and 92°06′E), Kumarghat (KUGT; 24°09′N and 92°02′E), Nalkata (NLKT; 24°03′N and 92°00′E), Manu (MANU; 23°59′N and 91°59′E), S.K. Para (SKAP; 23°58′N and 91°58′E), Jawaharnagar (JWNR; 23°55′N and 91°54′E), Ambassa (ABSA; 23°55′N and 91°51′E), Mungiakami (MGKM; 23°53′N and 91°42′E), Teliamura (TLMR; 23°51′N and 91°37′E), Jirania (JRNA; 23°49′N and 91°25′E), Jogendranagar (JGNR; 23°48′N and 91°18′E), Agartala (AGTL; 23°47′N and 91°16′E), Sekerkote (SKKE; 23°44'N and 91°16'E), Bishalgarh (BLGH; 23°40'N and 91°16'E), Bishramganj (BHRM; 23°35'N and 91°21'E), Udaipur (UDPU; 23°30'N and 91°28'E), Garjee (JRJE; 23°25'N and 91°29'E), Santirbazar (STRB; 23°19'N and 91°31'E), the Belonia railway station (BENA; 23°14'N and 91°29'E), Jolaibari (JLBRI; 23°11'N and 91°35'E), Thailik Twisa (THTW; 23°7'N and 91°36'E) and Manu Bazar (MUBR; 23°3'N and 91°38'E) (Fig. 1).

Presently, Tripura has only six freight stations: Dharmanagar, Kumarghat, Jirania, Udaipur, Belonia and Sabroom. Freight operations are conducted only through the Dharmanagar, Kumarghat, Jirania and Udaipur railway stations. Goods yards are under construction in the Belonia and Sabroom railway stations. Those rail freight stations provide freight services to answer demand by about 20 urban centres and 901 villages.

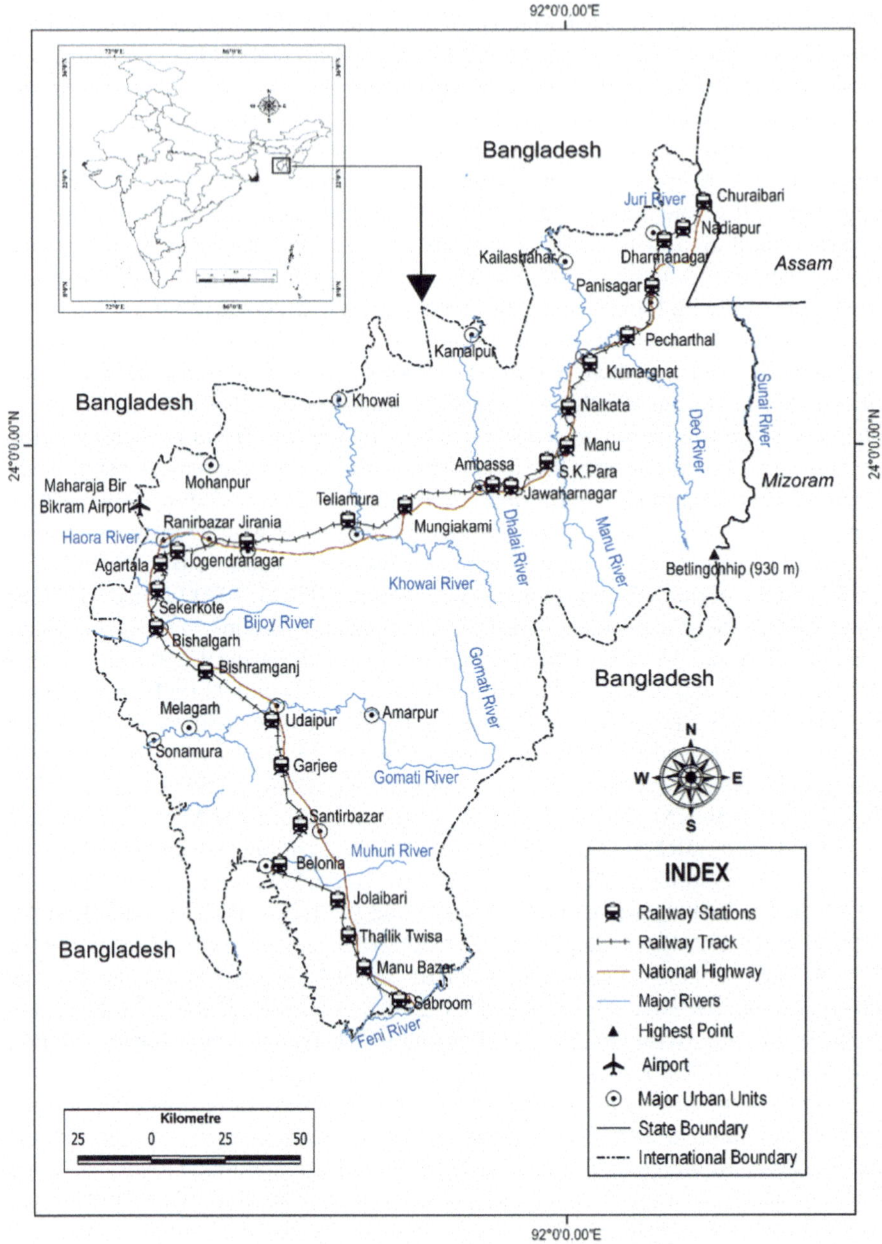

Fig. 1 Location map of the study area (*Source* Prepared by the Authors [data of railway stations and track have been collected by handheld GPS receiver])

3 Literature Review

Various kinds of literatures are available that are related to rail freight transportation. Literature has been reviewed and characterised based on three different perspectives: a geographical perspective, a functional perspective and a methodological perspective (Gale & Olsson, 1979).

About 13% of the reviewed literature on rail freight transport deals with geographical form (structural approach) and about 93.33% of the literature deals with the geographical relationships (man–environment relationship approach) in rail freight transport research. In functional perspectives, about 46.67% of research is thought (fundamental) oriented and 73.33% of active research oriented. As concerns methodology, about 73.33% of the literature reviewed is analytical (with different model-based solutions), whereas about 53.33% of the literature address the research problem in a phenomenological approach to policy measures (Table 1).

Barbour et al. (2020) proposed a method for performance optimisation of rail freight transport operations. They used data reconciliation for data-driven rail freight transport planning. They found that data reconciliation can reduce timing errors,

Table 1 Characterisation of the literatures

Authors	Geographical perspective		Functional perspective		Methodological perspective		Total number of entries
	Form	Relations	Thought	Action	Analytical	Phenomenological	
Barbour et al. (2020)	☐	☒	☒	☒	☒	☐	4
Merchan et al. (2020)	☐	☒	☒	☐	☒	☐	3
Khan and Khan (2020)	☐	☒	☒	☐	☒	☒	4
Zeybek (2019)	☐	☒	☐	☒	☒	☒	4
Anoop et al. (2018)	☐	☒	☒	☐	☒	☐	3
Islam (2018)	☐	☒	☐	☒	☐	☒	3
Zunder and Islam (2018)	☐	☒	☒	☒	☒	☒	5
Zhang and Pel (2016)	☐	☒	☒	☒	☒	☐	4
Drozdziel et al. (2015)	☐	☒	☐	☒	☐	☒	3
Amentae and Gebresenbet (2015)	☐	☒	☐	☒	☐	☒	3
Zitz and Matopoulos (2014)	☒	☐	☐	☒	☐	☒	3
Islam et al. (2013)	☒	☒	☒	☐	☒	☐	4
Bod and Havenga (2010)	☐	☒	☐	☒	☒	☐	3
Caris et al. (2008)	☐	☒	☐	☒	☒	☐	3
Zografos and Regan (2004)	☐	☒	☐	☒	☒	☒	4

Source Prepared by authors after Gale and Olsson (1979)

increase train frequency and overcome estimation errors. Their empirical study was undertaken between Nashville, TN and Chattanooga, TN, USA.

Merchan et al. (2020) conducted an empirical study in Belgium so as to determine environmental impacts on rail freight transport through the use of a Life Cycle Assessment (LCA) methodology. The authors divided rail freight transport into three subsystems: rail transport operation, rail equipment and rail infrastructure. Analysing LCA in four steps, they found that the performance of rail freight transport is more viable through electric trains.

Khan and Khan (2020) estimated the demand for rail freight transport in Pakistan using Johansen's co-integration and error correction model. The authors found that aggregate economic activity and freight rates are the most relevant factors when making effective decisions regarding rail freight planning and management.

Zeybek (2019) analysed the performance of freight modes, as well as the factors responsible for mode choice decisions. Both top-down and bottom-up approaches were considered. The study's sample was composed of 49 shippers and 47 forwarders. It revealed that most (78%) shippers and forwarders were involved in domestic freight transport, and 25% of stakeholders prefer rail freight transportation. The cost, quality and reliability of freight transport systems play a crucial role in decision-making.

Anoop et al. (2018) pointed out planning issues in rail freight transportation from a strategic, tactical and operational perspective. A multi-period Integer Non-Linear Programming (INLP) and optimum rake allocation algorithm were used to analyse the performance and planning perspectives of rail freight transport. The authors' alternative heuristic model was proposed as a solution (Table 1).

Islam (2018) examined the economic sustainability and developed transport infrastructure in Europe during the 2008 recession period. The study was conducted in eight European Union (EU) countries—France, Germany, Italy, Poland, Romania, Spain, Sweden and the United Kingdom. The author found that investment in transport infrastructures helps sustain economic growth.

Zunder and Islam (2018) assessed both existing and future rail freight services and technologies for Low-Density High-Volume (LDHV) goods. The authors conducted an online survey with 24 industry experts of Europe so as to evaluate the 'existing' situation for competitive rail freight service operation with LDHV goods (and proposed an alternative plan). The 'GAP' analysis observed three fundamental areas of rail freight transport: 'wagon', 'train and hubs' and 'business quality and planning'.

Zhang and Pel (2016) developed a model allowing for a comparative analysis of intermodal and synchromodal operations from social, economic and geo-environmental points of view. The model focuses on demand and supply dynamics of rail freight transport, as well as on flexible multimodal routing with transfers and transhipments facilities. The study revealed that transport services and capacity of operation have improved in Rotterdam's hinterland.

Drozdziel et al. (2015) focused on the broad-gauge lines in Europe and their impacts on international freight transport. The authors highlighted the heterogeneous line gauges in Europe, which are responsible for difficulties in freight transport from east to west. An inclusive analysis led to a suggestion to establish broad-gauge lines, developing employment, increasing the transport capacity and building transhipment stations capable of supporting Slovakia's position in international trade.

Amentae and Gebresenbet (2015) evaluated the performance of the intermodal import-export freight transport system in Ethiopia. The authors collected data from the field using a structured questionnaire. A comparative satisfaction model was employed to assess the performance of the rail freight transport. The study reveals that custom checking and wasteful custom inspections processes are the most severe problem in freight transport system in Ethiopia. In order to improve import-export freight movements, the authors suggest improving the intermodal freight transport system.

Zitz and Matopoulos (2014) analysed the development of rail freight transport system in Northern Germany using the Delphi technique. Authors found that existing logistics and freight infrastructure, as well as financial permanence, were somewhat lacking. They concluded that sustainability, environmental protection and working conditions represent the solution to deal with the forecasted growing freight volumes in the next years.

Islam et al. (2013) analysed the potential of three rail corridors: the Trans-Siberian, the Central Corridor and the Transport Corridor Europe-Caucasus-Asia (TRACECA) dedicated to rail freight transport between Central Europe and China. Authors also focused on the technical, operational and geopolitical environments of those corridors. Their research found that unreliable transit times, higher costs, complex border-crossings, lacking in infrastructure, rolling stock constraints, change of gauges, differing power supplies and signalling systems, as well as non-automated systems and freight damage and theft are the most persistent challenges to offering an efficient and integrated logistics and supply chain service along these corridors.

Bod and Havenga (2010) found out that cost reduction opportunities were possible through the densification of rail freight in sub-Saharan Africa (SSA). The research found that most countries in sub-Saharan Africa have a small freight transport market. In both regional and international trade, rail plays a significant role—especially in longer distances and with adequate density. Authors also found that rail is considerably more efficient than any other modes of transportation.

Caris et al. (2008) provided an overview of planning decisions in intermodal freight transport and proposed methods for the solution of several challenges. Intermodal freight transport suffers problems due to road congestion, environmental concerns and traffic safety. Authors classify planning problems based on the type of decision-makers and decision levels. Rail transport is recognised as the most growing and strategically significant supply chain and logistic service.

Zografos and Regan (2004) analysed situations and challenges in connection with the extensive execution of intermodal freight transport systems in both the USA and Europe. Authors examined the socio-economic, political, technological and environmental factors which influenced the development of supply chain managements (SCM). The study found that congestion and environmental factors create negative impacts and that intermodal transportation has an increasing importance.

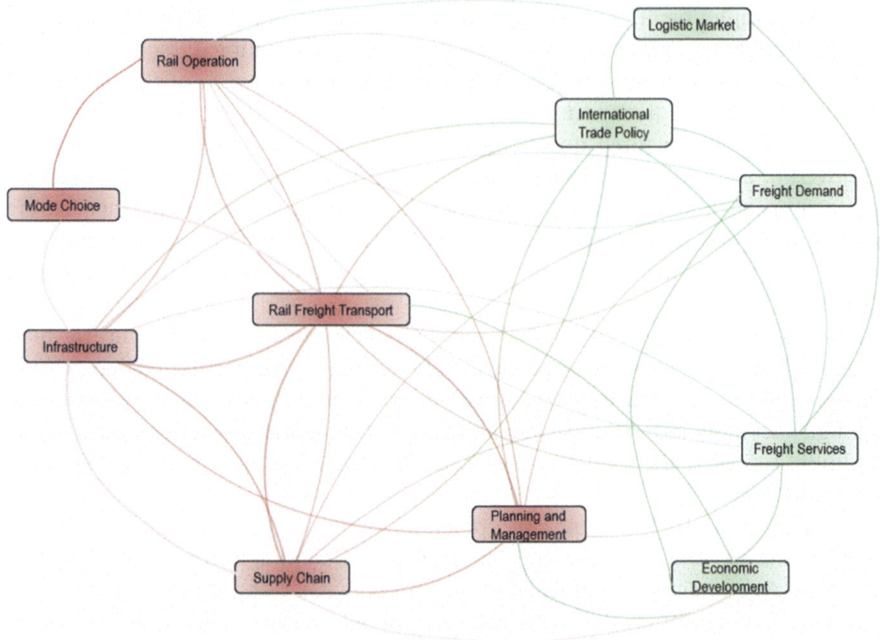

Fig. 2 Keywords—connectivity map of rail freight literature (*Source* Prepared by the Authors using the VOS viewer v. 1.6.12 [bibliometric data extracted from Google Scholar, Web of Science and Scopus database])

Bibliometric analysis was undertaken so as to identify research gaps by using a bibliometric cluster mapping (Oliveira et al., 2019). The concentration of keywords in the reviewed literature shows the threshold area of rail freight transportation (Fig. 2).

Metaphysical or conceptually based expositions were developed on the basis of this literature review. We found it to be widely held that rail freight is a key factor for economic development (Amentae and Gebresenbet, 2015; Islam, 2018; Stewart, 2013; Zitz & Matopoulos, 2014). Performance of rail freight transport system depends upon several factors: intra- and inter-regional demand-supply dynamics, trade policies, multi-modal freight infrastructures, regional stability (economic and political), social steadfastness and environmental susceptibility (Bod and Havenga, 2010; Caris et al., 2008; Ministry of Transport, 2014; Puri, 2017; Zografos & Regan, 2004). A few more internal aspects such as route design (direct links, static links and dynamic links), capricious transit times, higher establishment costs, trans-border permits, differential gauge structures, insufficient rail infrastructural facilities, rolling stock constraints, heterogeneous power supplies and different signalling mechanisms also influence the performance of rail freight transport (Islam et al., 2013; Janic & Vleugel, 2012; Woxenius, 2007). Different mathematical simulations (Monte Carlo simulation), econometrics models, freight models, disaggregate models and Johansen co-integration were used to address the research problems or provide tentative solutions on rail freight transport-related issues (Khan & Khan, 2020; Tian et al., 2020; Zgonic et al., 2019). A priori knowledge concerns previous research that does not focus on infrastructural conditions of fright stations and their performance. The efficiency of the rail freight stations has not been addressed by previous researchers.

4 Study's Objectives and Methodology

The main objective of the study is to find out infrastructural facilities of rail freight stations in Tripura. The paper also measures the performance and efficiency of rail freight stations, as well as prospects of rail freight transport in the state. The study is based on both primary and secondary data. Primary data was collected from the all six rail freight stations of Tripura: Dharmanagar, Kumarghat, Jirania, Udaipur, Belonia and Sabroom. Infrastructural data such as the length of the goods yards, goods holding areas and warehouse sizes was measured through the total station device (Model vide no. Leica TS105) and the handheld Global Positioning System (GPS) receiver (Model vide no. Garmin eTrex 30x). Secondary data such as year-wise commodity flows, earnings from goods, data related to wagon bookings, commodity-wise stakeholders' details and operational capacity was collected from the office of Area Manager, the Lumding Division, the Northeast Frontier Railway, Badarpur, Assam; the office of the Chief Commercial Inspector, Dharmanagar; and the office of the station supervisor of each rail freight stations in Tripura. Based on infrastructural data, the Dimension Index (DI) was calculated to understand comparative conditions in different goods stations in Tripura. In order to compare performance, each station's infrastructural data was standardised by calculating the 'Z Score'. The 'Z Score' has been calculated in the following manner: $Z_i = \frac{(x_i - \bar{x}_i)}{S_i}$,

where x_i = individual observation, $\bar{x}_i$ = arithmetic mean of the observation, S_i = standard deviation.

In the next stage, an aggregated synthetic indicator was prepared based on which performance of the rail freight stations was analysed (Jaroca & Glinska, 2017). DEA analysis was calculated for analysing the infrastructural efficiency of goods stations in Tripura. The Single Input (area of the goods yards) and Single Output (average amount of goods holding) methods were applied. The existing terminal structure was reviewed so as to understand the scope of future developments. The terminal model was designed in AutoCAD. The digital cartographic technique was used for the graphical representation of the spatial phenomenon. The prospects of rail freight transport in Tripura were articulated through Strengths, Challenges, Opportunities and Threats (SCOTs) and the Political, Economic, Social and Technological (PEST) methods.

5 Preliminary Data Analysis

Preliminary data shows that the longest goods station in Tripura is the Jirania railway station (1,246 m); the Dharmanagar railway station (554 m) is the smallest. The mean length of the goods station in Tripura is about 799 m. (Table 2). Preliminary data revealed that the largest goods holding area is found in the Sabroom railway station, whereas the Belonia railway station has the smallest goods holding area. The average goods holding areas of rail freight stations in Tripura is about 15,910.67 sq. m. (Table 2). The Food Corporation of India (FCI) has developed six warehouses in the Dharmanagar and one in the Kumarghat railway station. The total area of warehouses in Dharmanagar is about 32,012 sq. m. (Table 2). Indian Oil also set up a storage area with 13,829 sq. m. For multi-modal freight transit, Dharmanagar (30,306 sq. m.) has the largest parking area, whereas the Udaipur railway station does not even have a designated parking area. The average parking area of the rail freight stations in Tripura is 12,532.33 sq. m. (Table 2).

Sabroom and Belonia are newly developed railway stations in Tripura. Since October 2019, those stations are operated for passenger transport. Freight infrastructure has been developed at those stations; however, freight transport has not yet begun. As per preliminary data concerns, Tripura imports food grains (including rice, wheat, potato, pulses, oil, etc.,) stone chip/bolders, fertilisers, petroleum, bentonite powder, cement and salt. On average, about 255,912 metric tons of freight were imported through rail (Table 3). The maximum amount of food grains was received by the Dharmanagar railway station (61,600 MT). The Kumarghat and Udaipur railway stations only received food grains, especially rice. The Jirania railway station received a significant amount of stone chip/bolders along with food grains. The Jirania

Table 2 Descriptive statistics of rail freight stations infrastructure in Tripura

Name of the station	Length of the goods yards (m)	Goods holding area (Sq. m.)	FCI warehouse area (Sq. m.)	Indian Oil (Sq. m.)	Truck parking area (Sq. m.)
Dharmanagar	554	14,293	32,012	13,829	30,306
Kumarghat	1,073	11,301	3,038	0	3,774
Jirania	1,246	11,305	0	0	8,072
Udaipur	670	19,125	0	0	0
Belonia	631	7,094	0	0	10,856
Sabroom	620	32,346	0	0	22,186
Minimum	554	7,094	0	0	0
Maximum	1,246	32,346	32,012	13,829	30,306
Mean ($\overline{x}$)	799.00	15,910.67	5,841.67	2,304.83	12,532.33
Standard Deviation (σ)	286.99	8,980.69	12,878.25	5,645.67	11,537.50

Source Primary Surveys and calculations by the authors, 2020

Table 3 Preliminary data of average monthly inward commodities by rail freight in Tripura

Monthly inward commodities (metric tons)	Operational rail freight stations				Total
	Dharmanagar	Kumarghat	Jirania	Udaipur	
Food Grains	61,600	2,577	69,534	2,266	135,977
Stone Chip/Bolders	47,740	0	55,864	0	103,604
Fertilisers	2,643	0	2,673	0	5,316
Petroleum	7,726	0	0	0	7,726
Bentonite Powder	266	0	0	0	266
Cement	567	0	0	0	567
Salt	0	0	2,456	0	2,456
Total	120,542	2,577	130,527	2,266	255,912

Source Office of Chief Commercial Inspector (CCMI, Goods), Dharmanagar, 2020

station imports all salt for the entire state (Table 3). The Udaipur is newly operated freight station where a little amount of food grains was delivered. Bentonite powder was imported through the Dharmanagar railway station for the Oil and Natural Gas Corporation (ONGC).

6 Results

6.1 Comparative Status of Rail Freight Stations

According to the Average Composite Dimension Index (ACDI), the Dharmanagar railway station presents the best comparative conditions (Table 4). This railway

Table 4 Comparative status of rail freight in Tripura

Name of the stations	Dimension index					Average composite dimension index	Rank
	Length of the goods yards	Goods holding area	FCI warehouse area	Indian Oil storage area	Truck parking area		
Dharmanagar	0.00	0.29	1.00	1.00	1.00	0.66	1
Kumarghat	0.75	0.17	0.09	0.00	0.12	0.23	4
Jirania	1.00	0.17	0.00	0.00	0.27	0.29	3
Udaipur	0.17	0.48	0.00	0.00	0.00	0.13	5
Belonia	0.11	0.00	0.00	0.00	0.36	0.09	6
Sabroom	0.10	1.00	0.00	0.00	0.73	0.37	2

Source Computed by the Authors

station serves northern parts of the state, adjacent to neighbouring state of Assam. Sabroom is an under-operated freight station located in southern Tripura; it holds the second position. The Jirania has been developed so as to become an international trade with Bangladesh; it holds the third position in terms of rail freight infrastructure (Table 4). Jirania is located only 27 km away from Agartala, the state's capital city. It is the rail freight station nearest to the capital.

Rail freight infrastructure is in comparative poor conditions in Kumarghat, Belonia and Udaipur. Historiography of the stations and value of spatiality strongly influence the developmental dynamics of these rail freight infrastructures. The Kumarghat is one of the old stations of Tripura (developed in the 1990s). Since the beginning, this station has been an important freight station for the Dhalai and Unakoti districts of Tripura. At present, the Kumarghat railway station holds the fourth position (ACDI = 0.23) in Tripura (Table 4). Udaipur is the third-largest city after Agartala and Dharmanagar, as well as the headquarter of the Gomati district. Due to the spatial and economic significance of the area, rail freight infrastructure has been developed at the Udaipur station. Freight operation started in October 2019. According to the Station Supervisor (SS) of the Udaipur railway station, only food grains were imported through this station. Presently, it holds the fifth position in terms of rail freight infrastructure (Table 4). Belonia is a newly developed station located at the south bank of the River Muhuri (Fig. 1)—located only 3.50 km away from the centre of the Belonia station. A small freight (19,400 sq. m.) terminal has been established; however, freight services have not yet begun. At present means during the year 2018-2020 when data has been collected the Belonia are in atrocious conditions (ACDI = 0.37).

Rail freight stations were categorised based on infrastructural conditions. The synthetic indicator was proposed for grading the stations. The Kolmogorov–Smirnov (K-S) test was undertaken so as to understand the empirical distribution of data. The K-S test (0.8398) shows that data has been normally distributed and that the value of the Local Asymptotic Normality (LAN) is 0.5304.

A synthetic indicator shows that infrastructural conditions of the Dharmanagar railway station are far better than the case in other railway stations (Table 5). Poor

Table 5 Comparative performance of rail freight in Tripura

Name of the stations	Length of the goods yards	Goods holding area	FCI Warehouse area	Indian Oil storage area	Truck parking area	Synthetic indicator
Dharmanagar	−0.85	−0.18	2.03	2.04	1.54	4.58
Kumarghat	0.95	−0.51	−0.22	−0.41	−0.76	−0.94
Jirania	1.56	−0.51	−0.45	−0.41	−0.39	−0.20
Udaipur	−0.45	0.36	−0.45	−0.41	−1.09	−2.04
Belonia	−0.59	−0.98	−0.45	−0.41	−0.15	−2.57
Sabroom	−0.62	1.83	−0.45	−0.41	0.84	1.18

Source Computed by the Authors, 2020

Table 6 Classification of rail freight performance in Tripura

Station grade	Method of calculating class	Class range	Characteristic	Class of the state
A	$z_i \geq \bar{z} + s_z$	≥ 2.21	Very good performance	Dharmanagar
B	$\bar{z} \leq z_i \leq \bar{z} + s_z$	$0 - 2.21$	Good performance	Sabroom
C	$\bar{z} - s_z \leq z_i \leq \bar{z}$	$-2.21 \leq 0$	Poor performance	Kumarghat, Jirania and Udaipur
D	$z_i \leq \bar{z} - s_z$	≤ -2.21	Very poor performance	Belonia

Source Computed by the Authors

infrastructure in the Belonia railway station was reflected in the synthetic indicator.

Data based on the synthetic indicator gradation of the rail freight station was classified into four categories (Table 6). Only the Dharmanagar railway station falls under the 'A' category. The Dharmanagar is a very old railway station in Tripura, developed in the 1950s. Since the 1960s, the Dharmanagar railway station has been involved in goods mobility. After gauge conversion (2016), the Dharmanagar maintains the same legacy of freight transport. The Sabroom railway station was demarked as a 'B' category freight station. Still, developmental work has been going on in the Sabroom railway station for freight transport. The Ministry of Railways, Government of India, placed a special focus on the Sabroom due to its geo-strategic location. About 50% of the railway stations fall under the 'C' category, which indicates poor infrastructural conditions. Kumarghat, Jirania and Udaipur railway stations were all assigned this grade. Based on existing infrastructural conditions, the Belonia was graded as 'D', which epitomises 'very poor' infrastructural conditions.

Both the Belonia and Sabroom are newly developed railway stations, and freight transport here has not yet started. Their activation in near future will strongly impact the overall rail services' performance.

6.2 Performance and Efficiency of Rail Freight Stations

Performance and efficiency are highly dependent on infrastructural facilities (National Research Council, 1996). Only four freight operational stations in Tripura have been considered for performance and efficiency analysis. The study found that Jirania is the most efficient rail freight station in Tripura (Table 7).

The Jirania railway station acts as the efficient rail freight frontier in Tripura (Fig. 3). Relative efficiency of Dharmanagar and Kumarghat railway station is about 0.51 and 0.11, respectively.

The study reveals that the Udaipur railway station is the most inefficient railway station in Tripura. The efficiency of the Jirania railway station is higher than that of

Table 7 Performance of rail freight stations according to data envelopment analysis

Name of the stations (DMU)	Dharmanagar	Kumarghat	Jirania	Udaipur
Goods handling area (sq. m.)	14,293	11,301	11,305	19,125
Total amount of goods handled per month (tons)	15,399.08	2,576.51	23,902.17	2,265.64
Efficiency	1.08	0.23	**2.11**	0.12
Relative efficiency	0.51	0.11	1.00	0.06

Source Computed by the Authors

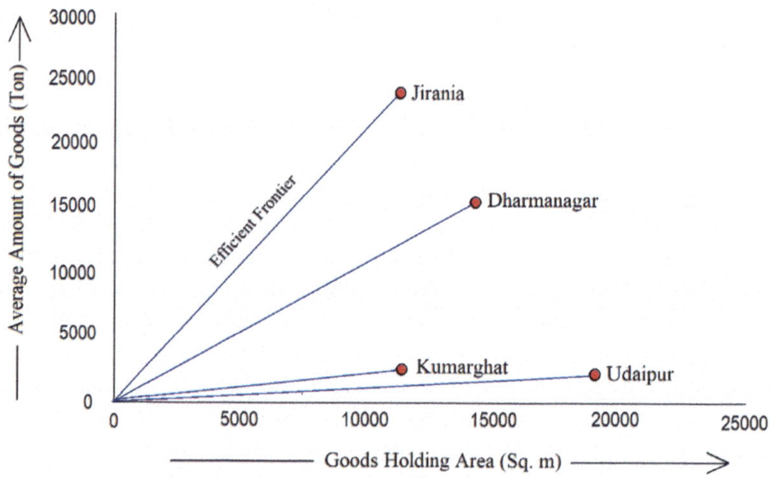

Fig. 3 Efficient frontier of rail stations in Tripura (*Source* Computed by the Authors)

other stations because this station supplies the demand by the Agartala city. This city is the fastest-growing and second-largest urban centre of Northeast India (Debbarma et al., 2018). Urban metamorphosis is higher than in other cities of the state. This strongly influences the Jirania station's performance.

6.3 *Prospect Analysis*

Spatial planning supported the significance of prospect and was more neutral about refuge. Studies related to natural environments provided evidence for the significance of both prospect and refuge (Dosen & Ostwald, 2016). Prospects of rail freight transport in Tripura were analysed based on two approaches: the in situ development model (IDM) and the ex situ development model (EDM).

6.3.1 In Situ Development Model (IDM)

In the in situ development model, three zones were demarcated, with an interval of 1 km from the centre of the rail freight station. The first kilometre was indicated as the inner core; the 1 to 2 km area was called outer core; and area between 2 to 3 km was named peripheral circuit. Land use characteristics of that zone are relevant in the analysis for future infrastructural development.

The Dharmanagar railway station is located in the eastern margin of the town. The western and north-eastern sides of the station have thick urban settlements (Fig. 4). As per 2011 census of India, the total population of the Dharmanagar town is about 40,595 persons. No fallow land has been found within the inner core area. The eastern portion of the station has about 30 hectares. Agricultural land is available which can be manoeuvred for future requirement. The Kakri River flowing from north to south in this area limited further expansion of the station. The outer core is mostly occupied by residential areas. In the south-east extension, about 80 hectares of fallow land was found surrounded by cultivable land. The peripheral circuit of the Dharmanagar railway station is mostly occupied by residential and agricultural land (Fig. 4).

The Kumarghat railway station is located in a low-density settlement area on the left bank of the Manu River. The Kumarghat town is located about 1.37 km to the north-east of the Kumarghat railway station. Its population is 13,054 (Census of

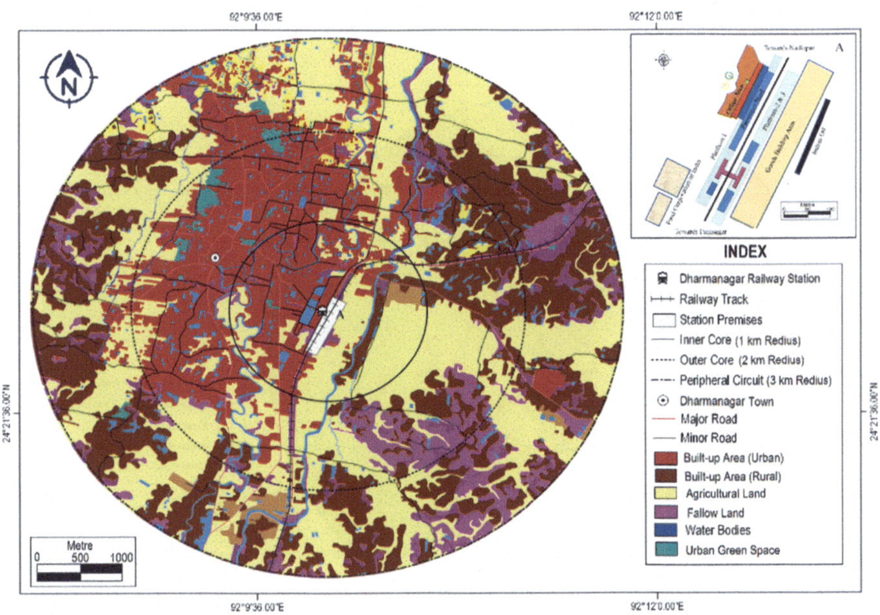

Fig. 4 Tri-zonal land use map of the Dharmanagar Railway Station (*Source* Prepared by the Authors)

India, 2011). This freight station serves not only the Kumarghat town but also the entire Unakoti district (with a total population of 298,574). Kumarghat is the only freight station in this district. The national highway travels about 364 m west of the Kumarghat railway station (Fig. 5). The area between the railway track and the national highway (NH) is the most developed area and not expendable. The proximity between the railway station and the national highway helps in the multimodal transit of goods and cursive door-to-door supplies. The inner core of the station is mostly occupied by semi-urban settlements. In the east portion of the Kumarghat railway station, an area of agricultural land (about 41 hectares) is available. Those areas are also occupied by scattered settlements. A total of 1367 hectares of forested land occupies the outer core and peripheral circuits. Forested land is mostly located in the eastern part of the Kumarghat railway station. However, elevation (105 m) in this area is much higher than other areas surrounding the Kumarghat railway station. The development of the station in the west is circumscribed by the Manu River (Fig. 5). Future expansion of the Kumarghat railway station will therefore occur eastwards—according to local physiography.

The Jirania railway station is the rail freight station nearest to Agartala, the state's capital city, which is located about 27 km north. Since 2008, this rail freight station serves the domestic freight demand of Agartala city and surrounding areas. The national highway travels about 0.96 km south of this station, which again helps in supply chain management. In the inner core area, the Jirania railway station is surrounded by low-density settlements (railway staff quarters, rural settlements such as Barjala and Bankimnagar, agricultural land, fallow land and brickfield). About

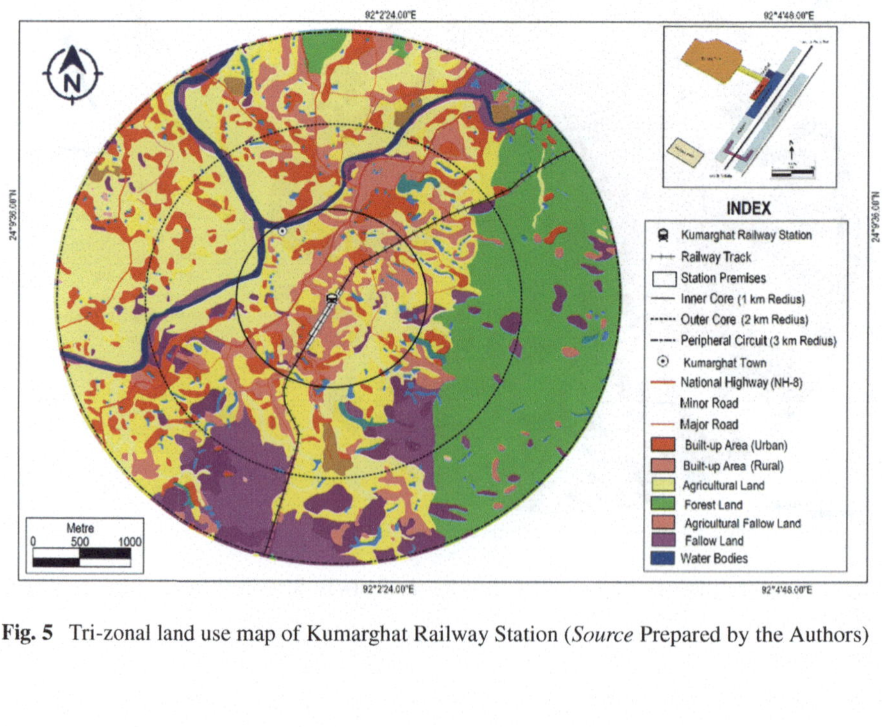

Fig. 5 Tri-zonal land use map of Kumarghat Railway Station (*Source* Prepared by the Authors)

26 hectares in the inner core of the Jirania railway station is used for residential purposes. About 33 hectares in the northern side of the inner core of the Jirania railway station has fallow land (Fig. 6). Expansion of the Jirania railway station can be undertaken northwards. The outer core of the Jirania railway station is mostly used for agricultural activity. The river has a total length of 61.2 km, of which 52 km flows within Indian Territory (Bandyopadhyay & De, 2018). Within this 52, 3.18 and 3.33 km fall under the outer core and peripheral circuits of the Jirania railway station, respectively (Fig. 6). A maximum built-up area (Rural types) has been developed in between the railway line and the national highway. Another built-up area was found in the northern part of the Jirania railway station. This is due to the existence of a 147.95 hectares-large campus of the National Institute of Technology, Agartala. This mixed land use and diversified cultural landscape both limit the future expansion of the Jirania railway station for rail freight transportation (Fig. 7).

Udaipur is the oldest city and the capital of the Manikya dynasty in Tripura. In 590 AD, the King of Tripura, 'Jhujaroofa', established this city. Earlier, the city was known as Rangamati. In the year 1567, Udai Manikya changed the name of the capital from Rangamati to Udaipur. In the year 1760, Maharaja Krishna Manikya moved the state capital from Udaipur to Old Agartala (Santra, 2017). According to the Census of India (2011), Udaipur is the third-largest city in Tripura. The total population of this city is 37,781 (Census of India, 2011). About 98% of the land in the inner core of the Udaipur railway station is used for agricultural purposes; the remaining two per cent is composed of fallow land and water bodies (Fig. 8). In the outer core, there are a few low dense settlement areas; agricultural land and water bodies were

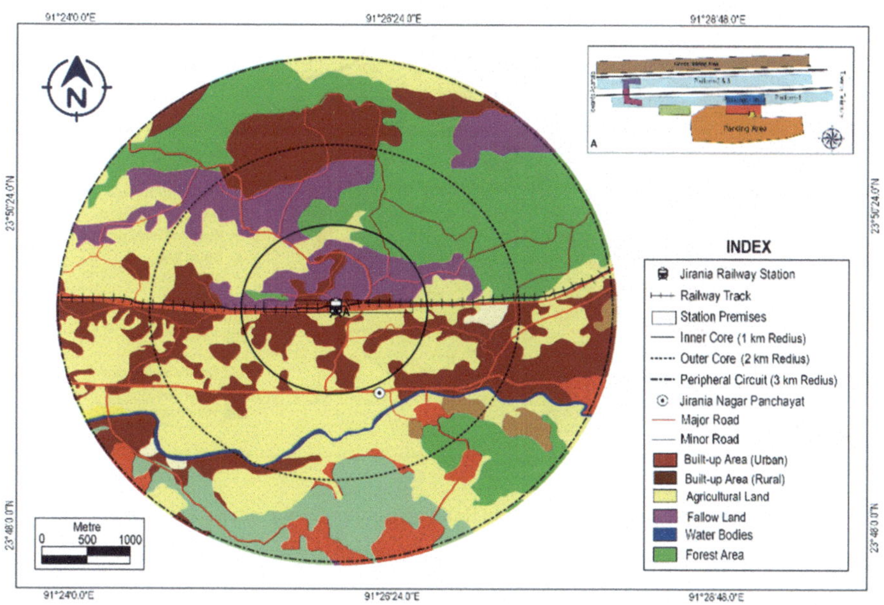

Fig. 6 Tri-zonal land use map of the Jirania Railway Station (*Source* Prepared by the Authors)

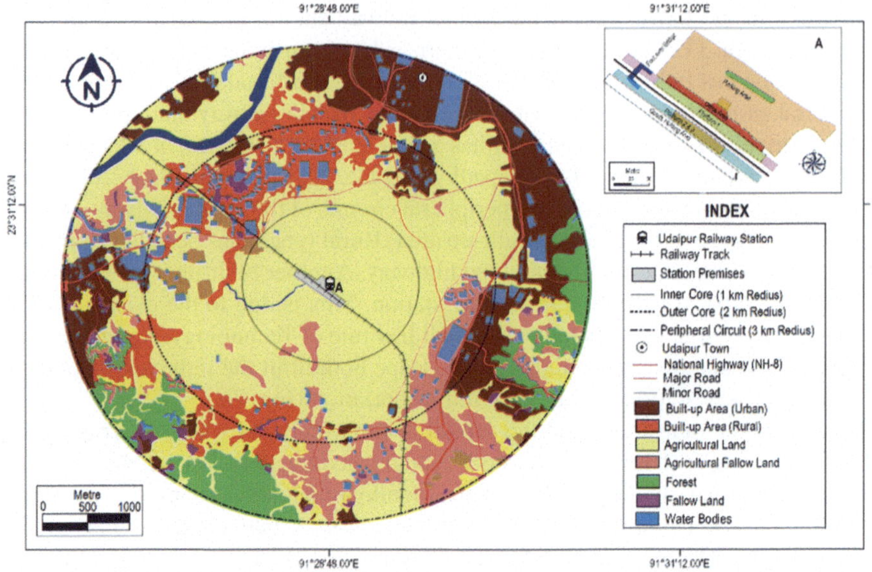

Fig. 7 Tri-zonal land use map of the Udaipur Railway Station (*Source* Prepared by the Authors)

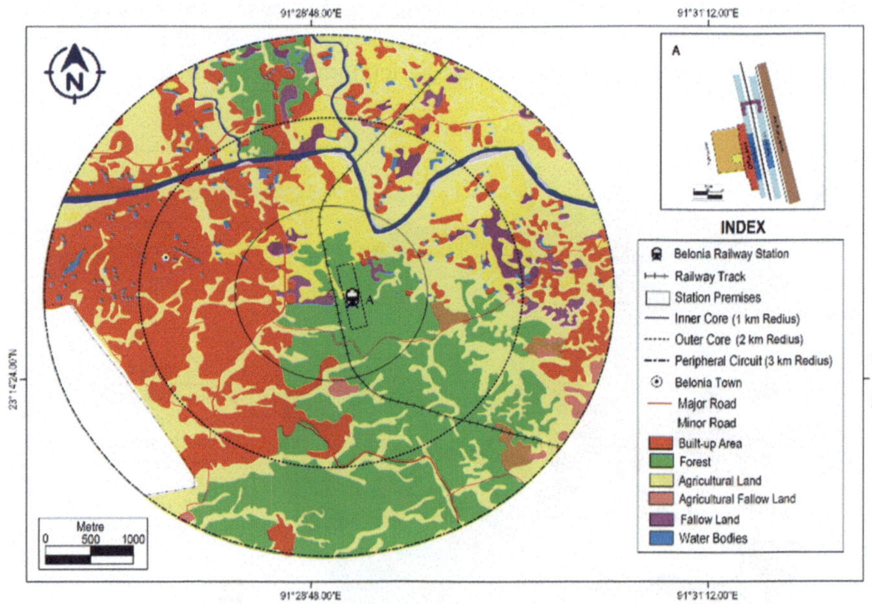

Fig. 8 Tri-zonal land use map of the Belonia Railway Station (*Source* Prepared by the Authors)

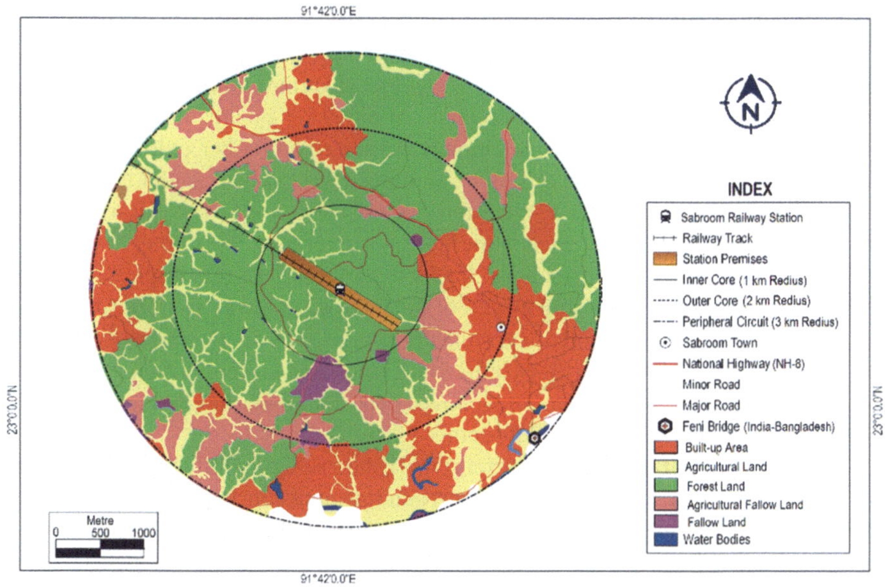

Fig. 9 Tri-zonal land use map of the Sabroom Railway Station (*Source* Prepared by the Authors)

also found. The Udaipur town and its lakes are located in the peripheral circuit. The Gomati River is the longest river in the state, flowing westward from the peripheral circuit of the Udaipur railway station. The national highway also travels along the peripheral circuit of the Udaipur railway station (Fig. 8). The study reveals that the Udaipur railway station's surrounding area contains large areas of homogeneous land with great potential for enhancing infrastructural establishment.

Belonia is a small town and the headquarter of the South Tripura District, located near the internationally significant Indo-Bangladesh border (about 72 km away from the capital city of the state Agartala; Fig. 9). As per the Census of India (2011), the total population of this town is 19,996. A large section of South Tripura depends upon the Belonia railway station. The first train reached Belonia on 9 February 2019. A huge goods holding area has been designed for future rail freight service. A small segment of the River Muhuri (0.43 km) flows westward from the inner core of the Belonia railway station. About 136 hectares of this inner core is covered by forest. The Belonia town starts from the western portion of the station. About 33 hectares of residential land is located at the inner core of the Belonia railway station. The Central Business District (CBD) of the Belonia town (1.78 km) is located in the outer core of the station. The outer core of the Belonia railway station is mostly occupied by settlement areas. In the eastern and north-eastern portions of the Belonia station, about 478 hectares of land is used for agricultural purposes. The international boundary with Bangladesh is located in the peripheral circuit of the Belonia railway station.

Sabroom is Tripura's southernmost railway station. It is located on the right bank of the River Feni. It is a newly developed railway station, which started operating passenger trains on 3 October 2019. Rail freight services have not yet started because construction works are not yet completed. About 94% of the inner core of the Sabroom railway station is forested (mostly dedicated to rubber cultivation). Few agricultural actives are found in this area. An industrial land bank under the Government of Tripura is available near the Sabroom railway station. The Sabroom town with its 7,142 population is located at the station's outer core. This outer core is also predominantly featured by agricultural land use patterns. The maximum settlement areas were found in the peripheral circuit of the Sabroom railway station. About 344 hectares of land in the outer core is combinedly occupied by residential and commercially settlements. The Feni bridge (India-Bangladesh) is located in this zone of the station, which is currently under construction. The Bangladeshi Chittagong Port is situated about 72 km south from Sabroom.

The future scope of rail freight transport in Tripura is highly influenced by current land use patterns of railway stations and surrounding areas. In Tripura, many railway stations were developed in the 1950s—a few between the 1980s and the first decades of the twenty-first century. The omnipresence of railway stations also influenced local land use patterns. Future scopes and potentials for current rail freight stations were determined based on the land use patterns present today. Land Acquisition, the Rehabilitation and Resettlement Act (LARR Act), 2013, has provided the Government of India with the power to acquire private land for industrialisation, development of infrastructural facilities or urbanisation (and to compensate affected landowners for their rehabilitation and resettlement; Hoda, 2018). Land acquisition always becomes a political issue in India (Sathe, 2017). In order to minimise political risks in land acquisition and analyse the prospects of spatial growth of rail freight stations, we started by identifying easily convertible land such as fallow land (weighted value = 0.40), forest land (weighted value = 0.30) and agricultural land (weighted value = 0.30) within the inner cores of the respective railway stations. We then proposed a weighted percentage index (Table 8).

The weighted percentage index reveals that the Sabroom railway station provides the greatest prospects in term of spatial expansion. Jirania and Belonia also enjoy a

Table 8 Weighted percentage index of Tripura's rail freight stations

Name of the stations	Fallow land	Forest	Agricultural land	WPI	Rank
Dharmanagar	6.98	0.00	3.43	10.41	5
Kumarghat	4.40	0.00	4.51	8.91	6
Jirania	19.18	0.00	3.47	22.65	2
Udaipur	3.24	0.00	13.48	16.72	4
Belonia	6.20	8.77	3.64	18.62	3
Sabroom	0.00	21.23	1.47	22.69	1

Source Computed by the Authors

significant land bank that can be used for future expansion. Kumarghat and Dharmanagar, two older railway stations, also contain some potential for further spatial expansion.

6.3.2 Ex Situ Development Model (EDM)

Presently, Tripura imports major food grains like rice, white, oil, potato, salt and pulses. It also imports petroleum, stone chips, cement, gravel, bolder, fertilisers, etc. Tripura exports only timber and once in a year at that. The Food Corporation of India (FCI) is the main food grain importer; apart from the Directorate of Food, Civil Supplies & Consumer Affairs, the Government of Tripura is also a major importer. The National Highway Authority of India (NHAI) and the Border Road Organisation (BRO) import stone chips and bolder for their constructional works (roads, guard walls, etc.) because hills in Tripura are composed of mud and landslides are common phenomena (Ghosh et al., 2016). Agartala is the fastest-growing city in the Northeast of India; other towns of Tripura arc also rapidly expanding; huge amounts of construction materials such as cement and stone chips are demanded and imported by rail by different stakeholders. Fertilisers are imported by the Agriculture Department, Government of Tripura. Indian oil imports petroleum from the Digboi of Assam (27°23'35"N and 95°37'14"E). Rice is imported from Barnala (BNN; 30°21'48"N and 75°32'35"E), Punjab; Sitarampur (STN; 23°43'25"N and 86°53'47"E), West Bengal; Dhulkot (DKT; 30°23'41"N and 76°47'09"E), Haryana; Dhuri (DUI; 30°22'22"N and 75°51'58"E), Punjab; Sirhind (SIR; 30°27'25"N and 76°22'56"E), Punjab; Kapurthala (KXH; 31°21'58"N and 75°22'56"E), Punjab; Kila Raipur (QRP; 30°45'31"N and 75°49'36."E), Punjab; Safidon (SFDE; 29°23'51"N and 76°39'42"E), Haryana; etc. (Figure 9). After rice, the most imported food grain is wheat. Wheat generally comes from Maur (MAUR; 30°3'46"N and 75°13'41"E), Punjab; Mullanpur (MLX;30°50'15"N and 75°39'54"E), Punjab; Phillaur (PHR; 31°1'0"N and 75°47'9"E), Punjab, Ajitwal (AJL; 30°48'28"N and 75°20'7"E), Punjab; Gauriganj (GNG; 26°12'14"N and 81°41'9"E), Uttar Pradesh; Mansa (MSZ; 29°59'16"N and 75°24'19"E), Punjab; Lalitpur (LAR; 24°41'18"N and 78°23'43"E), Uttar Pradesh; Hansi (HNS; 29°5'17"N and 75°56'50"E), Haryana; and the Chauri Chaura (CC; 26°38'41"N and 83°35'13"E) railway station in Uttar Pradesh. About 55% of wheat exporting railway stations are located in Punjab. Uttar Pradesh and Haryana contain 33.33% and 11.11% wheat exporting railway stations to Tripura, respectively. Potato is another important imported foodstuff; here, the state works with West Bengal. Potato has been imported from the Salbari (SXX; 26°32'43"N and 89°6'13"E), Dhupguri (DQG; 26°34'27"N and 88°59'37"E) and Changrabandha (CBD; 26°24'45"N and 88°55'11"E) railway stations in West Bengal (Fig. 9).

Bentonite powder is imported from the Koduru railway station (KOU; 13°56'40"N and 79°20'56"E) in Andhra Pradesh for the Oil and Natural Gas Corporation, India. The Ditokcherra (DTC; 25°5'33"N and 92°48'16"E), Assam, and Harnaut (HRT; 25°22'18"N and 85°32'27"E), Bihar, railway stations export bolder to Tripura. The state imports cement from the Kolaghat Thermal Power Station's railway station (KKPS; 22°26'0"N and 87°52'52"E), a railway station in West Bengal. The Panagarh railway station (PAN; 23°26'34"N and 87°27'31"E) of West Bengal is the main exporter of gravel to Tripura. Tripura used to frequently face fuel crises during the gauge conversion in 2015–2016. Now, the problem is almost eradicated because of efficient rail freight operation (Bose, 2016). Presently, petroleum is imported from the Digboi (DIGBOI; 27°22'59"N and 95°37'2"E) and Bongaigaon (BRGN; 26°28'21"N and 90°31'19.80"E) railway stations in Assam.

Tripura only exports timber to Nangloi (NNO; 28°41'11"N and 77° 3'26.12"E), New Delhi; Rajpura (RPJ; 30°29'10"N and 76°35'37"E), Punjab; Sanat Nagar (SNF; 17°27'42"N and 78°26'11"E), Telangana; and Kanakapura (KKU; 26°55'45"N and 75°42'8"E), Rajasthan; and Kolkata (CP; 22°36'4"N and 88°23'4"E), West Bengal.

Alternative (shortest) import-export routes can improve the performance and potentialities of rail freight transport in Tripura.

7 Discussion

7.1 SCOT Analysis

Presently, rail freight transport systems in Tripura present several Strengths, Challenges, Opportunities and Threats (SCOT). According to the analysis, rail is the primary mode of freight transport. Many railways now enjoy newly developed infrastructural facilities; these constitute the greatest strength of rail freight transport in Tripura. The majority of rail freight stations contain potential land banks for future development. The proximity to the national highway and urban centres strengthens the performance of rail freight transport systems (Fig. 3).

The main challenges faced by rail freight stations in Tripura concern the need to improve efficiency and performance. The improvement of rail freight infrastructures such as goods yards and warehouses is required so as to enhance service quality.

Stations have huge potentialities in both import and export of goods by railway because the unit cost transportation by road is much higher and condition of many roads is very poor due to rugged topography and geo-environmental phenomena such as landslides. Geo-strategic location can open up a new window for trade and commerce. Specifically, Sabroom's location can promote international, multi-modal freight gateways.

The major threat to Tripura's rail freight transport concerns the high performance of the recently upgraded national highways. Roadways provide doorstep freight services, and small-scale stakeholders tend to prefer roadways for freight transportation (Table 9).

Table 9 SCOT analysis of rail freight transport in Tripura

<table>
<tr><td rowspan="6">Positive Factors</td><td colspan="2">Internal Factors</td><td rowspan="6">Challenging Factors</td></tr>
<tr><td>Strengths</td><td>Challenges</td></tr>
<tr><td>1. Primary mode freight transport in Tripura
2. Newly developed train freight infrastructure
3. Land availability for future expansion
4. The proximity between rail freight stations and urban centres</td><td>1. Efficiency and performance of rail freight stations
2. Less frequent train services (depending upon booking)
3. Inadequate infrastructural set-ups like goods yards, warehouses, etc.</td></tr>
<tr><td colspan="2">External Factors</td></tr>
<tr><td>Opportunities</td><td>Threats</td></tr>
<tr><td>1. Stations with huge potentialities for import and export of goods by railway because the unit cost transportation by road is much higher
2. Geo-strategic location of Tripura in relation to the Chittagong Port, Bangladesh</td><td>1. Doorstep freight service through roadways
2. Increasing road traffic for small-scale supplies
3. High-performing national highway (NH-8)</td></tr>
</table>

Source Computed by the Authors

7.2 PEST Analysis

Risks and potentialities of rail freight transport were analysed through the PEST model. For further railway development, huge amounts of land are needed. However, land acquisition tends to create some political issues. The Indo-Bangla geopolitical relationship is the most dynamic relationship in South Asia and has been for the last 52 years. This relation can play a crucial role in the development of rail freight transport in Tripura (Table 10).

The supply chain of Tripura is highly influenced by railway with its comparative cheap services. International trade opportunity and rail transport corridor with Bangladesh can boost up the state's domestic economy as well as the regional economy of Northeast India.

Due to social demand for food supplies, goods carriage became efficiently operated by railways and even developed a social value. Many technological interventions were applied by the Ministry of Railway, Government of India, in wagon design—including bulk, cement and fly ash transportation, parcel wagon, etc. Modern technology in terminal design is still widely unavailable.

Table 10 PEST analysis of rail freight transport in Tripura

P	E	S	T
Political factors	***Economic factors***	***Social factors***	***Technological factors***
1. Land acquisition and related issues 2. Indo-Bangla geopolitical relationship	1. Low-cost freight transport mode 2. Domestic supply chain management 3. International trade opportunity	1. Food demand in Tripura 2. Social acceptance	1. The intervention of modern technology in goods loading and unloading 2. Developed on-line monitoring system 3. Digital marketing to promote Indian Railway Freight Service

Source Computed by the Authors

8 Conclusions and Recommendations

Rail freight infrastructure in Tripura is in an initial phase of development. The state imports rice from Punjab, wheat from Punjab and Uttar Pradesh, potatoes from West Bengal, stone chips from Bihar and oil from Assam; it exports timber to Delhi, Hyderabad, Kolkata, etc. For both import and export, rail freight stations require sufficient infrastructural support. However, the major problem in Tripura's rail freight services concerns exactly insufficient infrastructural facilities such as goods holding areas, loading and unloading areas, and goods yards. Much of the state's rail freight infrastructure has been newly developed or modified. Still, much infrastructural inequality was found. Railway infrastructure plays an important role in the performance of any station; however, demand-supply dynamics also play a crucial role. The huge convertible land bank creates a wide scope for rail freight infrastructural development in Tripura. Due to its favourable strategic location and positive geopolitical relations with Bangladesh, Tripura has great potentiality for rail freight transport through the successful implementation of the Agartala-Akhaura rail project connecting Sabroom with Bangladesh's Chittagong Port by both road and rail.

References

Amentae, T. K., & Gebresenbet, G. (2015). Evaluation of performances of intermodal import-export freight transport system in Ethiopia. *Journal of Service Science and Management, 1*, 57–70. https://doi.org/10.4236/jssm.2015.81007

Anoop, K. P., Panicker, V. V., Narayanan, M., & Kumar, C. T. (2018). A mathematical model and solution method for rail freight transportation planning in an Indian food grain supply Chain. *Sādhanā, 43*(12), 1–20. https://doi.org/10.1007/s12046-018-0958-z

Bandyopadhyay, S., & De, S. K. (2018). Anthropogenic impacts on the morphology of the Haora River, Tripura, India. *Géomorphologie: Relief, Processus, Environnement, 24*(2), 151–166. https://doi.org/10.4000/geomorphologie.12019

Barbour, W., Kuppa, S., & Work, D. B. (2020). Enhanced data reconciliation of freight rail dispatch data. *Journal of Rail Transport Planning and Management, 14*(2), 1–12. https://doi.org/10.1016/j.jrtpm.2020.100193

Bod, A. D., & Havenga, J. H. (2010). Sub-Saharan Africa's Rail freight transport system: Potential impact of densification on cost. *Journal of Transport and Supply. Chain*, 89–101. http://scholar.sun.ac.za/handle/10019.1/16053

Bose, P. R. (2016). *Tripura reeling under fuel crisis*. New Delhi: The Hindu.

Business Today. (2016). *These are the 5 most industrialised states in India*. New Delhi: BT Online.

Business Wire. (2019, September 19). Global rail freight transport market report 2019: Analyses & forecasts through 2017–2026. Dublin, Ireland. https://www.businesswire.com/news/home/20190919005354/en/Global-Rail-Freight-Transport-Market-Report-2019#:~:text=The%20Global%20Rail%20Freight%20Transport,3.9%25%20during%20the%20forecast%20period.&text=The%20large%20ability%20of%20rail,the%20oil%20and%2

Caris, A., Macharis, C., & Janssens, G. K. (2008). Planning problems in intermodal freight transport: Accomplishments and prospects. *Transportation Planning and Technology, 31*(3), 277–302. https://doi.org/10.1080/03081060802086397

Census of India. (2011). *Primary census abstract*. New Delhi: Census Commissioner of India and Registrar General of India.

Chandra, P., & Jain, N. (2007). The logistics sector in India: Overview and challenges. Ahmedabad: Indian Institute of Management, Ahmedabad. https://www.researchgate.net/publication/46437406_The_Logistics_Sector_in_India_Overview_and_Challenges

De, U. K. (2004). Food security and PDS in Tripura: A policy intervention. In A. Banerjee (Ed.), *Food security and public distribution system today: Failures and successes* (pp. 348–357). New Delhi: Kanishka Publishers. https://www.researchgate.net/publication/24114898_Food_Security_And_Pds_In_Tripura_A_Policy_Intervention

Debbarma, D., Roy, S., Santra, A., & Mitra, S. (2018). A spatial analysis of population distribution, density and growth in agartala city. *Asian Journal of Spatial Science, 6*(1), 24–36.

Department of Agriculture, Cooperation & Farmers Welfare. (2018). *Krishi Annual Report 2017–2018*. New Delhi: Government of India. http://agricoop.nic.in/sites/default/files/Krishi%20AR%202017-18-1%20for%20web.pdf

Dikshit, K. R., & Dikshit, J. K. (2014). *North-East India: Land, people and economy*. New York: Springer. https://doi.org/10.1007/978-94-007-7055-3

Dosen, A. S., & Ostwald, M. J. (2016). Evidence for prospect-refuge theory: A meta-analysis of the findings of environmental preference research. *City, Territory and Architecture, 3*(4), 1–14. https://doi.org/10.1186/s40410-016-0033-1

Drozdziel, P., Bukova, B., & Brumercikova, E. (2015). Prospects of international freight transport in the East-West direction. *Transport Problems, 10*(4), 5–13. https://doi.org/10.21307/tp-2015-043

Gale, S., & Olsson, G. (1979). *Philosophy in geography*. Dordrecht: D. Reidel Publishing Company. https://doi.org/10.1007/978-94-009-9394-5

Ghosh, K., Bandyopadhyay, S., & De, S. K. (2016). Geophysical investigation and management plan of a shallow landslide along the NH-44 in Atharamura Hill, Tripura, India. *International Journal of Geohazards and Environment, 2*(3), 119–130.https://doi.org/10.15273/ijge.2016.03.013

Ginevicius, R., & Sinkevicius, G. (2020). Quantitative assessment of the dynamics of rail freight transportation. *Economic Interferences, 22*(54), 579–591. https://doi.org/10.24818/ea/2020/54/579

Guha, A. (1968). A big push without a take-off: A case-study of Assam, 1871–1901. *The Indian Economic and Social History Review*, 199–221. https://doi.org/10.1177/001946466800500301

Hilaly, S. (2016). Railways and colonialism: The Kachugaon forest tramway in Assam. *Global Journal of Indian History and Culture, 1*(1), 45–57.

Hoda, A. (2018). *Land use and Land Acquisition Act in India*. New Delhi: Indian Council for Research on International Economic Relations. https://icrier.org/pdf/Working_Paper_361.pdf

Islam, D. M. (2018). Prospects for European sustainable rail freight transport during economic austerity. *Benchmarking: An International Journal, 25*(8), 2783–2805. https://doi.org/10.1108/bij-12-2016-0187/full/html

Islam, D. M., Jackson, R., Zunder, T. H., & Burgess, A. (2015). Assessing the impact of the 2011 EU Transport White Paper—A rail freight demand forecast up to 2050 for the EU27. *European Transport Research Review, 7*(22), 21–30. https://doi.org/10.1007/s12544-015-0171-7

Islam, D. M., Zunder, T., Jackson, R., Nesterova, N., & Burgess, A. (2013). Potential of alternative rail freight transport channels between Central Europe and China. *Transport Problems, 8*(4), 45–57.

Janic, M., & Vleugel, J. (2012). Estimating potential reductions in externalities from railroad substitution in Trans-European freight transport corridors. *Transportation Research Part D, 17,* 154–160. https://doi.org/10.1016/j.trd.2011.09.015

Jaroca, M., & Glinska, E. (2017). The state and prospects for development of railway transport in Eastern Poland—Secondary data analysis. *Procedia Engineering, 182*(3), 299–305. https://doi.org/10.1016/j.proeng.2017.03.198

Khan, M. Z., & Khan, F. N. (2020). Estimating the demand for rail freight Transport in Pakistan: A time series analysis. *Journal of Rail Transport Planning and Management, 14*(2), 1–13. https://doi.org/10.1016/j.jrtpm.2019.100176

Marinov, M., Giubilei, F., Gerhardt, M., Ozkan, T., Stergiou, E., Papadopol, M., & Cabecinha, L. (2013). Urban freight movement by rail. *Journal of Transport Literature, 7*(3), 87–116. https://doi.org/10.1590/s2238-10312013000300005

Merchan, A. L., Belboom, S., & Leonard, A. (2020). Life cycle assessment of rail freight transport in Belgium. *Clean Technologies and Environmental Policy, 22,* 1109–1131. https://doi.org/10.1007/s10098-020-01853-8

Ministry of Transport. (2014). *Contribution of transport to economic development: International literature review with New Zealand perspectives.* Wellington, New Zealand.

Mozumder, T. (2020). *An introduction to the Indian Railways.* Knowindia. http://www.knowindia.net/rail.html

National Research Council. (1996). *Measuring and Improving Infrastructure Performance.* Washington, DC: The National Academies Press.

Oliveira, O. J., Silva, F. F., Juliani, F., Barbosa, L. C., & Nunhes, T. V. (2019). Bibliometric method for mapping the state-of-the-art and identifying research gaps and trends in literature: An essential instrument to support the development of scientific projects. In M. Jibu and Y. Osabe (Eds.), *Scientometrics* (pp. 1–20). London: IntechOpen. https://doi.org/10.5772/intechopen.85856

P R Newswire. (2015, May 29). *India Freight Transport Market Analysis and Forecasts Report 2015–2020.* Research and Market. https://www.prnewswire.com/news-releases/india-freight-transport-market-analysis-and-forecasts-report-2015-2020-300090847.html

Pandey, N. N. (2008). *India's North East region: Insurgency, economic development and linkages with South-East Asia—New Delhi and Institute of South Asian Studies.* Singapore: Manohar Publishers.

Planning Commission. (1981). *Report on general issues relating to backward areas development.* New Delhi: Government of India.

Puri, B. N. (2017). Sustainable transport and inter modal mix. *Transport and Communications Bulletin for Asia and the Pacific*, 54–66.

Research and Innovative Technology Administration. (2010). *Freight transportation: Global highlights.* Washington: Bureau of Transportation Statistics. Retrieved July 2019, 25, from https://www.infrastructureusa.org/wp-content/uploads/2010/04/bts-freight.pdf

Rodrigue, J. P., Comtois, C., & Slack, B. (2013). *The geography of transport systems.* New York: Routledge.

Roy, S., & Mitra, S. (2015). Infrastructural status of railway transportation in Tripura, India: A geographical analysis. In *Proceedings of National Seminar on Infrastructural Development in India: Special Reference to North Eastern States* (pp. 61–73). Agartala: Women's College

Roy, S., & Mitra, S. (2016). Railway transport system in Tripura, India: A geographical analysis. *Geographical Review of India, 78*(1), 40–57.

Roy, S., & Mitra, S. (2020). Railway stations of Tripura, India: An assessment of infrastructural conditions. In S. Bandyopadhyay, C. R. Pathak, & T. P. Dentinho (Eds.), *Urbanization and regional sustainability in South Asia* (pp. 177–200). Belgium: Springer.

Saha, S. (2014). *A study of the geohydrological problems of the Khowai and Haora River Basins, West Tripura.* Tripura University, Geography and Disaster Management. Agartala: Shodhganga. https://shodhganga.inflibnet.ac.in/bitstream/10603/180835/7/07_chapter%202.pdf

Santra, A. (2017). *Changing land use pattern of the Agartala city, Tripura: A geographical perspective.* Agartala: Tripura University.

Sathe, D. (2017). *The political economy of land acquisition in India.* Singapore: Springer Nature. https://doi.org/10.1007/978-981-10-5326-9

Sen, S. K., Gupta, S., & Mukhopadhyay, I. (2013). Economic viability of alternative internal road network in Tripura: An application of shortest path algorithm. *Hill Geographer, 29*(1), 25–39. https://ssrn.com/abstract=2579580

Sen, S., Debbarma, C., De, S. K., & Mitra, S. (2015). Impact of faults on landslide in the Atharamura Hill (along the NH 44), Tripura. *Environmental Earth Sciences, 73*(19), 5289–5298. https://doi.org/10.1007/s12665-014-3778-4

Sharma, A. K. (2016). *Factors impacting railway freight traffic in India.* New Delhi: National Council of Applied Economic Research.

Singh, B. (2016, August 12). *Railways run flat-wagon goods train to Tripura transport fuel,* p. 1. https://economictimes.indiatimes.com/industry/transportation/railways/railways-run-flat-wagon-goods-train-to-tripura-transport-fuel/articleshow/53655299.cms?from=mdr

Singh, S. N., Narain, A., & Kumar, P. (2011). *Socio economic and political problems of tea garden workers: A study of Assam.* New Delhi: Mittal Publications.

Stewart, R. D. (2013). *Freight rail economic development.* Saint Paul: Minnesota Department of Transportation. http://www.dot.state.mn.us/ofrw/fred/PDF/final.pdf

Tian, Z., Sun, G., Chen, D., Qiu, Z., & Ma, Y. (2020). Method for determining the valid travel route of railways based on generalised cost under the syncretic railway network. *Journal of Advanced Transportation, 2020*(8287648), 12. https://doi.org/10.1155/2020/8287648

Tseng, Y. Y., Yue, W. L., & Taylor, M. A. (2005). The role of transportation in logistics chain. *Proceedings of the Eastern Asia Society for Transportation Studies, 5,* 1657–1672. https://www.researchgate.net/publication/281230908_The_role_of_transportation_in_logistics_chain

Tripura State Control Board. (2013). Surface Water Resources and Drainage System. Welcome to TSPCB site (trpenvis.nic.in).

Woxenius, J. (2007). Generic framework for transport network designs: Applications and treatment in intermodal freight transport literature. *Transport Reviews, 27*(6), 733–749. https://doi.org/10.1080/01441640701358796

Yi, S. (2018). *Principles of railway location and design.* London: Elsevier. https://doi.org/10.1016/b978-0-12-813487-0.00004-4

Zeybek, H. (2019). Analysis of freight mode choice decisions of shippers and forwarders: Pre-liberation survey. *World Review of Intermodal Transport Research, 8*(3), 209–221. https://doi.org/10.1504/writr.2019.102357

Zgonic, B., Tekavcic, M., & Jaksic, M. (2019). The impact of distance on model choice in freight transport. *European Transport Research Review, 11*(10), 1–18.

Zhang, M., & Pel, A. J. (2016). Synchromodal hinterland freight transport: Model study for the port of Rotterdam. *Journal of Transport Geography, 52,* 1–10. https://doi.org/10.1016/j.jtrangeo.2016.02.007

Zitz, G., & Matopoulos, A. (2014). Developments and prospects of freight railway transport in Northern Germany: A Delphi survey. *International Journal of Logistics Systems and Management, 18*(4), 500–516. https://doi.org/10.1504/ijlsm.2014.063982

Zografos, K. G., & Regan, A. C. (2004). Current challenges for intermodal freight transport and logistics in Europe and the United States. *Transportation Research Record: Journal of the Transportation Research Board, 1873*(1), 70–78. https://doi.org/10.3141/1873-09

Zunder, T. H., & Islam D. M. (2018). Assessment of existing and future rail freight services and technologies for low density high value goods in Europe. *European Transport Research Review, 10*(9), 1–12. https://doi.org/10.1007/s12544-017-0277-1

India's Public Transport Systems: The Role of Metro Rail

Paulose N. Kuriakose and Jayasmita Bhattacharjee

1 Introduction

Increasing urbanization and industrialization increase the demand for transport. Transport must be accessible to all in the community so that access to livelihood opportunities, economic functions, socio-cultural and political rights are not hindered (UCLG, 2008). Accessibility has various dimensions such as physical accessibility, financial accessibility and social accessibility. Socio-economic factors affect citizen mobility choices—and commuters are often influenced to take up private modes of travel. This forces policymakers to probe better travel demand management measures. Since there are many policy issues such as market failure, equity inaccessibility and financial and environmental sustainability, transport is one of the most heavily government-regulated sectors (Button & Gillingwater, 1983; Van Wee et al., 2013). Various negatives externalities are created by unregulated transport development. Transport is one of the major causes of Green House Gas (GHG) emissions; it dwarfs all other anthropogenic sources. Despite new modes of thinking and a greater awareness of the challenges facing transport as well as the need for technological change, the system is still far from adequate (Van Wee et al., 2013). Formalizing the negative externalities of transport is also a political process and governments in various countries and provinces concerned with the costs of these trends and are in a constant process to tackle the issue (Banister, 2005). So as to solve transport problems, various countries have started implementing public transport systems. Accessibility to and high frequency of public transport are necessary to a sustainable city. It is a proven fact and common knowledge among urban planners that increasing

P. N. Kuriakose (✉)
Department of Urban and Regional Planning, School of Planning and Architecture, Bhopal, India
e-mail: paulosenk@spabhopal.ac.in

J. Bhattacharjee
Architect and Urban Planner, North Eastern Space Application Centre (NESAC), Umiam, Meghalaya, India

© The Author(s), under exclusive license to Springer Nature Switzerland AG 2021

S. Mitra et al. (eds.), *Railway Transportation in South Asia*, Contemporary South Asian Studies, https://doi.org/10.1007/978-3-030-76878-2_8

modal share in public transport will eventually reduce the dependency on fossil fuels, as well as related ambient air quality deterioration and greenhouse gas emissions. Various types of public transport technologies have evolved over some time. Public transport ensures equity in accessibility, inclusiveness in road safety and welfare to the community. A city with higher rates of public transport also has a more democratic road space distribution among various types of users. Quality public transport services assure inclusiveness of lower-income citizens—thus to a certain extent, it reduces class conflict. Provision of Public transport ensures accrual of social benefits by providing affordable mobility options; it also induces community living. Provision of public transport also ensures lower per capita greenhouse gas emissions than private modes of transport. Increased dependency on the private modes of transport leads to high external cost—and most often the urban poor bear this burden through health issues, congestion, accidents and trauma. An efficient public transportation system that is integrated with the principles of smart urbanism and Transit-Oriented Development (TOD) offers a large number of benefits such as affordable housing and transport, physical and mental fitness, and reduced traffic crashes (Haque et al., 2013; Litman, 2010). Traffic accidents and congestion externalities of transportation inflict a massive pressure on both society and the environment. The Brasilia Convention held in 2015 in Brazil pushes for public transport infrastructure development so as to reduce road accidents. Public transport is not only sustainable, it also supports economic activities with minimum impact on the environment. Sustainable public transport development strategies can be widely classified into four categories: (a) integrated land use transport, (b) mass transit supply measures, (c) travel demand management measures and (d) inclusion of environment-friendly technologies (Haque et al., 2013). An efficient public transport system is essential to assist sustainable economic growth and sustainable urban development (Bachok et al., 2014). It is necessary to implement efficient public transport promotional strategies so as to allow the general public to identify with the policy and remove the so-called psychological barrier relating to public transport use. Factors leading to higher public transport usage are broadly divided into internal and external. External factors, which lie outside the purview of public transport soppier but influence the public transport system, include parking prices, subsidies to private vehicles, fuel prices, congestion charges, etc. Internal factors that influence public transport include (Rohani et al., 2013) ticket charges and fare policy, accessibility, availability, comfort, safety and reliability (National Academies of Sciences, Engineering, and Medicine, 2007; Paulley et al., 2006 Taylor & Fink, 2003).

2 Problem Statement: Urbanization and High Mobility Demand

Around 400 million people live in India's urban areas. According to the decadal census of 2011, 31% of the population live in urban areas. There are about 7,935

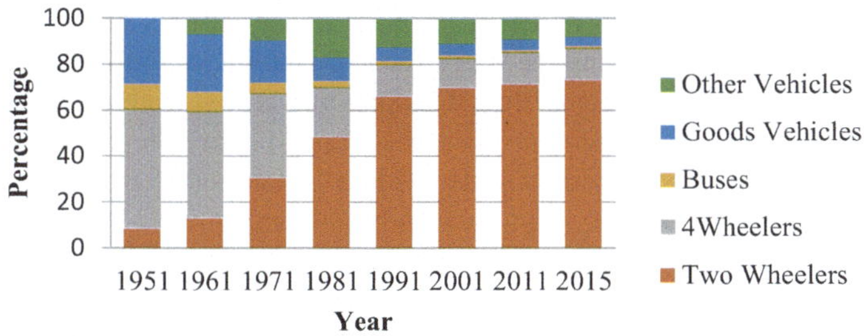

Fig. 1 Composition of registered motor vehicles in India: 1951–2015 (*Source* MoRTH [2016])

urban areas in India (Census of India, 2011). The number of urban centres with more than one million population in 2011 was 53. Post liberalization of the Indian economy in 1991, the country has grown at much greater speed, especially in industrialization and manufacturing and information technology. Cities are considered as the engines of growth; they themselves grow further through rural–urban migration, natural population increase, and the annexation of new areas to the urban boundary. With the explosive growth in the urban population, travel demand also grows and many transport issues emerge. The gross number of vehicles registered in India was 230 million in 2016. Two-wheelers (73.5%) constitute the highest share of registered vehicles, followed by four wheelers in the passenger vehicle segment (13.1%; Fig. 1). The number of registered buses is decreasing and the number of private passenger vehicles is fast-growing (MoRTH, 2017). Conscious of this evolving urban development and of the demand for travel the Government of India (GoI) launched the National Urban Transport Policy in 2006. It shows various categories of vehicles registered in India, around 85% of which are private motor vehicles (Fig. 1). The number of road accidents in India is growing at an alarming rate. Road accidents in the year of 2017 were 464,910 and road accident death was 146,377 (MoRTH, 2017). All these problems of the urban transport point to the need for better Public Transport infrastructure. In order to meet the derived demand, various measures have been taken up by the Government of India to improve public transport infrastructure. Metro rail projects are considered as one of the solutions for implementing sustainable transportation. But most cities that implemented metro rail projects are facing low ridership and huge financial loss.

3 Methodology

This study reviews the general policies, current status, barriers and opportunities faced by the metro rail development in India. This study relies on exploratory research methods such as secondary data available from various stakeholders and focus group discussions. The information collected was analysed in view of understanding the problems and issues involved and coming up with innovative solutions. This article is heavily based on existing published information by various Central, State and Municipal Governments. During the study, the team interacted with relevant government officials, experts in their affiliated fields, as well as academics, along with independent research organizations, so as to gain perspectives regarding the opportunities and growth of metro rail projects in India. A total of 76 people participated in the focus group discussion—from research and academia (24), private consultants (17), non-governmental organizations (14), municipal and state road transport corporations (8), and commuters (13). Out of the 76 participants in discussion, 32 were women. Three questions were posed to participants: (i) Do you think there is an enabling policy environment and programmes for improving the metro rail infrastructure in Indian cities? (ii) Are you satisfied with the metro rail infrastructure available in Indian cities? (iii) What strategies would you suggest for the further improvement of the metro rail infrastructure available in Indian cities? Each of the questions was analysed and corroborated with the quantitative data made available by Central, State and Municipal Governments. Data that is not available in the public domain was collected by using the provisions of the Right to Information Act, 2005.

4 Public Transport Development in India

During the early half of the eighteenth century, animal-driven carts were used as transport, but they were availed only by upper-class people. British rule in India brought trams service in the late nineteenth century. In early stages, trams were driven by horses; later they were converted to steam engines. In the early nineteenth century, horse-drawn carts were introduced to India. In 1895 the colonial administration introduced electric trams in Chennai (then Madras). Later, tram services were constructed in Kolkata (1900), Mumbai (1907), Kanpur (1907) and Delhi (1908; Rossman, 1998). But slowly, as road-based transport was improved, trams were removed from most of the Indian cities. Kolkata is the sole city in India where trams are still in use. By the early twentieth century, bus-based public transport facilities were shaped up by some visionary princely state kings and municipal administrations. At Present, there are 54 (24 State Road Transport Corporations [SRTCs], 8 Government Departmental Undertakings, 12 Companies, and 10 Municipal Undertakings) state road transport agencies offering public transport services in India (MoRTH, 2016). All three tiers of Governance have been working towards the common goal of providing quality public transport facilities to urban areas. Various types of public transport systems like the metro, monorail, BRTS and modern bus services are introduced or strengthened by following the objectives in the NUTP (MoUD, 2012). Today, there are over 600 kms of metro rail operating throughout the country in 11 cities; another 500 km plus tracks are under construction. Table 1 shows the cities having metro rail along with its daily average ridership and the length of operational tracks. Eight cities in India are going ahead with the construction of metro rail projects, and four cities are going ahead with the preparation of detailed project reports and a geotechnical

Table 1 Cities with operational metro rail system

City	Starting year of operation	Total operational route length (km)	Daily ridership	Projected ridership	Number of lines	Number of stations
Kolkata	1984	33.02	700,000	2,200,000	2	30
Delhi	2002	348.00	2,744,000	4,000,000	12	285
Bengaluru	2011	42.30	415,000	500,000	2	40
Mumbai	2014	11.40	450,000		1	12
Jaipur	2015	9.63	25,000	100,000	1	9
Chennai	2015	20.28	121,000	800,000	2	32
Kochi	2017	23.80	65,000	275,000	1	21
Lucknow	2017	22.87	60,000	644,659	1	22
Hyderabad	2017	69.00	255,000	2,200,000	3	56
Ahmedabad	2019	6.50	35,000	2,000,000	1	4
Nagpur	2019	22.29	22,000	363,000	2	16

Source Compiled from different metro rail project websites

survey. It has been decided to conduct a feasibility study for metro rail in all Indian cities bearing two million people or more (MoRTH, 2016).

The citizens of a city and their use of public transport usage generally maintain a positive connection; there is an over 40% public transport usage in Indian megacities (MoUD, 2008). There is a pull towards the use of personal vehicles along with growing urban populations. There has been a significant rise in the number of registered vehicles. Table 1 shows the current situation of the modal split in Indian cities. Before 2006 there were around 30 city bus services, 4 suburban rail transits, and two metro rail systems in India.

5 Existing Metro Rail Systems in India

The metro rail system has proved itself not only to be an efficient solution but also an environment-friendly answer to the problems of fast urbanization and rapid magnification of the cities. As the population is growing exponentially, so is the number of vehicles on the road, adding to the congestion of roads and pollution. A mass transit system such as the urban rail facilitates the easy and cheap movement of masses at a time, resulting in many other benefits such as less carbon footprint, less congestion, reduction in travel times and costs, less air and noise pollution, less road accidents, no parking cost, etc. It also has a significant impact on social diversity, as people from various economic and social classes access the same transit system. Due to the significant success of the metro rail system in cities like Kolkata and Delhi, its need and necessity are being equated by more Indian cities—today as many as 11 cities have metro rail system. Many cities like Pune took the initiative of launching the Urban Rail after reaching a saturated congestion phase which made it difficult to implement the project, whereas other cities such as Nagpur realized its importance, learnt from the mistakes made by others, and started working on the project way before its need was felt by citizens.

Kolkata is the pioneer city of introducing Urban Rail to India—and it has been performing significantly well since 1984. Constructed entirely underground, it has proved a relief for the thousands of daily commuters travelling kilometres for work. Delhi was the second city to introduce the metro rail, and today it has the longest operational route length and the highest number of daily riders. The Hyderabad Metro Rail is relatively new with a very high success rate. It attained the third-largest daily ridership in less than three years and has been performing very well. However, most cities are performing moderately, with some failing terribly. The Nagpur Metro has the least average per day ridership; the reason for this might be that it is relatively new and could not harness attraction yet. The Jaipur Metro project is considered a failure due to the decreasing daily ridership, which affects the coverage of operational expenses.

The National Urban Transport Policy in 2006 suggested that cities with a population of 20 lakh or more should construct a metro rail system. In 2014, the Union Government announced financial assistance for the establishment of metro rail in cities with a population over 10 lakhs. The Union Government approved the Union Urban Development Ministry's proposal for implementing metro rail in 50 cities. The projects were to be implemented by special purpose vehicles as a joint venture between the centre and the state. In a draft policy from 2017, the centre asked the State Government to consider the metro rail system as a last option for mass transit and only go forward with it after considering other, less costly options. The project for metro rail is on in 10 other Indian cities which are expected to be operational in future years, along with the completion of existing tracks in the cities mentioned above.

6 Metro as a Low Carbon Transport

India's contribution to the total greenhouse gases produced in 2012 was 5.73 (Carbon Brief, 2015). India has taken the Sa pledge in Paris Summit to bring down greenhouse gases emissions by 30–35% by 2030 (from the levels registered in 2005; INDC India, 2015). Improvement in public transport and especially the increase in the Mass Rapid Transport System (MRTS) can help India achieve the promised emission reduction. For example, a modal shift-based carbon emission reduction analysis done for the Mumbai Metro Rail revealed a reduction of 22.7 tonnes of CO_2 emission per day (Soni & Chandel, 2018). The United Nations have certified the Delhi Metro Rail Corporation as the first metro rail and rail-based system in the world to get carbon credits for reducing Green House Gas Emissions. This is because it has helped reduce global warming by reducing pollution levels in the city by 0.63 million tons every year. In 2012–2013, the Delhi Metro Rail Corporation (DMRC) has been successful in reducing emissions by 103,000 tonnes of carbon dioxide; in 2013–2014 emission reduction attained 157,000 tonnes. The agency ambitions to save 26.60 million tonnes by 2031. However, the full potential of the system can be achieved when a higher level of modal shift is ensured from personalized vehicles to metro rail systems (DMRC, 2017). This can be ensured through the integration of various modes of transport in terms of seamlessness.

7 Recent Enabling Environment

7.1 The National Urban Transport Policy (NUTP) of 2006

The NUTP recognizes the increasing trend of personal vehicles and their negative influence on both quality of life and the environment. It suggests a multipronged strategy to solve many issues and externalities such as lacunae in land use public transport integration, suboptimal parking charges, congestion, road fatalities, and declining trend of non-motorized transport. It also stresses the focus on citizens as the rightful owners of road space—all plans should point to their benefit. The various objectives of the policy include ensuring seamless travel through modal and fare integration, inducing public transport, democratic and inclusive road spaces, transport land use integration, and reducing travel time as well as distances. Apart from these, the NUTP recognizes the need for the adoption of Intelligent Transport Systems for traffic management and public–private partnerships in urban transport management. The policy makes recommendations for effective transport governance, as well as regulatory and enforcement mechanisms (MoUD, 2008). Special attention is given to public transport as a socially important sector for the whole population.

The strategies identified for achieving these objectives are multi-faceted and comprehensive. The government envisages to attain inclusive and sustainable public transport through various measures such as the Centrally Sponsored Scheme of Urban Transport Planning (CSSUTP). Under the CSSUTP, the government provides financial assistance up to 80% for taking up various urban transport studies/surveys, launching of awareness campaigns, etc. For devising a Detailed Project Report (DPR) for MRTs projects, the central financial assistance will be 50%—the remaining expense must be shared by the State Government and the Urban Local Body. The CSSUTP covers a wide scope of urban mobility matters. The goal is to promote comprehensive urban transport plans and studies, transit-oriented land use and transport planning, preparation of comprehensive mobility plans and Detailed Project Reports, Clean Development Mechanism (CDM) research and Intelligent Transport System (ITS) studies, Transport integration, etc. (MoUD, 2008). These schemes aim to guide and facilitate the NUPT's implementation.

7.2 The Transit-Oriented Development (TOD) Policy

TOD policy is prepared by taking the basic principles of smart urbanism and following the path of sustainable, public, transport-oriented, compact urban development A major objective of the policy is to promote public transport-friendly, high-density, mixed-use development. The TOD policy adopts the principles of new urbanism in expanding liveable communities. It highlights the need for promoting a safe environment for NMT users and pedestrians. Another aim of the policy is to rein the parking supply as well as private vehicle ownership. Controlling parking space supply in the transit influence area constitutes a salient objective to induce public transport ridership and optimal use of valuable land. It calls for a delineation of a 500- to 800-m area around the Transit Station as a TOD zone, depending upon transit capacity. The policy advises keeping a minimum of 30% or higher Floor area ratio for affordable housing in the transit's influence area. It also suggests that an upper limit for the size of dwelling units should be fixed in the transit influence area so as to facilitate transit-friendly housing development. The TOD policy demands the incorporation of value capture financing for the funding of transit and overall improvement of transport (MoHUA, 2017a).

7.3 The Metro Rail Policy

The Metro Rail Policy of 2017 opens the door for public–private partnerships in capital-intensive high-capacity metro projects, by making it a required component for receiving assistance from the centre for new metro projects. Metro rail project components would be unbundled so as to make way for private participation in automatic fare collection, operation and maintenance of services, etc. The policy recognizes the lack or inefficiency of the last-mile connectivity around existing metro corridors and seeks to improve it. The new policy envisages considering a five-km buffer area on both sides for metro-supportive, facilitative, integrated transport development. State Governments are required to add more feeder services and NMT infrastructure so as to bring more ridership to the metro. The policy also mandates an alternate scenario analysis so as to identify suitable transit modes such as BRTS, LRT, Tramways and metro rail. Setting up of Urban Metropolitan Transport Authority (UMTA) was one of the policy agendas of the NUTP 2006; however, it was not implemented by many States. The metro rail policy makes it mandatory to obtain approval for metro projects. Considering the absence of ridership in already established corridors, the metro policy makes it compulsory to conduct an appraisal of both metro projects and ridership projections (MoHUA, 2017b).

It also stipulated that instead of 8% Financial Internal Rate of Return, an Economic Internal Rate of Return of 14% should be the basis of approving investments in metro projects. Metro rail policy makes the connection with the TOD policy by reiterating the need for transit-facilitative, compact land use development. The policy asks State

Governments to use land value capture and betterment levies so as to create a fund for financing metro projects. It also entrusts States to make necessary rules and regulations for fixing fares and its revisions. State Governments are provided with three models of financing the metro project: (a) the PPP and Central Government viability gap funding, (b) the Grant by the Government of India, and (c) a 50:50 Equity sharing model between Central and State Governments (MoHUA, 2017b).

7.4 Financial Modalities of the Metro Rail System Development

Huge investments are required for the infrastructure of the Urban Rail System. The Union Government stressed in 2014 that all cities with 2 million or higher populations should begin planning for metro projects with immediate effect. For this purpose, the Union Government in 2016–2017 has earmarked a budget of 100 billion INR (UITP, 2020). However, urban transport systems are a state responsibility—planning, execution and development are all carried out by the states themselves in a joint venture with the Central Government. In some cases, Urban Local Bodies have also financed projects using debt financing from other multilateral funding agencies (Indian Infrastructure, 2019).

7.4.1 The Public–Private Partnership Model

In the metro rail policy of 2017, the Central Government promised assistance only if a Public–Private Partnership (PPP) model is adopted by the State Government for the execution of metro projects (MoHUA, 2017b). In this kind of model, there is a shareholding among the state, centre and private parties through a special purpose vehicle. To this day, Mumbai, Gurgaon and Hyderabad are the only cities to embrace this approach. The extent of centre versus state participation can vary. Most times, projects become privatized with almost zero engagement from the government (public).

7.4.2 Limitations to the PPP Model

Metro rail projects are capital-intensive and are often subjected to bitter experiences as a result of failed investment models or insufficient revenue generation. After the metro rail policy stated in 2017 that the centre would only aid projects endowed with private participation, the situation worsened as private companies seem unwilling to invest. E. Sreedharan, the metro man of India, said that the public–private concept will not work as the metro is not a profit-making deal for private companies (Sood, 2019).

Many officials have claimed that the policy would not be successful as it puts a double pressure on State Governments—which must both implement the project and ensure private investments in projects with negligible profit margins. There has also been a considerable drop in the growth of this sector in the years 2014–2016—after a success story since 2002. There have been many changes in policy over the years, wherein the centre kept diminishing its role in the development of urban rail infrastructure. In fact, one of the reasons for the dip in growth was the centre's trepidation to fund projects.

7.4.3 Value Capture Financing (VFC)

This is a public financing model that tries to capture the value of land and use it for infrastructure development. Many cities around the world have successfully used this model—including Kochi, Delhi, Bangalore, etc. in India under the metro rail project. In 2017, the Value Capture Finance Policy Framework was announced as a relief to Urban Local Bodies. The metro rail policy of 2017 also encourages the ULBs to use value capture finance as a tool to generate revenue for investing in metro projects (MoHUA, 2017b).

The VCF was used by some State Governments in different ways so as to generate revenue sustainably without depending on either private investments or the centre. For example, the Karnataka Government came up with the idea of creating a dedicated fund for mass transit projects investments in which funds would be generated by anybody defying the land use near the respective projects—and by notifying a premium floor space index. Similarly, the Delhi and Haryana Government has increased the Floor Space Index on the sides of transit corridors to finance the metro project through Transit-Oriented Development (MoHUA, 2017b).

7.4.4 Non-Fare Revenue Streams

The Non-fare Revenue policy was introduced by Indian Railways so as to decrease the dependencies on the conventional method of generating revenue (i.e. through tariff hikes). Here, the focus shifts from ridership to consumer-ship as revenue is generated from using the spaces around both the station and the metro. This is achieved, for example, by installing advertisement boards in the metro station, renting out spaces and corners of the station for retail outlets, ATMs, eateries, etc., branding the rail with popular brands, etc. (Arora, 2017). This is a very important and sustainable source of revenue generation for mass transit systems—and there has been a substantial increase in revenue generation since this policy was adopted (Meshram, 2018).

The details of the operational models of metro rail projects in India. It is clear that most metro systems are run by public sectors; however, in the future it is likely that more and more private companies start investing in metro projects if these are found financially viable (Table 2).

Table 2 Operational model of the metro rail

Metro system	State	Operated by
Kolkata Metro	West Bengal	Public
Delhi Metro	Delhi	Public
Noida Metro	Uttar Pradesh	Public
Gurgaon Metro	**Haryana**	**Private**
Bengaluru Metro	Karnataka	Public
Mumbai Metro	**Maharashtra**	**Private**
Jaipur Metro	Rajasthan	Public
Chennai Metro	Tamil Nadu	Public
Kochi Metro	Kerala	Public
Lucknow Metro	Uttar Pradesh	Public
Hyderabad Metro	**Telangana**	**Private**
Ahmedabad Metro	Gujarat	Public
Nagpur Metro	Maharashtra	Public

Source Compiled from different metro rail project websites

8 Existing Challenges and Prospects

8.1 Lack of Integration of Modes and Fares

There are multiple modes of public transport in larger Indian cities. Metro rail, BRT and bus services are the common modes among scheduled services—along with para-transit services, intermediate services, auto-rickshaws, etc. that cover shorter routes. Multimodal integration of transport systems permits a seamless transfer between different modes of transport. This integration has a major role to play in travellers' commute choices because the greater the convenience provided in travel and the more economical the fare system, more commuters will public transport attract. Close to 180 Cities have public transport facilities running. From these, 10 cities have multiple public transport modes such as metro, suburban rail, monorail, and BRTS or other bus services. The problem, however, lies in the lack of integration between the different modes as they all work independently and compete with one another. There are overlapping between transit routes with metro routes in many cities such as Mumbai, Delhi, Ahmedabad, and Kochi. For instance, in Kochi, there are around 800 bus routes with 1,400 buses, but almost 46% of all types of bus routes overlap with the metro (Fig. 2). This leads to problems such as muddled schedules, overlapping routes and higher interchange penalties. Because of the modal shift of passengers from city bus services to metro, many routes lost considerable ridership. All the cities have IPT and various Para Transit modes; however, since there is no proper route rationalization exercises conducted, many of the IPT Modes and Para Transit modes lie in the same route as BRT or metro services. Better last-mile connectivity is important to enhance access and egress trips; in order to achieve this,

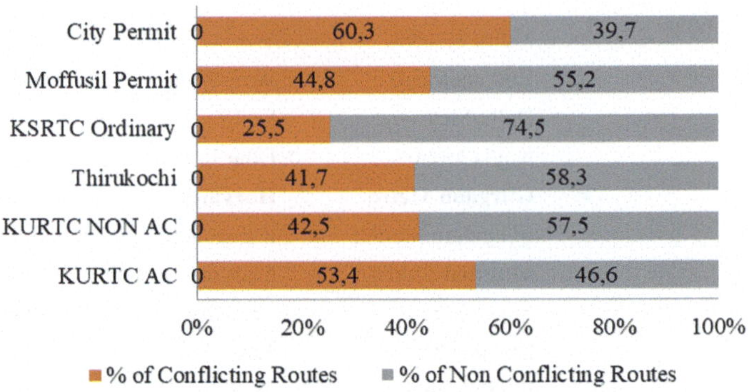

Fig. 2 Overlapping between Bus Routes and the Kochi Metro (*Source* Packirisamy and Kuriokose [2016])

IPT and para-transit mode routes must be reassigned. Along with this, integration in the fare system can create more support for public transport (Sharaby & Shiftan, 2012). Commuters have to pay higher fares because of the lack of fare integration. Absence of fare integration is compelling passengers to pay more for trips, thereby decreasing the overall effectiveness of the public transport system (Table 3).

Table 3 Ridership of a few Indian and International Metro Rail Systems

Metro	Route length	Average daily ridership (in thousands)	Ridership per km	Metro	Route length	Average daily ridership (in thousands)	Ridership per km
Delhi	217	2,700	12,442	Beijing	608	10,000	16,447
Bengaluru	42	330	7,857	Seoul	940	7,100	7,553
Kolkata	29	650	22,412	Moscow	326	9,000	27,607
Chennai	28	55	1,964	Guangzhou	260	8,000	30,770
Jaipur	10	18	1,800	New York	368	4,600	12,500
Kochi	14	45	3,214	Mexico	226	4,600	20,353
Mumbai	11	340	30,909	Hong Kong	175	4,400	25,143
Gurgaon	12	35	2,917	Paris	205	4,200	20,487
Mumbai monorail	9	15	1,667				
Hyderabad	30	100	3,333				
Lucknow	8.5	15	1,764				

Source (Agarwal, 2019)

8.2 *Lack of Transit-Oriented Development and Land Use Transport Integration*

Both coverage and ridership are low in most metro rail projects in India. More than 80% of the population lives beyond the one-kilometre buffer zone of metro corridors. The success of public transport systems depends upon the threshold of travel demand available in the transit influence corridor. Transit corridor planning and land use development regulations should be coordinated enough to ensure travel demand. The Transit-Oriented Development (TOD) and encouraging high-density growth along transit corridors constitute a proven strategy for ensuring high dependency on public transit. According to Cervero and Kockelman (1997), the objectives of the TOD include (1) reducing motorized trips, (2) escalating the frequency of non-motorized trips within the number of trips already produced, (3) increasing vehicle occupancy and reducing travel distances of motorized trips already being produced (Chen, 2012). The TOD provides equity inaccessibility for all sections of society (Cervero & Dai, 2014). Though lately a TOD policy was introduced, the implementation on the ground is negligible. Only a few metropolitan cities such as Ahmedabad, Delhi, Bengaluru, Hyderabad, Mumbai and Chennai have tried to incorporate TOD Principles in their master plans. Important aspects missing in the TOD zones include the lack of restrictions in the dwelling unit sizes that can be constructed in the TOD zone. Income, size of the dwelling unit and mode choice are interrelated. A study conducted in the Delhi and Kochi Metro Corridors shows that the higher the carpet area of houses, the less chance of using the Mass Transit. A study conducted in both Delhi and Kochi found that dwelling units with carpet area sizes ranging from 60 to 200 sqm have a higher chance of metro usage. Metro preference is highest among the respondents with 90 to 120 sqm dwelling units (Tables 4 and 5). But development control regulations are not modified to encourage construction of housing typologies that would be used by mass transit users. Making such modifications can avoid gradual gentrification of the TOD zone (Chava et al., 2018). Megacities have identified TOD zones and given higher FSI in their influence areas. There are almost 60 cities in India with more than one million people; however, less than 10 cities

Table 4 Dwelling unit sizes and metro preferences in Delhi

Dwelling unit size (Sq. mtr)	Metro preference				
	Punjabi Bagh	Kohat Enclave	Shastri Nagar	Mayur Vihar Extension	Dwarka Mor
< 30	13	0	0	12	0
30–60	47	12	14	31	53
60–120	20	47	36	38	41
120–200	13	41	43	19	6
> 200	7	0	7	0	0

Source Meesa (2017)

Table 5 Dwelling unit sizes and Mode Choice in Kochi

Dwelling unit size (Sq. mtr)	Walk	Two-wheeler	Car	Bus/Metro	Auto/Taxi	Cycle
< 30	40	40	0	17	0	3
31–90	11	37	22	25	5	0
91–120	6	29	18	42	5	0
121–200	1	25	26	39	9	0
201–250	0	36	47	11	6	0
251–300	0	33	51	6	10	0
> 300	0	0	70	5	25	0

Source Mukundan (2018)

have attempted to make an effort to bring TOD principles in their master plan. Many metro cities are progressing without an updated master plan. Higher FSI is provided; however, other elements of smart urbanism have not been followed.

8.3 *Affordability and Scope for Subsidies*

Travel fares in the Delhi Metro are probably one of the highest in the world—beating fare of many other prominent metro systems. The Indian metro charges Rs 35 to Rs 40 per trip (Fig. 3). Daily wage labourers are compelled to spend almost 20% of their earnings on travel, considering integrated journey costs (Table 6). This figure is higher

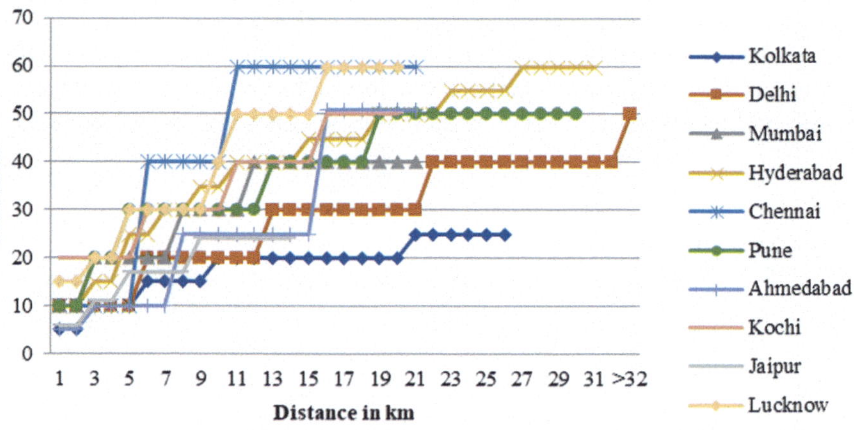

Fig. 3 Fares in major metro rail systems in India (*Source* Compiled from different metro rail project websites)

Table 6 Comparison of affordability among metro systems in India

City	Minimum wages (Rs per day)		Monthly income (Rs)		Average trip length (km)	Metro fare (Rs per trip	Monthly expenditure on public transport (Rs)	Percentage of transportation expenses over total income	
	Skilled	Unskilled	Skilled	Unskilled				Skilled	Unskilled
Delhi	648.00	534.00	16,848	13,884	12.90	40	3120	19	22
Bengaluru	565.54	471.95	14,704	12,271	11.90	35	2730	19	22
Mumbai	541.38	495.23	14,076	12,876	12.27	40	3120	22	24
Hyderabad	461.63	318.62	12,003	8284	900.0	35	2730	23	33

Source CSE (2019)

than the global benchmark of 10–15%, thus rendering the Indian metro unaffordable for the vast majority of the population belonging to lower-income groups (CSE 2019).

Automobile users often do not cover the average cost, let alone the marginal cost, of the service they consume; this leads to overconsumption of road infrastructure (Robinson, 2000). In turn, private automobile users enjoy under-priced road infrastructures, proliferating private vehicle ownership. Under-pricing of road infrastructure reduces the economic efficiency and productivity of bus services, thereby skewing the transport economy towards a sub-optimized allocation of transport resources. Tax rebates and incentives to private vehicles cause enormous transport market distortions and reduce the financial autonomy of transport operators. The prosperous increase in private counterparts catalyses the growth of perverse subsidies that are detrimental for both the economy and the environment. With the general reverse to the fair provision of transport, it is argued that there is no longer a subsidy for public transport but rather a payment for providing the service—which is opposite to its social mandate.

8.4 The Need for Innovative Financing of Metro Projects

There has been a drastic change in the way metro projects are funded. Inventive financing mechanisms raise funds from all the recipients of a system, rather than from its users alone (Agarwal et al., 2019). Beneficiaries of public transport include the real estate sector, business establishments, car users, etc. 'Versement Transport' a kind of transport tax in France imposed on all business establishments that employ more than a certain number of employees (Agarwal et al., 2019). The principle behind such a tax is that investments made in the transportation system are what allows employees to travel to work. An investment in mass transit increases land value and benefits the real estate sector. Some portion of this value addition can be captured for further improving PT. The TOD Policy emphasis value capture for the improvement of sustainable mobility. Vehicle taxation regimes should be revamped so as to develop a PT-supportive taxing system. It is time that India should think of applying a levy on fossil fuels based on the polluter-pays principle; the money received as levy should be reinvested on PT. Parking charges, depending on the land value of the property on which the car is parked, should be considered as a potential revenue option. Many municipal corporations have meagre revenues; the NUTP suggests charging parking fees based on land value. Congestion pricing in city cores can bring higher modal shifts and more revenue for investment in PT infrastructure (Mulley & Walters, 2014). No advance has been made in congestion pricing—even though optimal road pricing based on marginal social cost principles is the best way to proceed (Sen et al., 2010). PT systems usually possess valuable plots of land in urban areas. The pecuniary price of these lands is very high; however, they are under-utilized. Air rights above land parcels have allowed to commercially exploit the land and gain considerable revenue (Agarwal et al., 2019). Litman (2014) identified 18 funding alternatives (Table 7), some of which are widely used, whereas

Table 7 Potential PT funding options

Name	Description
Discounted bulk passes	Based on ridership, passes are sold
Property taxes	Increase local property taxes
Sales taxes	Special local sales tax
Fuel taxes	Extra fuel tax in the region
Vehicle fees	Extra fee for vehicles registered in the region
Utility levy	Levy on all utility accounts in the region
Employee levy	Levy on each employee within a designated area or jurisdiction
Road tolls	Levy on roads
Vehicle-Km tax	Distance-based fees on vehicles registered in the region
Parking taxes	On commercial parking transactions
Parking levy	Special property tax on parking spaces throughout the region
Expanded parking pricing	Increase when and where public parking facilities (such as on-street parking spaces) are priced
Development of transport impact fees	Fee on new developments to help finance infrastructure, including transit improvements
Land value capture	Special taxes on property that benefit from the transit service
Station rents	Collect revenues from public–private development at stations
Station air rights	Sell rights to build over transit stations
Advertising	Additional advertising on vehicles and stations
Fare Increases	Increase fares or change in fare structure so as to increase revenues

Source Litman (2014)

others, though innovative, are only used in particular jurisdictions.

8.5 Transport Governance and MaaS

The quality of urban mobility will be influenced by transport governance and intuitional arrangements for managing various transport modes (Lowe & Wright, 2018). Creation of UMTAs is occurring at snail's pace. Although 17 states have taken action to constitute UMTA, at present only seven out of 53 million cities are endowed with this institutional arrangement. A close examination of UMTA members has revealed that decisive power is concentrated at the State Government level. Coordinated urban transport development is slowed down due to ineffective institutional sets and deficient technical expertise. Though various policies such as the NUTP, the metro policy and the TOD are prepared by the Central Government, implementation of policy objectives at the ground is very meagre. Effective institutional integration would have helped articulate various modes in terms of both fares and physical integration. Cities such as Delhi, Kolkata, Mumbai, Chennai and Bangalore have multimodal public transportation; however, each of these PT service providers function independently. Urban transport governance is majorly concentrated in the hand of State and Central Governments. The creation of UMTAs can support coordinated route rationalization, physical and fare integration of modes, and fixing of parking charges—in tandem with Public Transport promotion principles.

Mobility-as-a-service (MaaS) surfaces as a most promising disruptive technology and service. It is a customer-focused service that provides easy access to a variety of choices of transport, hence changing the entire travelling industry. This system is in the limelight as it provides a wide range of 'better travel choices' and journey experiences for customers to choose from. MaaS is an intelligent mobility distribution model wherein a single digital platform provides aggregated mobility services (Mitashi, 2020).

The first Indian city to take tangible measures on multimodal integration was Kochi. The city is beautifully moving towards MaaS by integrating traditional transport modes such as water ferries and auto-rickshaws with the new metro rail in only one service platform. This new service has 'enabled dynamic transport governance, planning, and management from a redundant regulatory system' in Kochi. It can bring about multiple changes in governance, legislative, operational and functional simultaneously, involving many stakeholders and move the system from 'output' to 'outcome'. This new perspective has allowed Kochi to review its present institutional arrangement, identify its shortcomings and take measures to overcome the problems—progressing towards long-term effectiveness (Mitashi, 2020).

8.6 Open Data and Management Information Systems

General Transit Feed Specification (GTFS) is gaining popularity in developed cities. These are provided with transit agencies for storing and circulating PT schedules and other related geographic information (Devulapalli & Agrawal, 2017). It helps fuse information of transit to platforms like Google Maps—because of this, an astounding rise in the transit data sector has evolved in developed parts of the world. The Indian PT system data is not open to the public yet. Innovations are encouraged wherein open data ecosystems work with new ideas, techniques and approaches in the improvement of a transport mode (Devulapalli & Agrawal, 2017); however, not many transit agencies use or publish such data. Recently the Kochi Metro has started providing data. Opening up transit data could make private car aggregators provide last-mile connectivity. Within the closed environment of transit data, information is scattered in various agencies under different levels of governance—including Central, State, Para-statal, and urban local bodies.

8.7 Use Parking as a Travel Demand Management Strategy

Whether cars are fuel-efficient or not, a car on average spends about 95% of its life in a parking (Shoup, 2005) and takes up many parking spaces per week (Litman, 2013). Efficient parking strategies can help integrate land use and transit, reduce emissions from transport, induce high-density mixed land use, and reduce vehicle ownership and congestion. The link between transport and land use is seen in parking policy (Marsden, 2006). There is a need to amend parking standards by including parking supply caps, parking maximums, and flexible parking standards. PT accessibility, presence of mixed land uses, walkable streets, bicycle-sharing facilities, residential and employment opportunities, etc. should be contemplated in determining parking standards. The concept of flat standard for the whole city should be abolished—parking districts should be delineated. House registration procedures should be modified so as to unbundle parking; separate registry needs to be maintained for parking spaces. Putting a cap on parking supplies in city core areas and the introduction of parking maximums can much impact modal shift (Wang & Yuan, 2013). There is a need for strong legislative support to implement car parking free development in CBDs and residential areas within the accessible catchment area of PT nodes, thus allowing parking fees in private development, fines for parking fee violations, implementation of parking maximums, land value and congestion responsive parking pricing, and workplace levies. With the introduction of the GST parking space, managing vendors are paying taxes on parking fees. Parking fees and fines are can be a major source of revenue for the Municipal Corporation—and proper ring-fencing of this revenue can help urban local bodies push sustainable mobility. The role of parking charges in the average generalized cost of trips has to be given due importance by not only abolishing free parking in workspaces but also through designing adequate charges. It is an uncomplicated and relatively cheap method to reduce illegal parking in residential land. It can also help slow down private motor vehicle registration. The strategy is both spatially and dynamically efficient (Wang & Yuan, 2013). Similar strategies were implemented in Japan, which successfully decreased the demand on parking supplies as well as illegal parking in residential land uses (ADB, 2011; Morikawa et al., 2010).

9 Conclusions

Indian cities are adopting metro rail projects so as to improve public transport infrastructure. Many policies such as the NUTP, the metro rail policy and the TOD Policy, as well as legal support were provided for the sector to create an enabling environment. At present, there are 11 cities with metro rail facilities—and many more are going ahead with the implementation of feasibility studies. However, the current situation of metro coverage and usage is deficient in most Indian cities. Many a time metro fare is unaffordable to a major section of the society. The Indian metro rail implementation process has taken private stakeholders on board for the mobilization funds and is considered as one of the transport modes for bringing the GHG emissions down. Public Transport use must be encouraged so as to achieve quality living in urban areas. All PT infrastructures have to be perfectly integrated and need a TOD plan for the corridors involved. With the help of information technologies, in-vehicle travel time improvement methods can be achieved. Factors outside transit systems such as parking charges should be explored so as to bring ridership to the metro rail and bring negative externalities inside formal pricings. There is no panacea available to solve the problem of frequency and capacity; it can be achieved by allocating sufficient funds for more buses and coaches per million people as a feeder to the trunk routes of the metro. Indian metro rail management authorities should open up their data so as to foment demand-based travel and demand management. This can induce innovation and encourage a higher level of modal shift from private modes to the metro rail system. The creation of efficient transport governance infrastructures and a legal enabling environment will constitute the base for the introduction of MaaS. All these efforts can attract more patronage ridership in metro rail—and, eventually, a better quality of life in Indian cities.

References

ADB. (2011). *Parking policy in Asian cities*. Asian Development Bank.

Agarwal, O. P. (2019, January 29). Compulsion to choice: How can public transport in India be transformed? *Economic and Political Weekly*.

Agarwal, O. P., Kumar, A., & Zimmerman, S. (2019). *Emerging paradigms in urban mobility: Planning, finance and implementation* (pp. 139–158). Elsevier. https://doi.org/10.1016/B978-0-12-811434-6.00007-X

Arora, R. (2017, January 10). Railway launches non fare revenue policy. *The Economic Times*.

Bachok, S., Osman, M. M., & Ponrahono, Z. (2014). Passenger's aspiration towards sustainable public transportation system: Kerian District, Perak, Malaysia. *Procedia—Social and Behavioral Sciences, 153*, 553–565. https://doi.org/10.1016/j.sbspro.2014.10.088

Banister, D. (2005). *Unsustainable transportation. City transport in the new century*. Routledge.

Button, K. J., & Gillingwater, D. (1983). *Transport*. Ashgate Publishing.

Carbon Brief. (2015). *Carbon brief's Paris climate pledge tracker, Paris Summit 2015*. http://www.carbonbrief.org/paris-2015-tracking-country-climate-pledges

Census of India. (2011). *Rural urban distribution of population*. http://censusindia.gov.in/2011-prov-results/paper2/data_files/india/Rural_Urban_2011.pdf

Centre for Science and Environment. (2019). *The cost of urban commute: Balancing affordability and sustainability*.

Cervero, R., & Dai, D. (2014). BRT TOD: Leveraging transit-oriented development with bus rapid transit investments. *Transport Policy, 36*, 127–138. https://doi.org/10.1016/j.tranpol.2014.08.001

Cervero, R., & Kockelman, K. (1997). Travel demand and the 3Ds: density, diversity, and design. *Transportation Research Part D: Transport and Environment, 2*(3): 199–219. https://doi.org/10.1016/S1361-9209(97)00009-6

Chava, J., Newman, P., & Tiwari, R. (2018). Gentrification of station areas and its impact on transit ridership. *Case Studies on Transport Policy, 6*(1): 1–10. https://doi.org/10.1016/j.cstp.2018.01.007

Chen, Y. (2012). *Assessing climate value of cycling under different urban forms of Dutch cities* (Master Thesis). University of Twente.

Delhi Metro Rail Corporation. (2017). *UN body credits Delhi Metro- 6.3 lakh carbon credits for modal shift project*. http://www.delhimetrorail.com/press_reldetails.aspx?id=746xECETA6Qlld

Devulapalli, H., & Agrawal, G. (2017). Mapping bus transit services in Hyderabad—An illustrative example of the use of open geospatial data. *Transportation Research Procedia, 25*, 4196–4206. https://doi.org/10.1016/j.trpro.2017.05.369

Haque, M. M., Chin, H. C., & Debnath, A. K. (2013). Sustainable, safe, smart—Three key elements of Singapore's evolving transport policies. *Transport Policy, 27*, 20–31. https://doi.org/10.1016/j.tranpol.2012.11.017

Indian Infrastructure. (2019, November). *Innovative financing: Urban rail projects tapping new funding sources*. https://indianinfrastructure.com/2019/11/08/innovative-financing-2/. Accessed July 20, 2020.

Intended Nationally Determined Contribution India. (2015). *India's intended nationally determined contribution, working towards climate change*. https://www4.unfccc.int/sites/submissions/INDC/Published%20Documents/India/1/INDIA%20INDC%20TO%20UNFCCC.pdf

Litman, T. (2010). *Evaluating public transportation health benefits*. Victoria Transport Policy Institute. https://apta.com/wp-content/uploads/Resources/resources/reportsandpublications/Documents/APTA_Health_Benefits_Litman.pdf. Accessed October 19, 2013.

Litman, T. (2013). *Parking management: Strategies, evaluation and planning*. Victoria Transport Planning Institute. http://www.vtpi.org/park_man.pdf

Litman, T. (2014). Evaluating public transportation local funding options. *Journal of Public Transportation, 17*, 43–74. https://doi.org/10.5038/2375-0901.17.1.3

Lowe, C., & Wright, R. (2018). Institutional arrangements for greater inter-modality between bicycles and buses: A Melbourne case study. *Research in Transportation Economics, 69*, 369–376. https://doi.org/10.1016/j.retrec.2018.02.002

Marsden, G. (2006). The evidence base for parking policies—A review. *Transport Policy, 13*, 447–457. https://doi.org/10.1016/j.tranpol.2006.05.009

Meesa, S. H. (2017). *Housing typology and mode choice in a transit corridor* (Dissertation/Thesis). School of Panning and Architecture.

Meshram, R. (2018, December 5). *Opinion—There has been a 75% increase in non-fare revenue in 2016–17*. Live mint.

Ministry of Housing and Urban Affairs, Government of India. (2017a). *National transit oriented development policy*. http://mohua.gov.in/upload/whatsnew/59a4070e85256Transit_Oriented_Developoment_Policy.pdf

Ministry of Housing and Urban Affairs, Government of India. (2017b). *Metro rail policy 2017*. http://Mohua.Gov.In/Upload/Whatsnew/59a3f7f130eecmetro_Rail_Policy_2017.Pdf

Ministry of Road Transport and Highways. (2016). *Review of the performance of state road transport undertakings*. Ministry of Road Transport and Highways. https://morth.nic.in/sites/default/files/Review_2015_2016.pdf

Ministry of Road Transport and Highways. (2017). *Road accidents in India—2017*. Ministry of Road Transport and Highways. http://www.indiaenvironmentportal.org.in/content/459084/road-accidents-in-india-2017/

Ministry of Urban Development. (2008). *Notification to all states and union territories*. Government of India, Ministry of Urban Development. http://urbanindia.nic.in/programme/ut/guid_she_utp.pdf

Ministry of Urban Development. (2012). *Recommendations of working group on urban transport for 12th five year plan*. Government of India, Ministry of Urban Development. http://planningcommission.nic.in/aboutus/committee/wrkgrp12/hud/wg_%20urban%20Transport.pdf

Mitashi, S. (2020). India's shift from mass transit to MaaS transit: Insights from Kochi. *Transportation Research Part A: Policy and Practice, 131*, 219–227. https://doi.org/10.1016/j.tra.2019.09.037

Morikawa, S., Asao, K., Igo, T., & Kato, H. (2010). *Institutional system and current problems of car parking in Tokyo, Japan*. Internal Report for the Asian Cities Parking Study 2009, University of Tokyo. www.trip.t.u-zokyo.ac.jp/kato/WP/2010/2010wp_e2.pdf

Mukundan, P. (2018). *Regulating transit-oriented development for inducing transit friendly housing: A case study of Kochi* (Dissertation/Thesis). School of Panning and Architecture.

Mulley, C., & Walters, J. (2014). Workshop 7 Report: Innovative finance for innovative public transport. *Research in Transportation Economics, 48*, 389–392. https://doi.org/10.1016/j.retrec.2014.09.071

National Academies of Sciences, Engineering, and Medicine. (2007). *Elements needed to create high-ridership transit systems*. The National Academies Press. https://doi.org/10.17226/23175

Packirisamy, P., & Kuriokose, P. N. (2016). *Route rationalisation buses along the proposed metro corridor: A case of Kochi* (Unpublished Thesis). School of Planning and Architecture.

Paulley, N., Balcombe, R., Mackett, R., Titheridge, H., Preston, J., Wardman, M., Shires, J., & White, P. (2006). The demand for public transport: The effects of fares, quality of service, income and car ownership. *Transport Policy, 13*, 295–306. https://doi.org/10.1016/j.tranpol.2005.12.004

Robinson, N. (2000). *The politics of agenda setting: The car and the shaping of public policy*. Routledge.

Rohani, M. M., Wijeyesekera, D. C., & Karim, A. T. A. (2013). Bus operation, quality service and the role of bus provider and driver. *Procedia Engineering, 53*, 167–178. https://doi.org/10.1016/j.proeng.2013.02.022

Rossman, J. (1998). *Tram view of Asia, postcards and photographs of the early 20th century showing trams in the cities of the Middle East, The Far East, Southeast Asia and The Indian Subcontinent*. http://www.tramz.com/tva/tva.html

Sen, A., Tiwari, G., & Upadhyay, V. (2010). Estimating marginal external costs of transport in Delhi. *Transport Policy, 17*(1), 27–37. https://doi.org/10.1016/j.tranpol.2009.09.003

Sharaby, N., & Shiftan, Y. (2012). The impact of fare integration on travel behavior and transit ridership. *Transport Policy, 21*, 63–70. https://doi.org/10.1016/j.tranpol.2012.01.015

Shoup, D. C. (2005). *The high cost of free parking*. Planners Press.

Soni, A. R., & Chandel, M. K. (2018). Assessment of emission reduction potential of Mumbai metro rail. *Journal of Cleaner Production, 197*(1), 1579–1586. https://doi.org/10.1016/j.jclepro.2018.06.216

Sood, J. (2019, January 4). *Financial constraints may slow down India's metro rail story*. Live mint. https://www.livemint.com/Politics/LgLRo75WTlCTGaMo31xxnN/Financial-constraints-may-slow-down-nations-Metro-rail-stor.html

Taylor, B. D., & Fink, C. N. Y. (2003). *The factors influencing transit ridership: A review and analysis of the ridership literature*. University of California Transportation Center. https://escholarship.org/uc/item/3xk9j8m2

The International Association of Public Transport (UITP). (2020). *UITP in India*. https://india.uitp.org/articles/india-investing-metro-infrastructure

United Cities and Local Governments. (2008). *Agenda 21 for urban mobility*. https://www.uclg.org/sites/default/files/EN_480_agenda_21_for_urban_mobility_english.pdf

Van Wee, B., Annema, J. A., & Banister, D. (Eds.). (2013). *The transport system and transport policy: An introduction*. Edward Elgar.

Wang, R., & Yuan, Q. (2013). Parking practices and policies under rapid motorization: The case of China. *Transport Policy, 30*, 109–116. https://doi.org/10.1016/j.tranpol.2013.08.006

Intermodality—Towards Enhancing Rail Freight Transportation Prospects

Tanya Mittal and Paulose N. Kuriakose

1 Introduction

Logistics is a highly fragmented and unorganized sector which has always suffered from improper planning. As per the Economic Survey Report 2019–2020, the logistics sector contributes to 14% of GDP in India and is expected to grow 1.5 to 2 times at current GDP. The draft National Logistics Policy (NLP) prepared in 2019 by the Ministry of Commerce and Industry (MoCI), the Government of India Logistics' Division, set targets towards reducing logistics costs from 13–14% of GDP to 10% in the next four years—and increasing the share of railways as mode of transportation from 31% to 50–55%. The sector requires proper planning as well as coordinated management, with stringent regulations capable of enforcing an effective market supply chain. Measures to cut long road haulage and switch road freight to railway transportation are necessary to achieve these policy goals.

The intermodal approach—wherein goods are shipped using two or more successive transport modes without being handled during transhipment—proved very effective in intensifying rail freight transportation (Savelsberg, 2007). This concept promotes the use of multiple transport modes flexibly and dynamically. It implies a shift towards more environmentally friendly modes. Under this integrated service design, depending on specific delivery time requirements and availability of modes, the most appropriate mode is selected so as to improve service levels and increase utilization of transport means. These innovative technologies also allow a bundling of cargo flows, mode-free bookings, flexible switching between modalities, and synchronization with empty vehicles.

T. Mittal (✉)
Deutsche Gesellschaft Für Internationale Zusammenarbeit (GIZ) GmbH, New Delhi, India
e-mail: tanya.mittal@giz.de

P. N. Kuriakose
Department of Urban and Regional Planning, School of Planning and Architecture, Bhopal, India
e-mail: paulosenk@spabhopal.ac.in

© The Author(s), under exclusive license to Springer Nature Switzerland AG 2021
S. Mitra et al. (eds.), *Railway Transportation in South Asia*, Contemporary South Asian Studies, https://doi.org/10.1007/978-3-030-76878-2_9

In order to demonstrate our case, this study analyses current freight practices in both road and rail freight modes, along a stretch of 1504 km on the upcoming Western Freight Corridor (WFC). Tools intended to cut long-distance road haulage are selected through the use of primary surveys from various road and rail carriers. Average Generalized Costs (AGC) for freight in both modes are evaluated and compared. Cost influence on mode choice is assessed. Current challenges to rail freight are identified, and some modal shift measures are suggested. Using the stated preference approach, a prospective modal shift is also articulated. The study recommended suitable strategies to encourage rail freight transportation.

2 Problem Statement

In India, the highest share of freight movement occurs by road (60% by 2018). The share of railway transportation has steadily declined over the past decades (from 83% in 1950 to 31% in 2018; Juyal et al., 2018). Even as the costs of road transportation outweigh those of railway or waterway modes of transport, the first is most often preferred (National Transport Development Policy Committee, 2014). This is partly because these alternative modes have a hard time competing with door-to-door services, and partly because of poor efficiency in the whole supply chain. Even though the country is 4th largest rail freight carrier globally, and even though 65% of railway revenue emanates from freight transportation, the services involved are poor and deprived (Indian Railways, 2020). Current freight practices in Indian Railways (IR) are much neglected; this hampers both competitiveness and demand.

The demand for freight transport with lower logistic costs and dwell time is always high; nonetheless, in transportation by road the average vehicle load is low and numerous vehicles run empty on a daily basis (Juyal et al., 2018). Among all segments of logistics' costs, transportation accounts the highest share with 35%, followed by warehousing, packaging and losses, which account for 30% (Economic Division, Ministry of Finance, 2019). The sector is very fragmented. Moreover, freight transportation vehicles—which account for 4.4% of all vehicles (MoRTH, 2019)—generate almost 46%/month of $PM_{2.5}$ emissions (Sharma and Dikhsit, 2016). Sleep-deprived truck drivers are responsible for 40% of road accidents per annum (MoRTH, 2018). External costs are typically borne by the environment and society as a whole. Therefore, better fleet management practices, with the use of green vehicles to reduce truck movements and related emissions, are very much essential.

Since railways are considered environmentally friendly—both in use and as concerns the externalities involved—one could say that increasing its share within freight transportation holds the key for overall freight improvement in the country. In order to achieve this, the Government of India has proposed a Dedicated Freight Corridor (DFC) project in 2005—and directed the Ministry of Railways (MoR) to execute it. The project aims for a substantial increase in rail freight share, and it hopes to reduce total logistic costs as well as transit time for both Export–Import

(EXIM) and domestic cargo movements in railways. This would enhance overall efficiency and allow a modal shift towards greener modes of transportation. However, managing this task constitutes a great challenge in terms of infrastructure, coordination, and synchronization between requisite loads and targeted volumes. Thus, this study assesses current freight mode choice behaviours and suggests strategies for inducing intermodal rail freight transportation.

3 Research Objectives

In order 'To induce a modal shift towards rail freight transportation', four objectives are framed. (1) Identify freight modal shift strategies adopted across the globe. We aim to point out various tools adopted in different countries so as to shift freight transportation towards the rail mode, along with the tools more adapted to the Indian context. (2) Assess current freight practices on the Western Freight Corridor. This includes present freight behaviours and related impediments in both road and rail transportation modes. (3) Analyse total transportation costs associated with internal, external and time-related costs. We aim to appraise total transportation costs in both road and rail modes as well as analyse their influence on behaviours regarding mode choice. (4) Develop models of transport costs so as to deduce and recommend strategies that support intermodal rail freight. This final objective aims at the identification of an efficient mode that takes the current situation into account; we also endeavour to recommend strategies that can induce intermodal rail freight transportation.

4 Methodology

The study illustrates current freight practices and related impediments in two modes—road and rail—along a selected corridor. In order to grasp the present-day scenario, related plans and policies—existing or proposed—are thoroughly reviewed. Data regarding container traffic is collected and mapped out for additional analysis. Total Transport Costs (TTC) are evaluated for selected modes as well as compared so as to comprehend behaviours relating to mode choice in existing freight operations. Expert opinion surveys were carried out so as to identify tools able to minimize externalities caused by heavy freight vehicles. Lastly, suitable strategies for inducing intermodal rail freight transportation were endorsed. The detailed, stage-wise methodology is developed below.

Stage 1: Pertinent reports, plans, and proposals—along with rail freight enhancement tools and techniques adopted in various parts of the world—were reviewed. The results of such measures promoted in different cities were then assessed. Effective management approaches for freight transportation were analysed and possibilities regarding applications to the Indian context were identified.

Stage 2: The best available reports, papers, and statistics regarding container movements on the WFC were collected, compiled and reviewed. Existing freight practices were analysed and loopholes in current freight transportation were identified. The corridors with highest containerized traffic were explored and the factors discouraging the mode choice of rail were identified. Secondary information was collected through stakeholders involved in the functioning of the current freight industry—including the Transporters Associations (TA), the Custom Brokers Associations (CBA), the Container Shipping Line Associations (CSLA), the Federation of Freight Forwarders Associations of India (FFFAI), the Container Corporation of India (CONCOR), the Port Authorities (seaport and dry port), the MoCI, and several small players. The data collected incorporates the NLP, E-Way Bill Dashboard details, Logistics Data Bank analysis reports, Origin and Destination details, the Inland Container Report, the National Council of Applied Economic Research (NCAER) Logistics Cost Report, etc. Additionally, interview surveys aiming to collect detailed facts and figures regarding container movements (including chain-wise freight costs and transit times) and associated difficulties were carried out at each firm.

Stage 3: In order to evaluate internal costs (out-of-pocket expenditures), data regarding container size, number of Twenty-Foot Equivalent Units (TEUs), weight, type of mode, net km travelled, delays, losses, Terminal Handling Charges (THC), etc. were collected for a month. In order to deduce external costs, data on emissions factors, unit costs of air pollution, distances, inflation rates, etc. were collected. Time-based statistics were collected so as to evaluate total lead time—including running, delays and waiting times. Altogether, total logistics costs which include internal, external and time-based costs were estimated. Freight cost and time are taken at various stages in the supply chain and detailed out as concerns both road and rail freight modes, considering a 40′ non-hazardous general cargo container with a gross weight of 30 tonnes. Present road taxes (including Green Tax) and charges are also considered to come up with the final freight cost.

Stage 4: Potential modal shift strategies towards rail were selected from the literature reviewed and used for the 'Expert Opinion Surveys'. These surveys included interviews on the subject of which rail freight enhancement tools would best be implemented to the selected corridor. Survey responses were compiled and prospective tools for the modal shift were identified. The cost and time targets proposed in both the DFC and Delhi-Mumbai Industrial Corridor (DMIC) projects were reviewed. Modified freight costs and transit times in both modes—railways and roadways—were used so as to carry out a 'willingness to pay' survey. The responses collected were analysed using both disaggregate modelling, spreadsheet modelling, and intermodal TTC modelling. Besides, the average generalized cost for freight in both modes was evaluated and compared. An assessment of mode choice was carried out. Current challenges in rail freight transport were also identified and the impact of external costs internalization to the TTC on future mode choices was assessed.

Stage 5: After developing an understanding of current freight practices and related impediments, we suggest suitable strategies to shift road-based freight into rail-based—with a focus on enhancements in intermodal rail freight transportation systems. Emphasis is laid on future practices characterized by less emissions, a more efficient use of vehicles, cost-effective forms of transportation, and the availability of more flexible and competitive options.

5 Study Area

Out of six freight corridors proposed under the DFC (Fig. 1), we selected the WFC as the focus of our study's research. The corridor has a total length of 1504 km. It connects the Jawaharlal Nehru Port (JNPT) in Mumbai to Dadri, in Uttar Pradesh. The project proposed includes a double line, electric (2 × 25 kV) track via Vadodara-Ahmedabad-Palanpur-Phulera-and Rewari, thus joining the Eastern Freight Corridor (EFC) at Dadri (DFCCIL, 2019). The corridor crosses five states (Table 1).

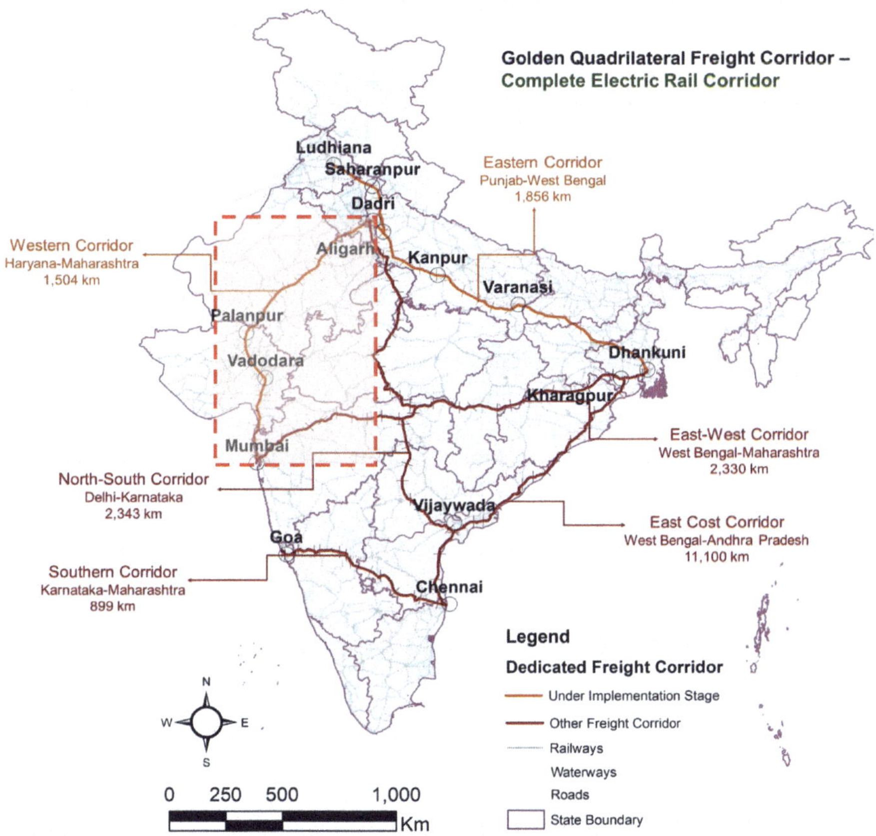

Fig. 1 Western freight corridor (*Source* DFCCIL [2019])

Table 1 Corridor length in different states

States	Length (km)	Percentage of total length
Uttar Pradesh	18	1.19
Haryana	177	11.77
Rajasthan	567	37.70
Gujarat	565	37.57
Maharashtra	177	11.77
Total	1504	100

Source DFCCIL (2019)

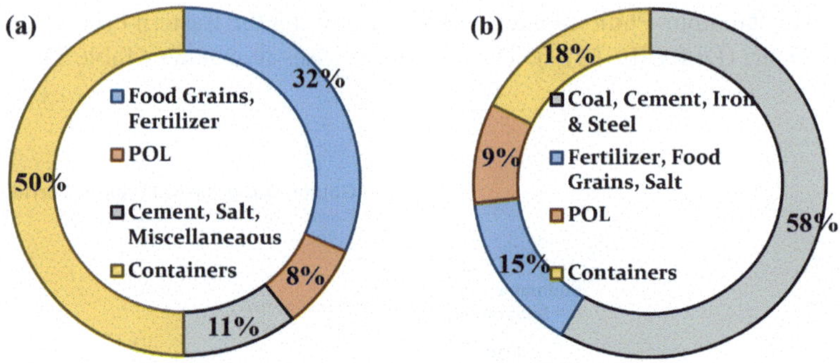

Fig. 2 Key commodities handled by the WFC in **a** an upward direction, and **b** a downward direction (*Source* DFCCIL [2019])

The length of the stretch implemented on the WFC by the Dedicated Freight Corridor Corporation of India Limited (DFCCIL) to date is 306 km—connecting Madar, in Ajmer, to New Rewari and then Kishangarh Balawas.

Traffic mainly comprises International Standard Organization (ISO) containers from the JNPT, the Mumbai Port in Maharashtra, as well as the Pipavav, Mundra and Kandla ports in Gujarat—destined for the Inland Container Depot (ICDs) located in northern India (especially at Tughlakabad, Dadri and Dandharikalan). Besides containers, other commodities moving on the corridor include Petroleum, Oil, & Lubricants (POL), fertilizers, food grains, salt, coal, iron & steel, and cement. These key commodities are handled in both upward and downward directions (Fig. 2).

Ports located in the north-west handle the highest number of containers in the country (Fig. 3).

6 Worldwide Strategies in Freight Modal Shift

As per the round table discussion regarding sustainable development held in the UK in 1996, urban freight carries a range of negative consequences: economic (congestion,

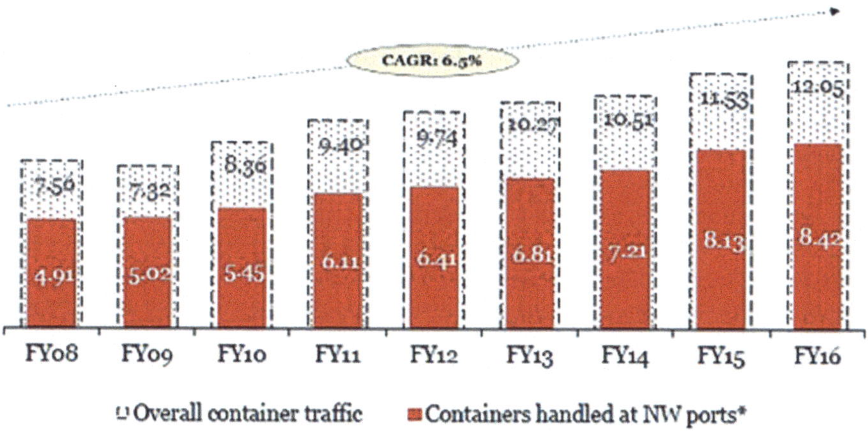

Fig. 3 Share of container traffic handled in north-west ports (*Source* PWC [2017])

inefficiency and resource waste), social (traffic accidents, noise & visual intrusion) and environmental (pollutant emissions, non-renewable fossil-fuel usage and waste products; Timilsina & Dulal, 2011). In order to curb these impacts, European countries in the last few decades promoted strategies so as to improve the efficiency of and conversion towards railway-based freight transportation. The planning involved ensured that prospective rail freight services offered end-point trains, with semi/fully automated loading/unloading equipment in hub terminals, as well as terminals at sidings so as to improve intermodal operations (UNECE, 2018). Figure 4 shows some characteristics of this intermodal transport chain.

In order to fulfil modal shift targets set by the EU White Paper 2011, an intermodal approach was adopted, as well as several tools for external cost formalization in road freight sector. The dominant concept applied was that of the 'Polluter Pays Principle' (Islam et al., 2016). Germany, Switzerland and Australia all introduced distance-based tolls for heavy trucks—based on the distance travelled, the number of axles involved, the vehicle's weight and emission class, etc. (Knorr et al., 2009). Switzerland introduced mileage-based charges—heavy-vehicle free-falling under a similar category (Kirk & Levinson, 2016). Time-based vignettes comprising accident and infrastructure costs were also developed. Area-based charges were implemented

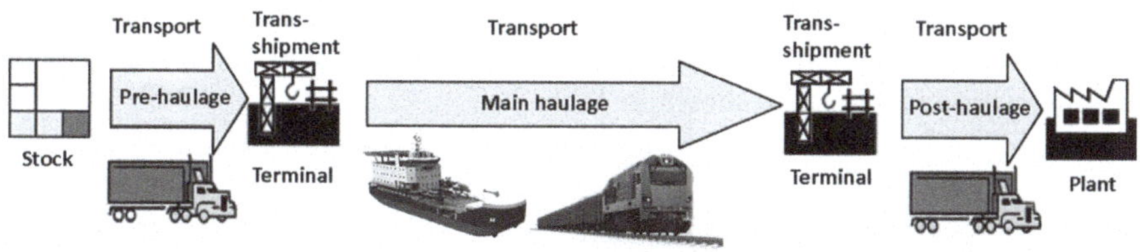

Fig. 4 Characteristics of the intermodal supply chain (*Source* Savelsberg [2007])

in Stockholm, London and Singapore according to day of the week, traffic flow, and time of travel—with the imposition of high tariffs during peak hours (Clement & Darmon, 2001). The Danish developed a fuel tax on larger or fuller trucks, as well as based on driving behaviour. A vehicle tax on operations such as registration and insurance was also implemented—with the provision of lower fees for 'cleaner' vehicles. In order to limit negative externalities, London imposed duty taxes on fuel, as well as congestion costs, noise costs, accident costs, and air pollution costs according to the tonnage carried by each vehicle category (Allen et al., 2008). Articulated with intermodal concepts, these strategies were able to cut long road haulages and shift a substantial share of transportation (up to 30%) towards railways (US Department of Transportation, 2015).

7 Current Freight Practices in Rail Freight Movements

The container, after lifting-off from the vessel, is taken to the Container Yard (CY) at the seaport—and then re-loaded to a rake for further transit to ICDs. Railway transit is provided by the CONCOR, a Public-Sector Undertaking (PSU) of Indian Railways. After customs clearance, the final distance to the factory is undertaken by transporters involved in first/last mile provisions. After de-stuffing of goods into the factory, the empty container needs to return to the assigned empty container depot according to new demand. The container is then used for either export or domestic goods transported through the rail carrier. Empty containers are also used by road carriers for repositioning at desired locations.

In order to analyse peak demand, container traffic (for a particular month) both to and from each port on the western corridor was mapped out (Fig. 5). Out of three major container handling seaports, namely the JN Port (JNPT), the Mundra Port (MDCC), and the Pipavav Port (PPSP), we noticed that a larger number of trips is undertaken towards both the MDCC and PPSP, while the demand for TEUs is higher at the JNPT.

The highest number of TEUs moved in March 2018 occurred towards the JNPT (39,955 TEUs), followed by the PPSP (26,823 TEUs) and the MDCC (23,074 TEUs). While MDCC imports huge amounts of TEUs, a higher number of exports are directed towards the JNPT. Irrespective of the similar amounts of TEUs imported from the JNPT and PPSP, inbound and outbound trips at the PPSP are very numerous. The total amount of TEUs—imported as well as exported—at all container handling seaports along the Western Corridor indicates a higher demand at the JNPT. Trips and composition in TEUs are shown in Table 2.

Around 99.4% of rail freight is moved in Boggy Loaded Container (BLC) rake types, which consist of 9 units of 5 wagons each. Each wagon can handle up to two TEUs, thus amounting to a full rake capacity of 90 TEUs. However, the carrier must wait for a full load so as to initiate the functioning of any BLC rake, which results in long-term storage of goods at CYs or Container Freight Stations (CFSs). At the two busiest ICDs (the Tughlakabad ICD, or TICD, and the Dadri ICD, or ICDD)

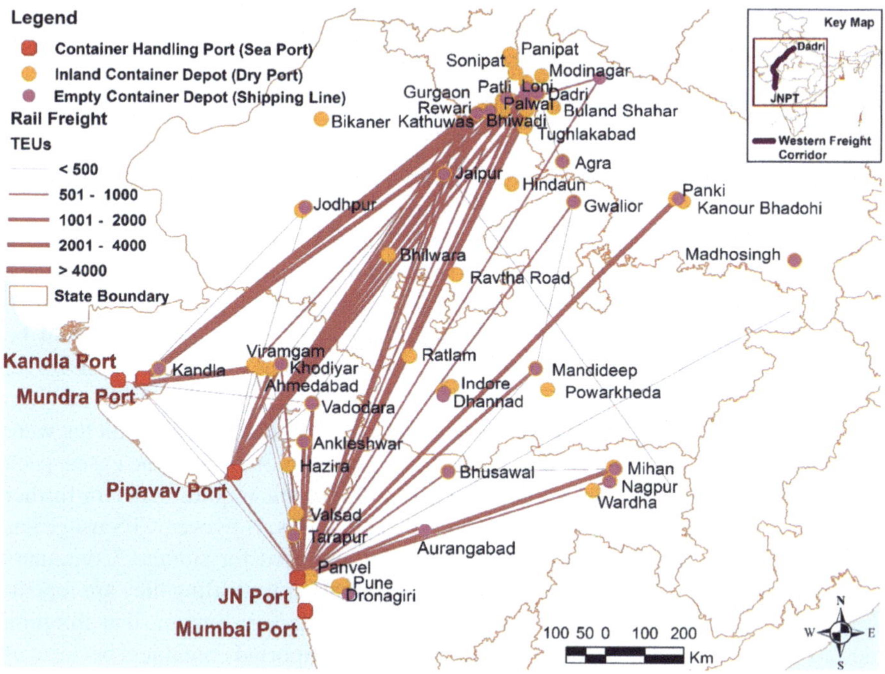

Fig. 5 Movement of containers through rail (*Source* CRIS [2018])

Table 2 Trips and TEUs composition of freight trains

Container handling port	Trips		TEUs	
	Import	Export	Import	Export
JNPT	151	171	16,199	23,756
MDCC	788	7	22,900	174
PPSP	356	290	16,941	9,882
	Total Trips—1915		Total TEUs—108,116	

Source CRIS (2018)

located at the northern end, only 1 train/day could be directly dispatched to the JNPT, MDCC and PPSP. Trains to Pipavav also cross the Kathuwas Junction, also known as the CMLK, where double-stack container rake systems are available between ICDs in both the north and various seaports in Gujarat. The train via the CMLK runs on Overhead Electric Traction (OHT), with a container height of 9.6′ required so as to maintain safe haulage. Apart from this, 70% of containers used in rail freight services are of 20′ size—40′ size containers are preferred for road-based transportation.

8 Current Freight Practices in Road Freight Movements

8.1 Through Containers

Road freight movement uses a variety of trailers, upon which shipping line containers are loaded and transported to the desired location. Trailers are available in three sizes, Single Axle (9ton), Double Axle (16ton), and Triple Axle (25ton). Only about 30% of these vehicles are privately owned—70% are rented trucks. Moreover, about 86% of road trips involves the movement of 40′ size containers.

As concerns the supply chain, Full Container Load (FCL) containers can be taken directly to the factory after arriving at the Container Yards through Direct Port Delivery (DPD) services. The share of loads handled through DPD services was 60% in Jan'2020. CFS inspections and customs clearance of all vehicles were previously mandatory, which led to wastes of time and money. In some cases, such as concerns Less than Container Loads (LCLs) and sporadic requirements for further customs examinations, containers must still travel to CFSs. However, CFS usage has strongly declined and most of these stations are now used for storage. Containers are then transported to factories through trucks, and after de-stuffing they are repositioned either towards empty depots or back towards a seaport. It seems that shipping lines are currently unable to make an effective use of imported containers because of imbalances in the nation's balance of trade. This is why about 99.9% of all containers are transported back to seaports.

In order to analyse road container traffic, we focused on data from 'Majha Transporters'. We observed the total numbers of trips involved, as well as TEUs carried in a particular month; this is represented in Fig. 6. The number of TEUs moved in Dec'2019 were greatest at the JNPT (925 TEUs) followed by the MDCC (453 TEUs) and the PPSP (104 TEUs). While there were more export trips directed to the MDCC (203 trips) than to the JNPT (99 trips), the TEUs moved to the JNPT (398 TEUs) and the MDCC (384 TEUs) are almost equivalent. The PPSP's transportation dynamics are characterized by much fewer road travels, both exports and imports. Trips and TEUs composition are provided in Table 3.

Since most containers are brought back to the seaport, they often carry illegal cabotage[1]—with a substantial discount to both shipping lines and domestic clients. Around 35% of these supposedly empty journeys were cabotaged by the carrier itself in this particular month Dec'2019—which resulted in a reduction of empty headings from 35 to 3%. Around 96% of all containers in the country belong to the Full Container Load category. Direct Port Delivery services involve about 60% of all loads. However, shipping line containers are only used in 1/4th of the trips made from the JNPT. The remaining 3/4th of transportations involves de-stuffing from EXIM containers. Over 50% of this cargo is transported in loose packages through the use of either 'BOX trucks' or containers leased from shipping lines.

[1] Cabotage is a practice wherein carriers use shipping line containers to deliver domestic goods.

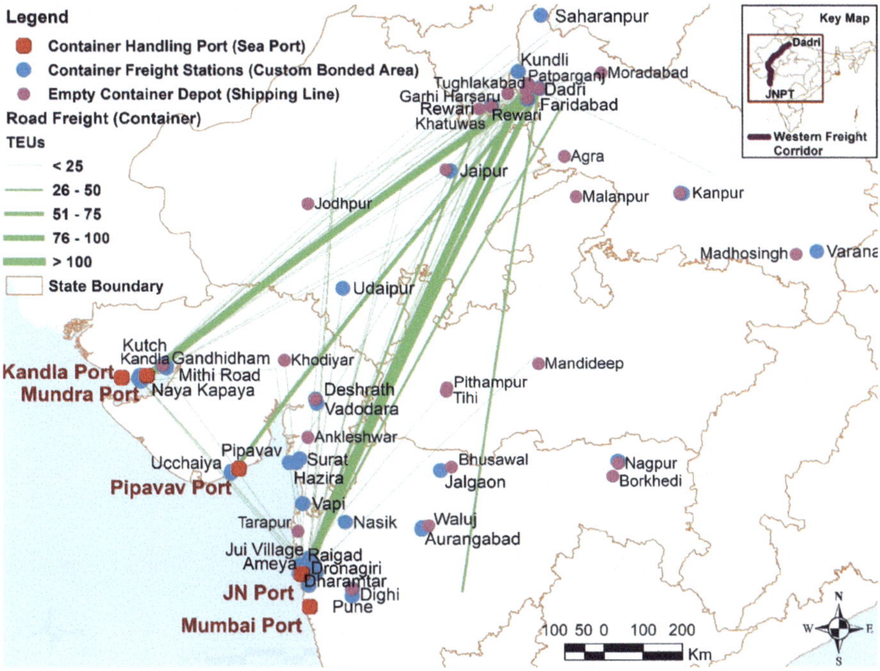

Fig. 6 Movement of containers through road (*Source* Majha Transporters [2019])

Table 3 Trips and TEUs composition of containers transported through trucks

Container handling port	Trips		TEUs	
	Import	Export	Import	Export
JNPT	165	99	527	398
MDCC	30	203	69	384
PPSP	25	34	42	62
	Total Trips—624		Total TEUs—1,127	

Source Majha Transporters (2019)

8.2 *Through Packages*

The movements of EXIM cargo in loose packages were analysed through monthly data supplied by the Supreme Corporation of India (SCI), bearer of the Bombay Goods Transport Association (BGTA). It seems that both the numbers of trips and tonnage carried are greater at the JNPT—with very few movements at the MDCC and almost no movement at the PPSP (Fig. 7).

The cargo is transported through BOX trucks available in four sizes: 18′ (Single Axle, 9 MT capacity), 22′ (Double Axle, 16 MT capacity), 25′ (Triple Axle, 21 MT

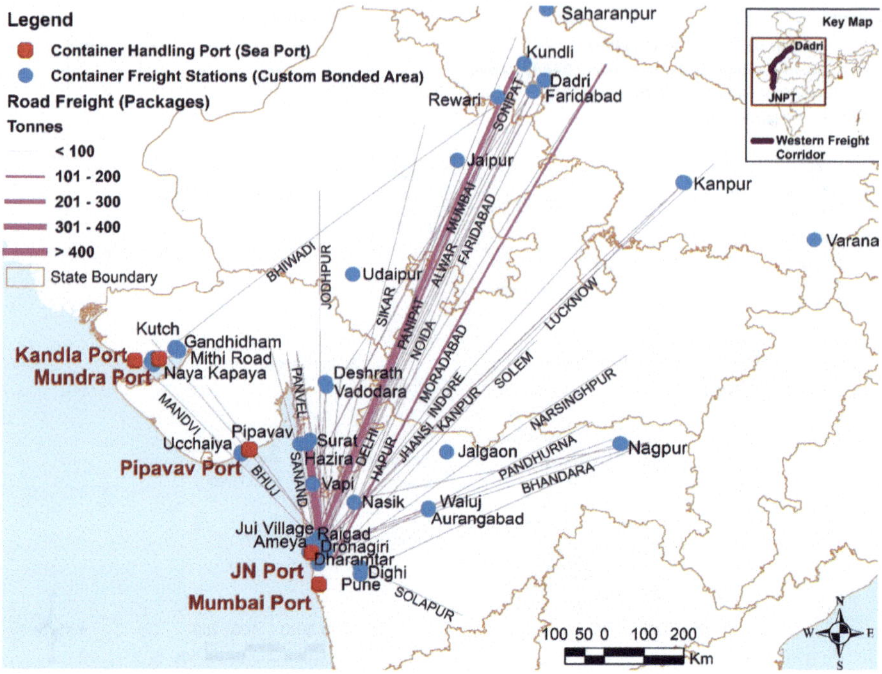

Fig. 7 Movement of packages through roads (*Source* SCI, BGTA, Jan'2020)

Table 4 Trips and tonnage composition of packages via trucks

Container handling port	Trips		Tonnage	
	Import	Export	Import	Export
JNPT	117	36	2,233	650
MDCC	4	1	49	21
PPSP	0	0	0	0
	Total Trips—1,915		Total Weight (MT)—108,116	

Source SCI, BGTA, Jan'2020

capacity) and 28′ (Triple Axle, 25 MT capacity). 22′ BOX trucks account for almost 50% of all movements. Again very few (about 20%) of these trucks are privately owned, while around 80% are hired. An advance payment of around 80% of the total cost is demanded when a vehicle is hired.

The volume of goods moved in packages from the JNPT in Jan'2020 was found to be very high, at 2,223 tonnes—compared to only 650 tonnes exported goods. Very low movements are observed at the MDCC while almost no movement occurred towards the PPSP by the Supreme Corporation of India. The composition of trips and tonnage are provided in Table 4.

Table 5 Empty container haulage from the TICD and the ICDD

O/D	Number (%)			Total TEUs Exported[a]
	JNPT	MDCC	PPSP	
TICD	6,000 (13.23)	639 (06.00)	16,000 (54.29)	85,454
ICDD	60 (00.35)	300 (00.46)	200 (01.00)	102,136

[a]TEUs exported could be of either category: relocated to the seaport or empty international export
Source CONCOR (TICD and ICDD)

Each loose package is transported to the Northern States via the Patparganj CFS (in 98% of the cases). This is due to time restrictions in the capital, which occur from 9 am to 9 pm. At present, vignette schemes only work efficiently in Delhi, while in Maharashtra, due to poor monitoring, restrictions in entries/exits from 8 to 11 am and 4 to 8 pm are mostly ignored—which results in 24*7 plying of trucks on city roads. The average rate of package movement along the corridor was found to be € 1.1 per kilogram. Only 60% of return journeys involve loads—even after waiting for 2 or 3 days near a delivery point.

9 Empty Container Haulage (Road + Rail)

Another concern arises during the backhaul of empty containers (either repositioned or exported) to seaports. In empty exports, empty containers are picked up from an empty depot (mostly in ICD premises) and taken to the CY at a seaport under the Direct Port Entry (DPE)—from where containers are lifted-on to the vessel and exported to the desired location. In railways, containers are exported either empty or filled with domestic goods, while in roadways, stuffing and de-stuffing of domestic goods occur in almost every backhaul trip. Only when the container is required to be stuffed with international cargo during exports does the movement necessarily cross CFSs/ICDs.

At present, a considerable number of empty containers are moved at longer distances. For example, in railways, around 25,000 empty TEUs per year are moved by the CONCOR from ICDs in the northern end (TICD and ICDD). Most of these travel to the PPSP; a few towards the MDCC. The Dadri, being a multi-modal logistics hub and second-highest ICD after Tughlakabad, provides loads to shipping line containers, and thus here we witness fewer empty trips towards any of the seaports. The detailed composition of empty TEUs moved by the CONCOR in 2019 is provided in Table 5.

In roadways, over 60% empty haulage occurs towards the JNPT—followed by un-authorized cabotage practices (Prabhu Logistics Pvt. Ltd.[2]). The carrier is offered a discount from the shipping line for empty repositionings, for example, a discount of € 77 for carrying a 20′ container and of € 165 for a 40′ container is provided

[2] Prabhu Logistics Pvt. Ltd is one of the carriers involved in empty repositioning via roadways.

from the base price of € 199 and € 361, respectively. Although the fact that the road sector involves many illegal practices may of course seem dubious, in practice private players much contribute to reducing empty headings—legally or otherwise.

10 Why not Rail?

The longest stretch on the Western Corridor, with a length of 1504 km and which connects the JNPT in the west and the ICDD in the north, is here thoroughly studied. We aim to capture detailed components of freight transportation. Focus group discussions were carried out involving several agencies: the FFFAI, CSLA, CBA, BGTA, TA; Traffic and Railway Managers at both the JN port and Mumbai ports; Managers at ICDs (Dadri and Tughlakabad); the CONCOR, Best Roadways, Majha Transporters, All Cargo Logistics, Road and Rail Carriers, the Logistics Division of MoCI; and, finally, Economists. Various segments of freight costs and transit times were reviewed and their impact on mode choice was assessed. Current challenges in rail freight transportation were identified and future preferences towards the rail assessed—considering the upcoming DFC, the DMIC and the internalization of road freight externalities.

Though container traffic at the JNPT has reached 5.13 million TEUs in 2019—with an annual growth rate of 18.22% (Fig. 8a)—the share of rail transportation has declined rapidly and fallen to 16% (Fig. 8b).

Among the various reasons suggesting why rail transportation is so underutilized we mention here the fact that railways only have access to 3 out of the 38 CFSs surrounding the JNPT. Lower flexibility in route selection (IR prerogative) constitutes another damage to railway competitiveness. Road travel is of course highly flexible and journeys can be organized according to which route brings higher profits for those involved. For example, out of two routes available on the Delhi-Mumbai stretch (via Gujarat or via Madhya Pradesh) the route via Madhya Pradesh is often preferred so as to avoid Weight in Motion (WIM), a load detection IT application in Gujarat.

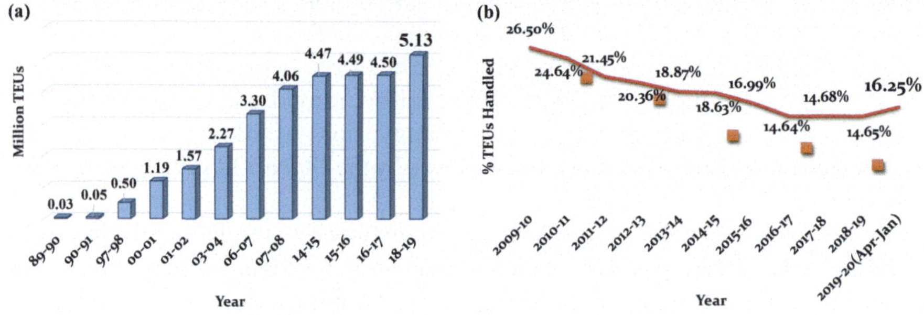

Fig. 8 **a** Container traffic at JNPT. **b** TUEs handled through rail at JNPT (*Source* Traffic Manager, JNPT)

Avoiding this scrutiny is important for those who overload and/or are not interested in paying fines. Road transportation is sometimes cheaper as well—and less restricted as concerns illegal practices.

Survey responses concerning current rail freight transportation issues stressed out various other factors making railway transportation less attractive to clients. It involves long waiting times for a full load, often a poor availability of wagons/rakes, shaky schedules, high freight tariffs, improper intermodal terminal locations, a lot of documentation and lack of door-to-door services from a single party. The much reduced opportunities for bargaining and negotiation in the THCs imposed by shipping lines is another barrier. The results of survey responses are shown in Fig. 9.

Waiting for 90 TEUs is always a big challenge—it often takes 4–5 days to get the requisite load, resulting in a levy of CFS charges of around € 110–132 per day. Additionally, demurrage charges of € 1.65/hr./wagon apply for late loads to the rake. Further, during each stop as the train lets a passenger train cross by, halting charges are paid to the IR for the use of their infrastructure. Besides, a THC is levied by the shipping line on each handling of a loaded container—at rates almost doubling those of the CONCOR inland haulage tariff. All these factors squarely make railway transportation expensive, uncompetitive, and inflexible. This is why, out of 100% containers coming from the JNPT (with 16% railway share), only 30% returns through the same mode (either empty or loaded).

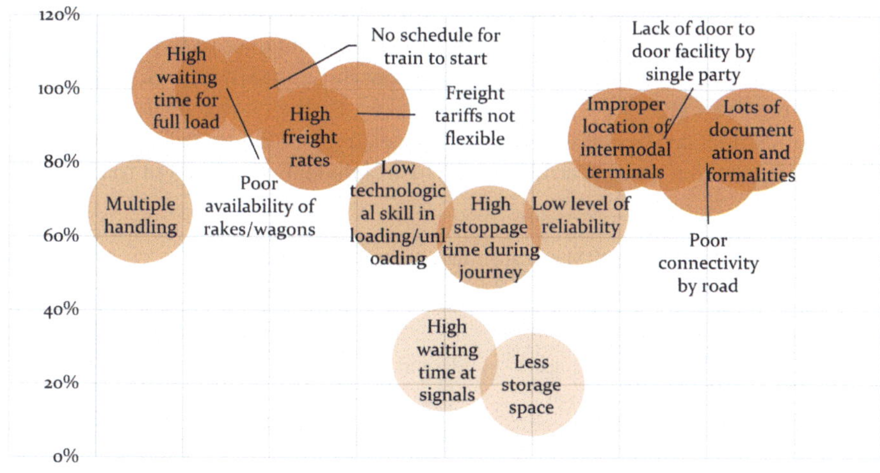

Fig. 9 Challenges in current rail freight transportation (*Source* Primary Survey from Carriers and Managers in both Road and Rail Freight Transportation)

11 Mode Choice Behaviour

Freight costs and transit times have a great influence on mode choice behaviour. TTC in EXIM cargo movements involves four major segments, i.e. documentation, customs clearance, terminal handlings and inland haulages. Terminal handling and inland haulage charges vary with mode used.

11.1 Freight Costs

The cost of moving an EXIM container is fully dependent on weight class and distance covered—irrespective of the type of cargo moved. Though the average inland haulage charges on the Delhi-Mumbai stretch are cheaper in railways, roadways are much more efficient. Table 6 discuss detailed tariff rates for EXIM container movements via road and rail.

However, while roadway costs involve the total transit costs from port to the factory, costs in railways are limited to the transit between the port and the ICD. Table 7 provides the costs for the first and last miles from the ICD to the factory.

Charges for empty container repositioning are also imposed on importers by the shipping line. Table 8 shows the comparison in tariff rates for repositioning from the factory to the nearby ICD and from the factory to the JNPT.

It can be stated that, in order to transport an EXIM container from port to factory and from factory to port, inland haulage is cheaper through railways. However, after incorporating the THC into the inland haulage costs, the choice of travel by road becomes most inexpensive in this country. For example, total transit costs for a 40′ FCL container with a non-hazardous cargo of weight slab 30 tonnes in roadways is

Table 6 Rail transit tariffs in EXIM container movements

	Rail transit tariff			Road transit tariff		
Container size	Weight class (Tonnes)	Tariff (€)		Weight class (Tonnes)	Tariff (€)	
		Import	Export		Import	Export
20′	≤ 10	353	309	6 T	550	341
	> 10–20	427	387	15 T	825	605
	> 20–26	524	476	20 T	935	660
	> 26–31	548	532	30 T	1375	1,045
	> 31	594	578	–	–	–
40′	≤ 20	604	550	7 T	715	495
	> 20	714	679	15 T	825	605
	–	–	–	20 T	935	660
	–	–	–	30 T	1375	1045

Source CONCOR and Best Roadways

Table 7 Transit costs of first/last mile deliveries in railways

Container size	Container handling Cost[a]	Transit cost[b]	Toll charges	Green tax[c]
20′	9	66	6	29
40′	17	77	13	14

[a]The CONCOR imposes handling charges for every loaded container, wherein free services are provided for handling unloaded containers
[b]The costs are imposed by the carriers providing first/last mile services
[c]The Delhi government levies a green tax of €29 on trailers with containers (loaded or empty) and €14 on trailers carrying no container during each entry/exit of heavy vehicles in the city
Note All costs are in EURO
Source Ashutosh Logistics Private Ltd. (Carriers are involved in first/last mile transit between the ICD and the factory)

Table 8 Tariffs for empty container repositioning

Container size	Factory to ICD (€)	Factory to JNPT (€)
20′	88	199
40′	110	361

Source Inland Container Market Report (2019), CONCOR (2019), and CSLA

€ 1,190—as compared to € 1,446 in railways (considering first and last miles within a 500 km radius).

11.1.1 Break-Up of Freight Costs

Among various components of road transportation costs (Fig. 10), over 50% of expenditures concern fuel—around 500 L of fuel are consumed by a 40′ container with a gross weight of 30 tonnes in each travel direction. Tolls and driver salaries hold a 15% share each—a driver's salary amounts to € 275/month (including local expenses) with an extra payment of € 11–33 under express services. Remaining costs include maintenance, green taxes, container handlings, insurance, bribes, etc. Cost components in rail include crew, fuel, halts, infrastructure, equipment, maintenance, and THC.

11.2 Transit Times

Transit times constitute another factor in selecting a suitable model for freight transportation. On the Delhi-Mumbai stretch, the time taken by a rake to reach the ICD from the port is much longer than that taken by a truck travelling from the port to the factory. For a single journey, the average transit time through railways is 96 h during import and 85 h during export—thus much higher than 60 h involved in roadway

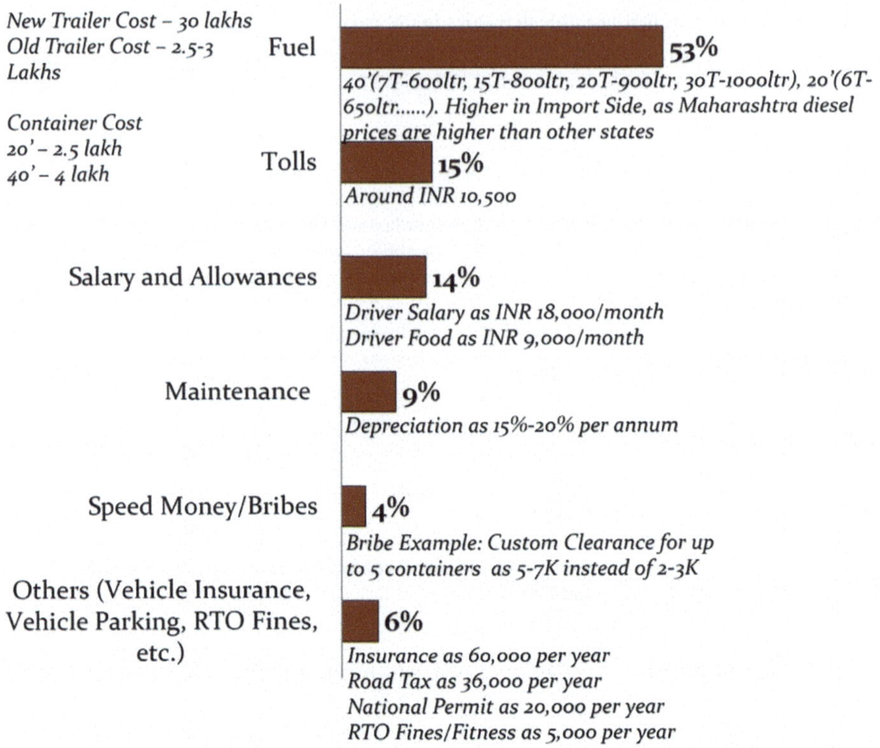

Fig. 10 Break-up of road transportation costs (*Source* National Council of Applied Economic Research [2019])

travel (this sector also provides an express service of 36 h only). In railways, out of 96 h and 85 h, the time during which a rake actually moves is only 67 h and 56 h, respectively. This is due to the waste of nearly one and half days for sidings, arranging wagons, and waiting times for a full load.

The average speed of a rake is 20–25 km/hr. Perishable goods are moved through express trains, carrying one goods container in the middle and one at both its ends. In roadways, the average speed of a trailer with a 40′ container lies around 40–45 km/hr. However, because express services allow goods to be transported within 36 h, sleep-deprived trucks drivers are responsible for almost 40% of all road accidents on the stretch.

11.3 Internalization of External Costs to the TTC

In order to incorporate negative externalities from the road freight sector to the TTC, modal shift strategies[3] are suggested by freight experts. A rank-based 'Expert Opinion Survey' with a sample size of 50 responders—involving the FFFAI, CSLA, CBA, Traffic and Railway Managers, the CONCOR, Best Roadways, Majha Transporters, All Cargo Logistics, and small-scale Road and Rail freight carriers and several Economists—was conducted. Experts were invited to express their views on the adoption of road pricing strategies, the imposition of road haulage taxes and regulations, and the introduction of service quality improvement measures in the rail freight system. Table 9 highlights the share opined by respondents under each category in a given rank: R1 denotes 'very low potential', R2 'low potential', R3 'moderate potential', R4 'high potential' and R5 'very high potential'.

Among the options provided, the imposition of an 'emission tax' drew the greatest levels of agreement. In order to analyse its impacts, air pollution costs for the Delhi-Mumbai stretch were derived. Air pollution costs in India at 2019–2020 prices were evaluated using the inflation rate method applied on costs calculated by 'Sengupta and Mandal at 2004–2005 prices' in 2005. The final cost is deduced by considering BSIV emission factors for Heavy Commercial Vehicle (HCV) diesel trucks with a combustion capacity of more than 6,000. Table 10 depicts the steps taken for evaluating these final costs.

Air pollution costs are estimated at € 0.024 per km, with total costs on the stretch amounting to € 35.32 for each travel direction.

11.4 Average Generalized Costs (AGC)

Cost parameters altogether—internal, external and time-based—were considered and an AGC was evaluated for both road and rail. The calculation is based on the formula suggested in the 17th European Automobile Manufacturers Association (ACEA) Safety Advisor Group Meeting in 2011 (Tavasszy & Meijeren, 2011).

$$GCm, g = VOTg{*}Tm + Pm, \tag{1}$$

where GC = Generalized Costs, VOT = Value of Time, T = Time, P = Price and Subscripts, m = Mode of Transport, g = Good.

The AGC for moving an EXIM container of 40′ size with a gross weight of 30 tonnes on the Delhi-Mumbai stretch is estimated as much lower in roadways than in railways—with a difference of € 517 in imports and € 748 in exports (Table 11).

[3] Freight modal shift tools were identified through a global perspective in the literature review.

Table 9 Share of respondents' views regarding modal shift strategies

Rank	The imposition of road haulage tax/regulation					Introduction of road pricing				Rail service quality improvements				
	FT (%)	VT (%)	ET (%)	DBT (%)	TBV (%)	MBC (%)	ABC (%)	NAI (%)	TS (%)	LFR (%)	ITP (%)	ITA (%)	STF (%)	IT (%)
R1	13	7	–	7	–	–	–	–	–	7	–	–	–	–
R2	–	20	–	13	20	27	27	–	–	7	–	–	–	–
R3	40	20	–	13	7	27	27	13	–	7	7	7	–	–
R4	20	33	7	33	27	20	20	13	7	7	7	40	20	13
R5	27	27	93	33	47	27	27	73	93	73	87	53	80	87

Notes Modal Shift Strategies: FT-Fuel Tax, VT-Vehicle Tax, ET-Emission Tax, DBT-Distance-Based Toll, TBC-Time-Based Vignettes, MBC-Mileage-Based Charges, ABC-Area-Based Charges, NAI-Network Access Improvements, TS-Time Scheduling, LFR-Low Freight Rates, ITP-Intermodal Terminal Provision, ITA-Innovative Technology Adoption, STF-Single Party Transit Facility, TT-Transparency in Tariff

Table 10 Air pollution cost for Delhi-Mumbai stretch

	Air pollution costs for India for 2020 at 2019–2020 prices in €/g **(A)**	Average emission factors (BSIV) for HCV Diesel trucks (>6000 cc) in (g/km) **(B)**	Total pollution costs for HCV Diesel trucks in €/Km **(A*B)**
CO	0.00002	4.345	0.00007
HC	0.00022	0.26	0.00007
NOx	0.00352	6.04	0.02157
PM	0.02981	0.07	0.00211

Source Shekhar (2015) and Sen et al. (2010)

Table 11 Average generalized costs on the Delhi-Mumbai stretch

	Import (€)	Export (€)
Rail	2,131	2,155
Road	1,618	1,407

Notes inputs received from freight service providers in railways and roadways
Source Author Generated

Low-cost roadways have a direct impact on mode choice behaviour for the majority of freight transit. Therefore, developing inexpensive and competitive rail freight services is essential to achieve future NLP's mode share targets.

12 Potential Shift to Rail

The potential modal shift to rail is analysed through a few hypothetical assumptions such as the full implementation of the Dedicated Freight Corridor and the Delhi-Mumbai Industrial Corridor as well as the levy of air pollution costs to the road freight sector. Three scenarios with variable transport costs and times are developed: in line with governmental targets from the DFC (40% reduction in rail freight costs and 22 h in rail transit times), in line with the DMIC's ambitions (6 h[4] reduction in road transit time), and through addition of air pollution costs to road freight costs.

Air pollution costs were internalized with 15, 30 and 50% shares in consecutive scenarios—and road transit times were considered as concerns the express service with a relay system.[5] Internalizations of air pollution costs increased total freight costs in roadways, while the DFC targets decreased total transit times and freight

[4] The DMIC proposes a new highway of 1,200 km in length (in comparison to 1,463 km at present) with aims to reduce overall transit time by 6 h.

[5] Relay trucking is a model wherein drivers rotate every few hundred kilometres through a network of relay pit-stops before travelling back to their home base. The model 'Relay-as-a-Service (RaaS')' was first introduced by Rivigo, a technology-enabled logistics company in Gurugram, in 2019.

cost in railways. Both costs and time in each mode are assessed for a 40′ general FCL container with a non-hazardous cargo of gross weight slab 30 tonnes travelling from the JNPT to the factory (within a 500 km radius from either the TICD or the ICDD) and vice a versa. Table 12 represents scenarios built for mode choice surveys.

A stated preference approach is adopted so as to collect responses regarding future mode choice. The survey was conducted with carriers involved in both road and rail freight transportation. Since so many carriers are involved in road freight transit and approaching them all would be unfeasible, we used a sample size of 50 responders.

The survey results emphasized that most preferences refer a 'mixed usage' from the JNPT to the Dadri, while road travel is favoured in the opposite direction. The results collected are shown in Fig. 11.

In import journeys, when freight by road becomes expensive (although still much faster), 60–67% of responders showed a preference towards 'using both modes'—while 'the only rail' option covers less than 20% of responses. Again, high preferences towards roadways are due to timely delivery and to the fact that goods can

Table 12 Scenario building for import–export journeys

Scenario	Mode	Freight cost (€)		Transit time (Hrs.)	
		Import	Export	Import	Export
I	Road	1,195	789	30	30
	Rail	1,161	886	45	35
II	Road	1,200	794	30	30
	Rail	1,161	886	45	35
III	Road	1,207	801	30	30
	Rail	1,161	886	45	35

Notes values defined by considering existing costs & time as well as several hypothetical assumptions, as explained above
Source Author Generated

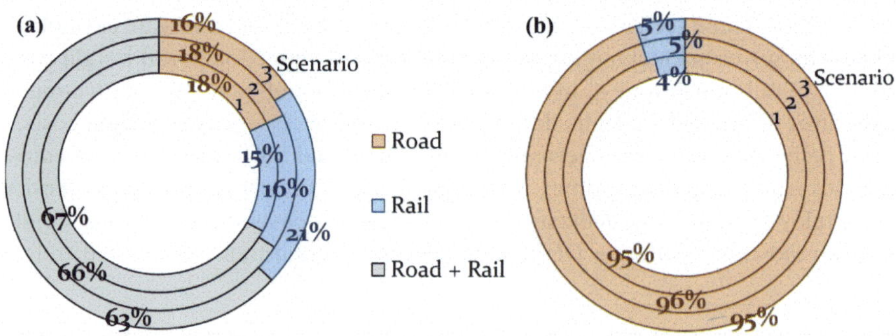

Fig. 11 Responses regarding mode choice in **a** import journeys and **b** export journeys (*Source* Primary Surveys from the FFAI, TA, CBA, CSLA, Traffic Managers, Railway Managers, CONCOR, Economists)

be shifted towards the cheaper rail option after transporting a substantial amount of cargo by road.. In the export journey, when road freight becomes both cheaper and faster, only 5% preferences mention the rail. Henceforward, the rail sector might not achieve the targeted volume even after the DFC is implemented. Suitable strategies are indispensable so as to discourage long road haulage and augment rail freight efficiency and promote intermodality.

13 Conclusion and Recommendations

India is a fast-developing country and a competitive economy wherein freight transportation plays an important role. In order to overcome challenges in the current freight supply chain, it is important that that the sector is well organized. Making rail efficient and competitive is crucial so as to increase its freight share in an environment of strong competitiveness of road travel. The generalized cost of green modes must decrease so that their use increases. Green practices must be encouraged—and their related challenges dealt with efficiently. Regulatory and incentive-based strategies are sine qua non.

The cost of negative externalities must be charged directly to the users creating it. The incorporation of a substantial amount of emission costs in terms of air pollution, noise pollution, congestion and accidents to the road freight sector must be accomplished. Proper implementation of IT applications so as to restrict illegal road activities and develop cost-effective rail services must be promoted. Containerization through the intermodal approach is focused on enhancing overall rail freight efficiency; it too must be applied. Railways must be the dominant haulage option, while the road sector may be limited for drayage (first and last mile goods delivery). Intermodal rail freight transportation must be fortified so as to achieve the rail freight share targets of the National Logistics Policy.

References

Allen, J., Piecyk, M., & McKinnon, A. (2008). *Internalising the external costs of light and heavy goods vehicle transport in London*. Heriot-Watt University.

Behera, S. K., & Oruganti, R. (2019). *Inland container market report*. New Delhi. Maritime Gateway. http://containersindia.in/pdf/INDIAN%20CONTAINER%20MARKET%20REPORT-2019.pdf

Bombay Goods Transport Association. (2020). *BGTA road freight movement through packages* (Mittal, T., Interviewer).

Centre for Railway Information Systems. (2018). *CRIS rail freight movement* (Mittal, T., Interviewer).

Clement, L., & Darmon, F. (2001). Transalpine transportation policies in Switzerland and Austria. *Japan Railway and Transport Review, 26*, 28–37. http://www.ejrcf.or.jp/jrtr/jrtr26/pdf/f28_cle.pdf

Container Corporation of India. (2019). CONCOR empty container haulage (Mittal, T., Interviewer).

Container Corporation of India & Best Roadways. (2020). *Rail transit tariff in EXIM container movement* (Mittal, T., Interviewer).
Container Shipping Line Association. (2020). *CSLA tariff for empty container repositioning* (Mittal, T., Interviewer).
Dedicated Freight Corridor Corporation of India. (2019). *DFCCIL details on western freight corridor*. https://dfccil.com/Home/DynemicPages?MenuId=77
Indian Railways. (2020). *Share in rail freight revenue*. http://www.rbs.indianrail.gov.in/ShortPath/
Islam, D. M., Ricci, S., & Nelldal, B. L. (2016). How to make modal shift from road to rail possible in the European transport market, as aspired to in the EU Transport White Paper 2011. *European Transport Research Review, 14*.
Jawahar Lal Nehru Port Trust. (2020). *JNPT container traffic at JN Port* (Mittal, T., Interviewer).
Juyal, S., Saxena, A., Sharma, S., & Srivastava, A. (2018). *Transforming India's Mobility*. NITI Aayog. New Delhi. http://www.niti.gov.in/writereaddata/files/document_publication/BCG.pdf
Kirk, R. S., & Levinson, M. (2016). *Mileage-based road user charges*. Congressional Research Service. https://fas.org/sgp/crs/misc/R44540.pdf
Knorr, A., Heinemann, A. W., & Eisenkopf, A. (2009). *Germany's Autobahn toll for heavy vehicles after four years: Experiences and perspectives*. FÖV Discussion Papers (53). https://dopus.uni-speyer.de/frontdoor/deliver/index/docId/554/file/DP-053.pdf
Majha Transporters. (2019). *Road freight movement through containers* (Mittal, T., Interviewer).
Ministry of Commerce and Industry Logistics Division. (2019). *MoCI draft national logistics policy*. New Delhi
Ministry of Finance Economic Division. (2019). *Economic survey report 2018–19*. New Delhi. https://www.indiabudget.gov.in/economicsurvey/
Ministry of Road Transport and Highways. (2018). *MoRTH road accidents in India*. New Delhi. https://morth.nic.in/road-accidents-india
Ministry of Road Transport and Highways. (2019). *MoRTH annual report 2018–19*. New Delhi. https://morth.nic.in/annual-report-2018-19
National Council of Applied Economic Research. (2019). *NCAER analysis of India's logistics costs*. New Delhi. Ministry of Commerce
National Transport Development Policy Committee. (2014). *NTDPC India transport report: Moving India to 2032*. http://www.nair.indianrailways.gov.in/uploads/files/1411707071818-Moving%20INDIA%20to%202032%20Railway%20Sector%20Report%20of%20Indian%20Transport%20Report(Highlights)%20-%20Rajnish%20Kumar.pdf
PricewaterhouseCoopers. (2017). *PWC multi modal logistics hub Uttar Pradesh*. New Delhi
Savelsberg, E. (2007). *Innovation in European freight transportation: Basics, methodology and case studies for the European markets*. Springer-Verlag, Berlin Heidelberg.
Sen, A. K., Tiwari, G., & Upadhyay, V. (2010). Estimating marginal external costs of transport in Delhi. *Transport Policy, 17*(1), 27–37. https://doi.org/10.1016/j.tranpol.2009.09.003
Sharma, M., & Dikhsit, O. (2016). *Comprehensive study on air pollution and Green House Gases (GHGs) in Delhi*. New Delhi. Indian Institute of Technology Kanpur. http://web.delhi.gov.in/wps/wcm/connect/c791f4804acf7f498ee18f09c683c810/Final+Report09Jan2016.pdf?MOD=AJPERES&lmod=-1601641899
Shekhar, S. (2015). *Status of pollution generated from road transport in six mega cities*. Retrieved from Central Pollution Control Board, MoEFCC. https://cpcb.nic.in/
Tavasszy, L. A., & Meijeren, J. V. (2011). Modal shift target for freight transport above 300 km: An assessment. 17th ACEA SAG Meeting (p. 20). Netherlands. https://www.acea.be/uploads/publications/SAG_17.pdf
Timilsina, G. R., & Dulal, H. B. (2011).Urban road transportation externalities: Costs and choice of policy instruments. *World Bank Research Observer, 26*(1), 62–191. https://doi.org/10.1093/wbro/lkq005
United Nations Economic Commission for Europe. (2018). *UNECE railways role in intermodality and the digitalization of transport documents*. New Yord and Geneva. United Nations. https://unece.org/fileadmin/DAM/trans/main/wp24/ECE_TRANS_262_E_Web_Optimized.pdf

United States Department of Transportation. (2015). *Modal shift comparative analysis technical report*. United States. https://ops.fhwa.dot.gov/freight/sw/map21tswstudy/technical_rpts/mscanalysis/modal_shift_comp.pdf

Passengers Mobility and Passengers' Perception on Railway Transport System

Assessing Perceptions of Railway Service Quality: A Compendium of Literature Studies

Laura Eboli and Gabriella Mazzulla

1 Introduction

Rail-based transportation experienced a rapid decline in the second half of the twentieth century, mainly due to developments in both road and air modes of transportation. At the same time, railway transportation must be considered a strategic sector—one on which the success of all efforts to shift the balance from private to public preferences will depend. In order to achieve this aim, rail services should offer adequate quality standards to users. The importance of 'user perceptions' regarding service quality was highlighted and prioritized by numerous researchers.

Service quality could be improved and service use increased by examining customers' experiences with trains and stations. The conduction of passenger satisfaction surveys regarding perceptions and expectations is most useful in this respect. Developing an understanding customer service is essential for an increase in user preferences towards railways (Bruhn & Grund, 2000). Rail service quality is influenced by several characteristics, including accessibility, safety, comfort, service, information availability and personnel. In order to understand users' perceptions of the services involved, it is fundamental to collect users' perceptions regarding these various characteristics.

As mentioned above, we focus here on user perceptions as regards railways services in northern Italy through two different methods. The methods diverge in terms of convenience, potential purposes and type of conclusive findings.

The first method is the Classification and Regression Tree Approach (CART). This technique allows us to identify the characteristics that most influence overall service quality and investigate perceptions by different user groups. The second method

L. Eboli (✉) · G. Mazzulla
University of Calabria, Rende, Italy
e-mail: laura.eboli@unical.it

G. Mazzulla
e-mail: gabriella.mazzulla@unical.it

© The Author(s), under exclusive license to Springer Nature Switzerland AG 2021
S. Mitra et al. (eds.), *Railway Transportation in South Asia*, Contemporary South Asian Studies, https://doi.org/10.1007/978-3-030-76878-2_10

is the Structural Equation Modelling (SEM) approach. This method is particularly appropriate for describing complex phenomena such as transit passenger perceptions of used services. This is because, as regards the traditional regression methods, it allows latent variables to be introduced for a better description of the relationship among service factors and overall service quality.

2 Context of the Study Area

The case study supporting this research concerns a railway service operating in the North of Italy, specifically in the city of Milan. The analysed data refers to the years 2011, 2012 and 2014, when the service offered 32 regional lines and 9 suburban lines connecting towns of the local hinterland of Milan, as well as 2 express lines connecting Milan to the Malpensa airport. A total of about 570,000 passengers per day travelled on these lines, which offered a between 35 and 83 runs per day—thus providing a service frequency of 2–4 runs per hour.

Face-to-face interviews were realized between the end of June and the middle of July 2011, with a sample of 16,718 passengers. 16,647 users were sampled in May 2012. The interviews were conducted on board during the whole week (weekday, before holidays and during holidays) in a time slot between 6.00 a.m. and 10.00 p.m.

Users were to answer a questionnaire structured into two main sections. The first section included: general information (e.g., time period of the interview, train, line, station and operator); socio-economic characteristics (e.g., gender, age, qualification, professional condition and income); travel habits (e.g., trip scope and frequency, ticket). The second section specifically centred on passengers' perceptions of used services: users expressed importance and satisfaction rates, on a scale from 1 to 10, regarding 27 service quality factors including safety, cleanliness, comfort, service, information availability, personnel, etc.

Over 50% of respondents were interviewed on trains of regional lines, over 40% on suburban lines and the remaining percentage on the Malpensa express service. Most users were interviewed on weekdays. Most respondents travel by train for the purposes of working and studying.

The following constitute the main characteristics of those sampled. Most were aged between 16 and 40. Most were employed, although a considerable share was still studying. Most respondents obtained a diploma, and almost one third held a degree. Over half respondents travelled by train every day (Table 1).

3 Literature Review

The literature is rich in studies analysing different modes of public transportation (services offered by buses, railway services, airport and airline services). Independently of the transport modes analysed, studies differ in terms of the approaches

Table 1 Sample characteristics

Characteristics	Categories	Percentages (2011)	Percentages (2012)
1. Gender	Male	45.5	45.5
	Female	54.5	54.5
2. Age	≤40 years old	70.5	74.5
	>40 years old	29.5	25.5
3. Professional condition	Employed	58.7	48.0
	Student	30.5	41.4
	Other (unemployed, housewife, pensioner, etc.)	10.8	10.6
5. Qualification	Degree	32.9	31.5
	Diploma	67.1	68.5
8. Travel scope	Work	42.3	36.8
	Studying	21.3	36.2
	Other (bureaucratic and/or personal activities, tourism)	36.4	27.0
9. Travel frequency	Habitually (daily or weekly)	64.8	73.7
	Occasionally	35.2	26.3

Source Prepared by the Authors

adopted for investigating service quality. Generally, the objective of these studies was to determine the service characteristics mostly affecting overall service quality—as based on passengers' points-of-view. The quality of rail services has been investigated by many researchers. Interesting findings were discovered thanks to the development and application of different methodologies, all valid. For example, Cavana et al. (2007) found, through a regression analysis, that assurance, responsiveness and empathy had significant effects on perceptions of overall service quality. From a study by Nathanail (2008), we learned that itinerary accuracy and system safety had been attributed the highest grades. Brons et al. (2009) applied a principal component analysis and found that satisfaction with the quality of the access to the station itself is an important dimension of the rail journey. Cantwell et al. (2009) discovered through a multinomial logit model that rail service passengers would benefit from an improvement in service reliability and a reduction in crowding. Rahaman and Rahaman (2009) found that the quality of security inside the train often dominates service satisfaction. Geetika (2010) identified through a factor analysis a number of determinants of service quality—such as availability and quality of refreshments, the effectiveness of information systems, behaviours by railway staff, basic amenities provided on platforms, as well as safety and security. Prasad and Shekhar (2010) applied a model based on SERVQUAL, the well-known service quality model developed by Zeithaml, Parasuraman and Berry in 1988 and concluded that service delivery was the most important factor involved, and social responsibility the least. Chou et al. (2011) found that ease in access to a station and personal space within the train were top-priority quality indicators. Eboli and Mazzulla (2012, 2015) applied a Structural Equation Model (SEM) and found that service characteristics such as punctuality, regularity and run frequency, as well as cleanliness, have the highest positive effects on service quality. De Oña et al. (2015) found through the Classification and Regression Tree Approach (CART) approach that perceptions regarding service differ among distinct user groups. More recently, Allen et al. (2020) proposed a Structural Equation Multiple Cause Multiple Indicator (SEM-MIMIC) model, which corrects for heterogeneity in user perceptions regarding satisfaction with various service attributes, overall service and loyalty. Service aspects such as reliability, added-value services are primary service attributes have a positive effect on satisfaction with the overall service and, thus, on loyalty.

4 Study Objectives and Methodology

This work's objective is to describe and compare methodologies applied to the same case study concerning rail service quality as assessed in users' perceptions. The methods differ for various reasons, especially in terms of proposed purposes and type of final findings.

We'll now present a brief theoretical framework regarding the two methodologies applied to our case study.

4.1 *The CART Methodology*

The development of a CART model begins by gathering all the data in the root node, which is the node located at the top of the tree. This root node is divided into two 'child nodes' based on an independent variable (splitter) that maximizes their "purity". Then, each child node is recursively split until all are "pure" (all the cases are of the same class) or until their "purity" cannot be increased. The most famous splitting index is the Gini Index (Gini, 1912), which measures mode impurity.

The measure of impurity at a node t, designated as $I(t)$, may be defined as follows (SAS Institute Inc., 2004):

$$I(t) = 1 - \sum_{i=1}^{i} \left(\frac{n_i}{n}\right)^2 \tag{1}$$

where i is the number of classes in the target variable, n_i is the number of cases belonging to the class i, and n is the total number of cases.

If a node is "pure", all observations in the node belong to one and the same class, and the Gini Index or Impurity (node) will equal zero. We can then define the split criterion based on the Gini Index as the Gini reduction criterion, which measures the "worth" of each split in terms of its contribution towards maximizing homogeneity through the resulting split. A set of candidate split rules are evaluated and ranked as the tree grows. If the split is obtained from splitting one parent node into B branches (in this case two branches, because the CART model generates binaries trees), the "worth" of that split may be measured as follows (SAS Institute Inc., 2004):

$$\text{Worth} = I(P) - \sum_{b=1}^{B} P(b) * I(b) \tag{2}$$

where $I(P)$ denotes the impurity of the parent node, $P(b)$ denotes the proportion of observations in the node assigned to a branch b, and $I(b)$ denotes the impurity of the node b.

Following this procedure, a maximal tree which overfits the data is created. In order to decrease complexity and create simpler trees, pruning is undertaken according to a cost-complexity algorithm (Breiman et al., 1984) based on removing the branches that add little to the tree's predictive value. If the increase in misclassification costs is significantly lower than the decrease in complexity costs, the branch will be pruned, and a new tree will be created.

The last step is to select an optimal tree from these pruned trees. Using misclassification costs on the testing dataset (or an independent dataset), the optimal tree to be selected is the one with the least misclassification costs.

One of the methods for developing this model randomly divides the sample used in the training phase into k sets (k-fold cross-validation). Sequentially, each subset is kept so as to be used as a testing set against the tree model generated by the remaining $k - 1$ subsets. Thus, different k models are obtained, in which the accuracy of the classifications in the training set ($k - 1$) and the testing subsets (k) can be evaluated and the optimal tree can be selected.

A more detailed description of the CART analysis and its applications can be found in Breiman et al. (1984).

One of the main advantages of decision trees, as opposed to other modelling methods, is that they are presented in easily understandable visual branching images which provide effective "If–Then" rules. Every leaf of the decision tree corresponds to a decision rule that extracts very useful information regarding the data. It is a logic conditional structure starting in the root node with "If", continuous with every variable that takes part in the tree growing—thus making an "If" of the rule. It ends in the child nodes with "Then", in which the class of the target variable showing the highest number of cases in the analysed child node is associated.

Another valuable outcome provided by the CART analysis concerns the values of the standardized importance of independent variables, which reflects the impact of such predictor variables on the model.

4.2 SEM Methodology

SEM is a specific type of regression analysis. It uncovers relationships between independent (exogenous) and dependent (endogenous) variables. It is composed of up to three sets of simultaneous equations, which are estimated at the same time: (i) a measurement model for the endogenous variables, (ii) a measurement model for the exogenous variables and (iii) a structural model. This full model is known as "SEM with latent variables". Latent variables are constructs which cannot be directly observed—they must be defined in terms of underlying observed variables called indicators. Each latent variable is defined by a measurement model, whereas the structural model represents relationships between exogenous and endogenous variables. The basic equation of the latent variable model is the following (Bollen, 1989):

$$\eta = B\eta + \Gamma\xi + \zeta \tag{3}$$

where η (eta) is an $(m \times 1)$ vector of the latent endogenous variables, ξ (xi) is an $(n \times 1)$ vector of the latent exogenous variables, and ζ (zeta) is an $(m \times 1)$ vector of random variables. The elements of the B (beta) and Γ (gamma) matrices are the model's structural coefficients; the B matrix is an $(m \times m)$ coefficient matrix for latent endogenous variables; the Γ matrix is an $(m \times n)$ coefficient matrix for latent exogenous variables.

The basic equation of the measurement model (4) concerns exogenous variables, and Eq. (5) concerns endogenous variables:

$$x = \Lambda_x\xi + \delta \tag{4}$$

$$y = \Lambda_y\eta + \varepsilon \tag{5}$$

where x and δ (delta) are column q-vectors related to the observed exogenous variables and errors, respectively; Λ_x (lambda) is a $(q \times n)$ structural coefficient matrix for the effects of latent exogenous variables on observed variables; y and ε (epsilon) are column p-vectors related to the observed endogenous variables and errors, respectively; Λ_y is a $(p \times m)$ structural coefficient matrix for the effects of the latent endogenous variables on those observed.

The structural equation system is generally estimated through the use of the Maximum Likelihood method (ML). In other cases, structural equation model parameters can be estimated through the use of other estimation methods such as Unweighted Least Squares (ULS), Weighted Least Squares (WLS), Generalized Least Squares (GLS) and so on. These estimation methods are described in Bollen (1989) and Washington et al. (2003)—where useful information about goodness-of-fit measures and their statistical interpretation is also provided. For a more detailed discussion on structural equation models, please refer to Bagozzi (1994), Bollen (1989), Golob (2003) and Joreskog (1973).

5 Preliminary Data Analysis

A preliminary evaluation of service quality was made by calculating average importance and satisfaction rates based on judgements expressed by respondents, on a scale from 1 to 10, regarding the various service attributes (Table 2). We can observe a similar trend by comparing the two different years in which surveys were undertaken. For this reason, the considerations reported below are valid for both years.

Average importance rates hovered around values of 8 and 9. It can be stated that users consider all attributes as very important; only the attribute linked to the possibility of transporting bicycles on board presents a lower average importance rate of 7.3. More specifically, the attributes considered as most important were the three service aspects linked to travel safety and personal security, which showed an average importance rate higher than 9.

On the other hand, the observation of average satisfaction rates led us to believe that people judge most service characteristics as deficient; in fact, only nine attributes were bestowed an average rate higher than 6. We can observe that service characteristics considered as most satisfying regarded safety, the personnel and the integration with the other modes of public transport, as well as station localization. By contrast, characteristics judged as the less satisfying concerned cleanliness, comfort and information availability.

6 Results

6.1 Results from Application of the CART

The CART was applied to the data collected in 2012. The 27 service attributes shown in Table 2 were used as independent variables of the model and named from 'Item1' to 'Item27'. In order to find out more applicable decision rules, the target variable (overall SQ) and the independent variables were re-coded according to a reduced semantic scale—specifically, a three semantic scale comprising rates from 1 to 4 as POOR, from 5 to 7 as FAIR and from 8 to 10 as GOOD. The CART used tenfold cross-validations of the sample, which provided a precision ratio in the categorization of the variable class of 77.43%. This value was higher than the values obtained in other studies wherein decision trees were applied with similar objectives (e.g. de Oña et al., 2014; Wong & Chung, 2007).

The results obtained through the proposed model which allowed us to reach the objective of this work—investigating service attributes affecting service quality—were mainly related to the importance of the variables. This is achieved through the use of the importance index (Kashani & Mohaymany, 2011), of which a standardized form was applied in this paper.

Table 3 shows the standardized importance rates stated by users as well as the standardized importance rates deduced through the model (de Oña et al., 2014).

Table 2 Importance and satisfaction rates regarding service quality

Service quality attribute	Importance rate (2011)	Satisfaction rate (2011)	Importance rate (2012)	Satisfaction rate (2012)
Travel safety	9.2	7.4	9.2	7.4
Personal security on board	9.1	6.7	9.2	6.6
Personal security at the station	9.1	6.5	9.2	6.3
Vehicle cleanliness	8.9	5.0	8.7	5.1
Seat cleanliness	8.9	4.8	8.8	5.0
Seat maintenance	8.6	5.1		
Restroom cleanliness	8.8	4.4	8.7	4.2
Cleanliness of the stations	8.5	5.3	8.3	5.4
Station maintenance	8.3	5.4	8.1	5.3
Crowding on board	8.4	5.4	8.3	5.3
Air-conditioning on board	8.7	5.1	8.4	5.3
Comfort on board	8.4	5.6	8.3	5.7
Fare/service ratio	8.8	5.1	8.8	5.0
Run frequency	8.9	5.9	8.8	5.9
Run punctuality	9.0	5.4	9.0	5.6
Run regularity	9.0	5.7	8.9	5.9
Integration with PT	8.7	6.0	8.3	5.9
Station localization	8.6	6.5	8.3	6.6
Parking	8.0	5.7	7.9	5.5
Bicycle transport on board	7.3	5.8	7.3	5.8
Facilities for the disabled	8.8	5.2	8.9	5.1
Substitute services	8.4	5.4		
Information availability at stations	8.7	5.9	8.5	5.7

(continued)

Table 2 (continued)

Service quality attribute	Importance rate (2011)	Satisfaction rate (2011)	Importance rate (2012)	Satisfaction rate (2012)
Information availability on board	8.5	5.5	8.4	5.4
Info timeliness at stations	8.7	5.5		
Info timeliness on board	8.6	5.3		
Complaints	8.5	5.0	8.3	5.1
Communications to the office	8.3	5.1		
Info connections with PT	8.5	5.4	8.2	5.2
Kindness on board	8.5	6.6	8.3	6.6
Competence on board	8.7	6.6	8.3	6.6
Ticket inspection	8.3	6.3	8.1	6.2
Kindness at the station	8.6	6.4	8.3	6.3
Overall service		5.8		5.7

Source Prepared by the Authors

The most important items identified by the model concern aspects such as comfort, personnel, information availability and service. The items with the highest importance deduced by the model are Item11 'Windows and Doors Working', Item25 'Courtesy and Competence on Board", Item21 'Clear and fast info in the stations', Item14 'Train Punctuality', Item27 'Courtesy and Competence in Station' and Item 15 'Regularity of Runs'—all with standardized importance values exceeding 64.6%. Little importance was deduced for the items related to safety, cleanliness or other.

On the other hand, the most important factors stated by users concern safety (Item1, Item2 and Item3), which are not relevant for the model (standardized importance values lower than 41%). However, when users are asked to rate the importance of each attribute, they consider all said attributes as highly important—as we can observe by analysing importance rates reported in the preliminary analysis in Table 2. This is one of the serious drawbacks encountered when studying the importance of variables based on stated passenger opinions (Weinstein, 2000).

Table 3 CART results

Stated importance			Derived importance		
Item3	Personal security at station	100%	Item11	Windows and doors working	100%
Item2	Personal security on board	99.9%	Item25	Courtesy and competence on board	88.5%
Item1	Travel safety	99.7%	Item21	Information at stations	81.0%
Item14	Run punctuality	97.8%	Item14	Run punctuality	73.5%
Item20	Facilities for the disabled	96.8%	Item27	Courtesy and competence in station	69.3%
Item15	Run regularity	96.4%	Item15	Run regularity	64.6%
Item12	Fare/service ratio	95.5%	Item12	Fare/service ratio	55.4%
Item5	Seat Cleanliness	95.3%	Item13	Run frequency	54.8%
Item13	Run frequency	95.1%	Item1	Travel safety	41.0%
Item6	Restroom cleanliness	94.8%	Item17	Localization of stations	40.9%
Item4	Vehicle cleanliness	94.7%	Item19	Bicycle transport on board	39.1%
Item21	Information availability at stations	92.0%	Item10	Air-conditioning on Board	37.6%
Item22	Information availability on board	91.3%	Item16	Price integration with PT	37.6%
Item10	Air-conditioning on board	90.9%	Item22	Information availability on board	35.4%
Item16	Price integration with PT	90.1%	Item4	Vehicle cleanliness	35.2%
Item27	Courtesy and competence at the station	89.9%	Item5	Seat cleanliness	33.4%
Item25	Courtesy and competence on board	89.9%	Item2	Personal security on board	28.3%
Item17	Localization of stations	89.9%	Item3	Personal security at the station	28.1%
Item11	Windows and doors working	89.9%	Item7	Station cleanliness	25.8%
Item7	Station cleanliness	89.7%	Item8	Station maintenance	25.4%

(continued)

Table 3 (continued)

Stated importance			Derived importance		
Item9	Crowding on board	89.7%	Item23	Complaints	19.3%
Item23	Complaints	89.6%	Item24	Info connections with PT	19.3%
Item24	Info connections with PT	88.5%	Item26	Ticket inspection	15.8%
Item8	Station maintenance	88.0%	Item6	Cleanliness of restroom facilities	10.6%
Item26	Ticket inspection	87.6%	Item20	Facilities for the disabled	9.1%
Item18	Parking	85.9%	Item18	Parking	7.1%
Item19	Bicycle transport on board	79.1%	Item9	Crowding on board	6.8%

Source Prepared by the Authors

6.2 Results from the Application of the SEM

A structural equation model was estimated through the use of the ML method—based on the data collected during the 2011 survey (Eboli & Mazzulla, 2015). We assumed the presence of seven latent exogenous constructs representing seven main characteristics of a railway service: safety, cleanliness, comfort, service, additional services, information and personnel. These latent factors are assumed to be linked to a latent construct representing overall service quality, represented by a latent endogenous variable named as service quality. Each latent construct is explained through 33 observed service quality factors reported in Table 2, which represents the latent endogenous variables. Each latent endogenous variable is linked to three observed indicators of global service quality. The first is simply the satisfaction rate expressed by each user as regards the overall service. The second indicator is the Customer Satisfaction Index (CSI; Hill et al., 2003). It constitutes a measure of service quality based on users' perceptions on some service aspects—expressed in terms of satisfaction rates—compared with users' expectations expressed in terms of importance rates. The third indicator is represented by the number of factors in which the user has experienced problems in the 30 days previous to the interview (Transportation Research Board [TRB], 1999).

The model was calibrated through the use of the AMOS 4.0 package from the SmallWaters Corporation (Arbuckle & Wothke, 1995). Table 4 shows the model results reported. The first and second columns report the model's variables. The third column shows the values of the coefficient regression weights (R.W.). The fourth and fifth columns report the values of the coefficient standard error (S.E.) and the probability levels (P)—so that the estimated coefficient is significantly different from zero. Finally, the last column shows the values of standardized regression weights (st. R.W.) representing the weights assigned by users to the various attributes, according

Table 4 SEM results

			R.W	S.E	P	st. R.W
Latent endogenous variable		*Latent exogenous variable*				
Service quality (η_1)	$\Leftarrow$	Safety (ξ_1)	0.080	0.003	0.000	0.191
Service quality (η_1)	$\Leftarrow$	Cleanliness (ξ_1)	0.154	0.003	0.000	0.525
Service quality (η_1)	$\Leftarrow$	Comfort (ξ_1)	0.111	0.003	0.000	0.287
Service quality (η_1)	$\Leftarrow$	Service (ξ_1)	0.200	0.004	0.000	0.488
Service quality (η_1)	$\Leftarrow$	Additional services (ξ_1)	0.136	0.004	0.000	0.289
Service quality (η_1)	$\Leftarrow$	Information (ξ_1)	0.240	0.005	0.000	0.560
Service quality (η_1)	$\Leftarrow$	Personnel (ξ_1)	0.114	0.003	0.000	0.280
Observed endogenous variable		*Latent exogenous variable*				
F1 (x_1) Travel safety	$\Leftarrow$	Safety (ξ_1)	1	–	–	0.737
F2 (x_2) Personal security on board	$\Leftarrow$	Safety (ξ_1)	1.354	0.012	0.000	0.962
F3 (x_3) Personal security at station	$\Leftarrow$	Safety (ξ_1)	1.225	0.011	0.000	0.858
F4 (x_4) Seat cleanliness	$\Leftarrow$	Cleanliness (ξ_1)	1	–	–	0.941
F5 (x_5) Seat cleanliness	$\Leftarrow$	Cleanliness (ξ_1)	1.018	0.004	0.000	0.952
F6 (x_6) Seat maintenance	$\Leftarrow$	Cleanliness (ξ_1)	0.967	0.004	0.000	0.911
F7 (x_7) Cleanliness of restroom facilities	$\Leftarrow$	Cleanliness (ξ_1)	0.817	0.005	0.000	0.808
F8(x_8) Station cleanliness	$\Leftarrow$	Cleanliness (ξ_1)	0.737	0.006	0.000	0.736
F9 (x_9) Station maintenance	$\Leftarrow$	Cleanliness (ξ_1)	0.691	0.006	0.000	0.697
F10 (x_{10}) Crowding on board	$\Leftarrow$	Comfort (ξ_1)	1	–	–	0.715
F11 (x_{11}) Air-conditioning on board	$\Leftarrow$	Comfort (ξ_1)	1.268	0.013	0.000	0.872
F12 (x_{12}) Comfort on board	$\Leftarrow$	Comfort (ξ_1)	1.134	0.012	0.000	0.845
F13 (x_{13}) Fare/service ratio	$\Leftarrow$	Service (ξ_1)	1	–	–	0.682
F14 (x_{14}) Run frequency	$\Leftarrow$	Service (ξ_1)	1.094	0.012	0.000	0.775
F15 (x_{15}) Run punctuality	$\Leftarrow$	Service (ξ_1)	1.235	0.013	0.000	0.820

(continued)

Table 4 (continued)

			R.W	S.E	P	st. R.W
F16 (x_{16}) Run regularity	⇐	Service (ξ_1)	1.168	0.012	0.000	0.836
F17 (x_{17}) Integration with PT	⇐	Service (ξ_1)	0.983	0.011	0.000	0.758
F18 (x_{18}) Station localization	⇐	Service (ξ_1)	0.899	0.011	0.000	0.713
F19 (x_{19}) Parking	⇐	Additional services (ξ_1)	0.899	0.013	0.000	0.587
F20 (x_{20}) Bicycle transport on board	⇐	Additional services (ξ_1)	0.848	0.011	0.000	0.699
F21 (x_{21}) Facilities for the disabled	⇐	Additional services (ξ_1)	1.091	0.013	0.000	0.754
F22 (x_{22}) Substitute services	⇐	Additional services (ξ_1)	1	–	–	0.767
F23 (x_{23}) Information availability at stations	⇐	Information (ξ_1)	1.137	0.010	0.000	0.811
F24 (x_{24}) Information availability on board	⇐	Information (ξ_1)	1.254	0.011	0.000	0.864
F25 (x_{25}) Info timeliness at stations	⇐	Information (ξ_1)	1.258	0.010	0.000	0.876
F26 (x_{26}) Info timeliness on board	⇐	Information (ξ_1)	1.269	0.010	0.000	0.879
F27 (x_{27}) Complaints	⇐	Information (ξ_1)	1.055	0.010	0.000	0.762
F28 (x_{28}) Communications to the office	⇐	Information (ξ_1)	1.020	0.010	0.000	0.764
F29 (x_{29}) Info connections with PT	⇐	Information (ξ_1)	1	–	–	0.757
F30 (x_{30}) Kindness on board	⇐	Personnel (ξ_1)	1.164	0.009	0.000	0.923
F31 (x_{31}) Competence on board	⇐	Personnel (ξ_1)	1.158	0.009	0.000	0.941
F32 (x_{32}) Ticket inspection	⇐	Personnel (ξ_1)	0.967	0.011	0.000	0.694
F33 (x_{33}) Kindness at the station	⇐	Personnel (ξ_1)	1	–	–	0.745
Observed endogenous variable		*Latent endogenous variable*				
Satisfaction (y_1)	⇐	Service quality (η_1)	1	–	–	0.423
CSI (y_2)	⇐	Service quality (η_1)	1.115	0.020	0.000	0.748
Critical event (y_3)	⇐	Service quality (η_1)	−3.110	0.095	0.000	−0.287

Source Prepared by the Authors

to the aims of our study—that is, the assessment of the influence of service attributes on overall service quality.

All parameters have a correct sign and are statistically significant. The minimum value of the discrepancy function is statistically significant according to the chi-squared test. The Goodness of Fit Index (GFI) is 0.651, the Adjusted Goodness of Fit Index (AGFI) is 0.605, and the Comparative Fit Index (CFI) is 0.785. The Root Mean Square Error of Approximation (RMSEA) has a value of 0.105; its lower and upper confidence interval boundaries are 0.105 and 0.106, respectively. Besides, the Root Mean Residual (RMR) is lower than 2.1. All these tests are quite satisfactory.

The latent exogenous variables mostly affecting service quality are information (0.560), cleanliness (0.525) and service (0.488). Another group of three service aspects present similar coefficients, indicating a certain level of importance for users: additional services (0.289), comfort (0.287) and personnel (0.280). On the other hand, safety has the lowest effect on service quality (0.191).

Other important findings can be highlighted by analysing the relationship between the latent exogenous variables and their observed indicators. First, we can easily observe that comparative weights of the observed indicators for each latent variable are quite similar, maybe because service attributes included within the same macro-factor registered very similar satisfaction rates. However, interesting distinctions can be underlined. For example, coefficient values for the indicators explaining Safety vary from 0.962 for the most important indicator, which is personal security on board, to 0.737 for the least important, which is travel safety. Cleanliness mostly results from the cleanliness of seats and vehicles, as well as seat maintenance; the other indicators have lower weights. On the other hand, comfort is often interpreted by passengers in terms of air-conditioning and degree of comfort on board—but less as concerns crowding on board, which often is the indicator mainly adopted by researchers for representing comfort factor (Eboli et al., 2016). Service characteristics, instead, are clearly related to regularity and punctuality. Additional services are moreover understood as the substitution of irregular services, while information availability is mostly understood as the timeliness of available information both on board and at stations—as well as regards information availability concerning services. Finally, the latent variable representing Personnel characteristics is best explained by competence and kindness of the personnel on board. On the other hand, concerning the most important attributes, the results diverge more highly between the two methods. The CART suggests that aspects linked to windows and doors working, courtesy and competence of personnel both on board and at the station, information availability at the station, train punctuality, and run regularity are the most important.

7 Discussion

The first important observation that emerged from the analysis of these results surely derives from the comparison between the importance directly stated by users and the importance calculated through the methodologies CART and SEM. The results from the models suggest that the stated and the calculated levels of importance are very dissimilar—and the results very contrasting. For example, service attributes linked to safety are considered as the most important by users when they must express a level of importance directly; however, the same attributes are the less important if we derive importance levels from the models wherein service attributes are related with overall service quality. In other words, users consider safety an important aspect of service if they think of this aspect without considering others—while in fact, they tend to give importance to all attributes. On the contrary, when we relate the various service attributes with the overall service through the proposed models, we can verify which attributes are correlated to overall user satisfaction—and understand which attributes users think of when they judge the service. For example, the SEM suggests that information and cleanliness are the aspects mostly affecting overall satisfaction and therefore the most important for users. This means that when users judge the overall service, they think more about aspects such as information and cleanliness, and less about safety—which therefore came up as the factor least affecting overall satisfaction.

The second important issue concerns comparisons between the two methodologies. We can state that both methods seek the same objective: deriving the importance that users ascertain to the various service attributes. More specifically, the CART directly provides a derived importance, while through the coefficients calibrated by the SEM we obtain calculated importance represented from the values of the model coefficients. The most convergent result emerging from the two methods is that attributes concerning safety are less important to users, confirming the considerations mentioned above.

8 Conclusions and Recommendations

In this paper, we propose an analysis of railway service quality based on users' perceptions expressed in terms of satisfaction and importance rates. The considerations emerging from this analysis suggest and confirm that it is fundamental to derive the importance of the various service attributes through modelling, and specifically by relating the various service attributes with the overall service, so as to determine the attributes that mostly affect the service. This kind of analyses can be very useful to determine which are the attributes mostly considered by users and on which transit operators have to concentrate their efforts so as to improve the service and offer satisfying quality levels. The recommendations to researchers and practitioners in the sector are to further the study of methodologies for deriving the importance of service attributes based on users' perceptions. This point is crucial because of the variety of aspects characterizing these services and because users' perceptions are fundamental for analysing transit service quality.

References

Allen, J., Eboli, L., Mazzulla, G., & Ortuzar, J. D. (2020). Effect of critical incidents on public transport satisfaction and loyalty: An Ordinal Probit SEM-MIMIC approach. *Transportation, 47*(2), 827–863. https://doi.org/10.1007/s11116-018-9921-4.

Arbuckle, J. L., & Wothke, W. (1995). *AMOS 4.0 user's guide*. SmallWaters Corporation.

Bagozzi, R. P. (1994). Structural equations models in marketing: basic principles. In R. P. Bagozzi (Ed.), *Principles of marketing research* (pp 317–385). Blackwell.

Bollen, K. A. (1989). *Structural equations with latent variables*. Wiley.

Breiman, L, Friedman, J. H., Olshen, R. A., & Stone, C. J. (1984). *Classification and regression trees* (368 pp.) Chapman and Hall/CRC.

Brons, M., Givoni, M., & Rietveld, P. (2009). Access to railway stations and its potential in increasing rail use. *Transportation Research Part A, 43,* 136–149.

Bruhn, M., & Grund, M. A. (2000). Theory, development and implementation of national customer satisfaction indices: The Swiss Index of Customer Satisfaction (SWICS). *Total Quality Management, 11,* 1017–1028.

Cantwell, M., Caulfield, B., & O'Mahony, M. (2009). Examining the factors that impact public transport commuting satisfaction. *Journal of Public Transportation, 12*(2), 1–21.

Cavana, R. Y., Corbett, L. M., & Lo, Y. L. (2007). Developing zones of tolerance for managing passenger rail service quality. *International Journal of Quality & Reliability Management, 24*(1), 7–31. https://doi.org/10.1108/02656710710720303.

Chou, J. S., Kim, C., Kuo, Y. C., & Ou, N. C. (2011). Deploying effective service strategy in the operations stage of high-speed rail. *Transportation Research Part E, 47,* 507–519.

de Oña, J., de Oña, R., Eboli, L., & Mazzulla, G. (2015). Heterogeneity in perceptions of service quality among groups of railway passengers. *International Journal of Sustainable Transportation, 9,* 612–626. https://doi.org/10.1080/15568318.2013.849318.

de Oña, R., Eboli, L., & Mazzulla, G. (2014). Key factors affecting rail service quality in the Northern Italy: A decision tree approach. *Transport, 29*(1), 75–83. https://doi.org/10.3846/16484142.2014.898216.

Eboli, L., & Mazzulla, G. (2012). Structural equation modelling for analysing passengers' perceptions about railway services. *Procedia—Social and Behavioral Sciences, 54,* 96–106. https://doi.org/10.1016/j.sbspro.2012.09.729.

Eboli, L., & Mazzulla, G. (2015). Relationships between rail passengers' satisfaction and service quality: A framework for identifying key service factors. *Public Transport, 7,* 185–201. https://doi.org/10.1007/s12469-014-0096-x.

Eboli, L., Mazzulla, G., & Pungillo, G. (2016). Measuring bus comfort levels by using acceleration instantaneous values. *Transportation Research Procedia, 18,* 27–34.

Geetika, S. N. (2010). Determinants of customer satisfaction on service quality: A study of railway platforms in India. *Journal of Public Transportation, 13*(1), 97–113.

Gini, C. (1912). *Variabilità e mutabilità* (in Italian) (p. 158). Tipogr. di P. Cuppini.

Golob, T. F. (2003). Structural equation modelling. In K. G. Goulias (Ed.), *Transportation systems planning* (pp. 1–23). CRC Press.

Hill, N., Brierley, G., & MacDougall, R. (2003). *How to measure customer satisfaction*. Gower Publishing.

Joreskog, K. G. (1973). Analysis of covariance structures. In P. R. Krishnaiah (Ed.), *Multivariate analysis*-III (pp 263–285). Academic Press.

Kashani, A. T., & Mohaymany, A. S. (2011). Analysis of the traffic injury severity on two-lane, two-way rural roads based on classification tree models. *Safety Science, 49*(10), 1314–1320. https://doi.org/10.1016/j.ssci.2011.04.019.

Nathanail, E. (2008). Measuring the quality of service for passengers on the Hellenic railways. *Transportation Research Part a: Policy and Practice, 42*(1), 48–66. https://doi.org/10.1016/j.tra.2007.06.006.

Prasad, M. D., & Shekhar, B. R. (2010). Impact of Service Quality Management (SQM) practices on Indian railways—A study of South-Central Railways. *International Journal of Business and Management, 5*(9), 139–146.

Rahaman, K. R., & Rahaman, A. (2009). Service quality attributes affecting the satisfaction of railway passengers of selective route in southwestern part of Bangladesh. *Theoretical and Empirical Researches in Urban Management, 3*(12), 115–125.

SAS Institute Inc. (2004). *SAS/STAT® 9.1 user's guide*. SAS Publishing. http://support.sas.com/documentation/onlinedoc/91pdf/sasdoc_91/stat_ug_7313.pdf.

Transportation Research Board. (1999). *A handbook for measuring customer satisfaction and service quality* (TCRP Report, 47). National Academy Press.

Washington, S. P., Karlaftis, M. G., & Mannering, F. L. (2003). *Statistical and econometric methods for transportation data analysis*. Chapman and Hall/CRC Press.

Weinstein, A. (2000). Customer satisfaction among transit riders: How customers rank the relative importance of various service attributes. *Transportation Research Record, 1735*, 123–132. https://doi.org/10.3141/1735-15.

Wong, J., & Chung, P. (2007). Managing valuable Taiwanese airline passengers using knowledge discovery in database techniques. *Journal of Air Transportation Management, 13*(6), 362–370.

The Role of Railways in Rural Development

Ana Ferreira

1 Introduction

As mentioned above, rural areas are generally distinctive for their low population densities—and small scale of human building. They are usually associated with low investment levels and employment rates below the national average. There are often shortfalls in terms of services and qualifications of the resident population (Ferreira, 2016). Rural areas were once defined by the main economic function of producing food of plant origin, raising livestock and other activities related to the forestry sector and beekeeping. This was usually translated into a set of distinct social characteristics typifying farming families. The dysfunctional (or simply lacking) adaptability of these ancient lifeways to the new urban–rural contexts led to a subalternation of rural areas—where we now find a concentration of older, less qualified populations with a lower incomes. Problems arose such as increase in unemployment, underemployment and social exclusion, which also led to a decrease in rural populations—a decrease we cannot dissociate from population ageing. Rural areas often have problems related to accessibility to (distance from) specialized services (Hoggart et al., 1995). The lack of opportunities and services makes rural areas unattractive for emigration. Thus, rural areas are depopulated as urban areas become overpopulated—a problem which has been debated worldwide. However, in order to face this problem, it is necessary to implement political measures capable of attracting populations. This is yet to take place. Rural areas present several areas of potential development, such as the agricultural sector, biodiversity, as well as pluriactivity and multifunctionality as concerns the energy sector and natural resources (Covas, 2006). An innovative and efficient agricultural system and sustainable multifunctionality constitute two important paths for rural development. This development can also be fostered through improved accessibility, the spread of Information and Communication Technologies (ICT), an

A. Ferreira (✉)
Northern Regional Development and Coordination Commission, Porto, Portugal

Centre for the Study of Geography and Spatial Planning, Coimbra, Portugal

© The Author(s), under exclusive license to Springer Nature Switzerland AG 2021

S. Mitra et al. (eds.), *Railway Transportation in South Asia*, Contemporary South Asian Studies, https://doi.org/10.1007/978-3-030-76878-2_11

increase in qualifications and an improvement of living conditions among resident populations. In the twentieth century, several scientific advances are occurring in areas such as physics and chemistry—with potential repercussions in the agricultural sector. For example, emerging new terminologies such as "agro rural and rural eco, marked by the replacement of biodiversity, the pluriactivity of energy activities, the multifunctionality of activities, the sustainability of processes and natural resources, the quality and safety of food and the cycle of agro-rural entrepreneurship, which will give rise to the post-modern multifunctional order, the second modernity of agriculture or the eco-rural cycle" (Covas, 2006, pp. 110–111). With the option of teleworking, it is possible to equate an increase in rural populations, provided that incentives are provided that can attract new populations, thus reducing population asymmetries between rural and urban areas.

The concept of rural development has gained importance in the last two decades. It is defined as the set of processes that focus on improving living conditions among rural populations—comprising the principles of "economic efficiency, social and territorial equity, heritage quality and environmental, sustainability, democratic participation and civic responsibility" (Ferreira, 2016).

For "rural development" to become an applicable concept, it is necessary to invest in decentralization and above all in cooperation between public and private entities. In some regions, given their peripheral position, development may involve enhancing the local natural and environmental heritage—on top of the cultural and social heritage which so often already characterize rural areas. There is a need to invest in the agricultural sector, combined with other activities and services, while focusing on multifunctionality and pluriactivity. In this way, it will be possible to take advantage of endogenous resources and attract investment—as well as the establishment of companies.

Multifunctionality is one means to support rural development. It identifies activities that benefit rural areas and studies the ways in which these can be combined. One of the models relates the agricultural sector to recreation and leisure, often associated with Rural Tourism and similar activities. Another model distinguishes nature conservation and the particularity of natural resources. Another defends the preservation of cultural identities. The last model defends the implementation of agro-industries and the production of bioenergy (Pinto Correia, 2007).

Another way of promoting the development of rural areas involves betting on innovation—or the combination of tradition and innovation. Rural populations are often more connected to traditions and customs, and these must be preserved. Thus, traditions represent an idea of continuity and not just part of history. Several success cases exist in which tradition is associated with innovation—to the benefit of rural areas, which thus attract tourism-related initiatives and become able to fix both resident populations and projects.

There are also examples of private initiatives that promote rural development—such as citizens who acquire uninhabited villages or towns and create dynamic projects centred on the region's cultural or natural heritages.

Other rural development initiatives are often developed by the government itself in conjunction with private companies or other communities. For example, the ASEAN

(Association of Southeast Asian Nations) has been developing a project for over 20 years now which aims to promote rural development, employment and inclusive growth throughout the region. In order to achieve these objectives, the programme aims not only to complement the industrial sector but also to make the whole economy more resilient—through the promotion of better connections to urban areas and the improvement of living conditions among rural populations (Thanh & Duong, 2016).

2 Some Context on the Study Area

2.1 Rural Areas in South Asia

South Asia used to be markedly destitute. In the year 2000, its still largely rural populations represented 40% of the world's total poverty. As mentioned above, the area known as the Asia Pacific (which includes parts of East Asia as well as South and Southeast Asia) concentrates 87% of all small agricultural farms worldwide, and agriculture provides income to approximately 2.2 million people. However, as urbanization has advanced, rural areas have been neglected and remain poorly connected to the new metropolises. Improving connectivity is therefore crucial (UNCRD, 2017).

Since South Asia is an area with such a high concentration of poverty, a programme based on the millennium goals, was created to reduce poverty in rural areas in this region. The goal of this and other programmes was to halving poverty by 2015 (United Nations, 2015).

In order for this to happen, measures were implemented to promote opportunities, facilitate local empowerment and increase security.

These measures aimed to develop less-favoured areas; instigate growth and reduce poverty; improve agricultural production capacities, especially female-led so as to promote social transformation; enhance capacities by indigenous and other marginalized populations; improve access by the most disadvantaged to productive resources and sustainable agricultural technologies; improve access to financial resources; and promote non-agricultural employment (Thapa, 2005).

According to Trading Economics, rural populations have significantly decreased in South Asia—from approximately 73% of the total population in 1996 to 66% in 2019. Urbanization is becoming ever more intensive. According to the United Nations (UN 2018), currently, 55% of the population lives in urban areas. A projection predicts that in 30 years, 68% of the population will be urban. On the Asian continent, approximately half of the population lived in urban areas in 2018 (United Nations, 2018).

In 1996, over 60% of the population residing in South Asia was employed in the agricultural sector; in 2019, this sector concentrated approximately 41% of the employed population. However, despite this significant decrease and after a sharp decline between 1996 and 2010, the percentage of the agricultural areas is currently gradually increasing. As an example of this increase, some areas of permanent cultivation of crops such as cocoa, coffee, rubber, flowering shrubs, fruit trees, walnut and vines have significantly increased since 1996—from 2.2% of the total area to 3.4% in 2017 (Trading Economics, 2020).

In India, poverty reduction has long constituted a major goal. When India became independent in 1947, several five-year plans were designed which aimed to gradually eradicate poverty. However, even with the implementation of these plans and over sixty years later, the failures and lacking are many—and a large section of the Indian population still lives in poverty. In recent years several government initiatives in infrastructure and rural development have led to an increase in rural wages, which grew considerably between 1983 and 2004. Other were investments targeted the education system. Yet urban and rural areas still present a huge disparity in terms of economic and human development (Singh et al., 2015).

According to the 2011 censuses, 70% of the Indian population lived in rural areas, that is, approximately 800 million people (*Business Standard*, 2013) Agriculture remains the largest source of income for rural families; however, in recent years investments in non-agricultural activities have increased. The bet on non-agricultural and multifunctional activities is of course a positive measure; however, it remains ill applied. Measures are yet to significantly favour populations; in fact, they seem to have increased economic and social inequalities (Elbers & Lanjouw, 2019).

Connectivity between South and Southeast Asia is essential for rural development in the region. Ideally, a connectivity system would be composed of an intermodal transport system that includes highways, railways, waterways and airports. Because the area includes mountainous regions, efficient transport systems are both more difficult to develop and much more essential (Florento & Corpuz, 2014).

South Asia is of course a region with numerous potentials—including in the agricultural sector, which remains the most important. However, it is also a region running at different speeds, with significant economic asymmetries. On the one hand, some countries are fast developing and present themselves as the region's engines of development—for example Singapore, which is classified as one of the most competitive economies in the world, or Bangladesh and its booming textile sector. On the other hand, Sri Lanka continues to struggle so as to develop key development factors such as infrastructure, transport and energy.

2.2 Railways on South Asia

The bet on railroads, implemented since the nineteenth century, could be a way to increase development in rural areas. As mentioned above, rail transport has numerous advantages such as safety, sustainability, speed, load capacity (Asim & Qanita, 2014).

For decades, travelling by train in South Asia was slow and extremely tiring, part of the lines ran through dry and arid areas with no air conditioning. A railway line with these characteristics is the line between Yangon and Mandalay, in Myanmar. Such is the lack of conditions, in fact, that locals prefer long bus trips, as these are more comfortable and are equipped with air conditioning. Myanmar Railways has been investing on the modernization of this line since 2017—the end of this endeavour is scheduled for the year 2023. This change will reduce travel time by 2/3 (from the previous 12 h). The company is also planning to modernize the line between the Yangon and Pyay, the oldest route in the country, which will serve rural areas north of the city of Yangon. Investment in more remote routes is also planned so as to attract adventure tourists to the scenic routes in mountainous areas (*Nikkei Asia*, 2020).

Lately, the possibility has been raised of an extension of the railway network from China to South Asia, this line being more modern, of large dimensions and allowing faster trains. However, several railway line points in Southeast Asia started to be rehabilitated after the airline crisis resulting from the COVID-19 pandemic. This change would serve much of the region and would constitute a much less expensive option than developing a whole new network—especially a whole new, high-speed network.

In countries such as Thailand, there were also major investments in rail transport, mainly on peripheral routes to Bangkok city. Electrification and the placement of a standard signalling system so as to increase safety and increase speed are planned. The investment in a dual path has allowed the commercialization of services—as well as the commercial and touristic marketing of a system which was in decline for almost four decades.

In Cambodia, the Royal Railway is winning over freight transport services, especially in transport to the port of the country's capital. This happened after the line was modernized and allowed to reduce freight rates by 8% in 2016. In 2018, the railway line between Phnom Penh and Poipet on the border between Cambodia and Thailand was reopened. The resurgence of railways has been extremely important for the development of companies and the citizen movement in Southeast Asia, especially for those most disadvantaged and most in need of a cheap and reliable means of transport. The focus on rail transport is also beneficial in the fight against climate change, as this means of transport is less polluting than road and air transport.

Although a controversial process, the implementation of the railway in India was a very important milestone—the proof of this fact being that almost no line has ceased being used since construction.

In India, according to the International Public Transport Association, trains transport 3.5% of the population to work—and is only used in commuting over 50 km. In rural areas, the use of rail transport decreased to 2.6% (*Nikkei*, 2020).

Indian Railways is one of the largest railway companies in the world and has all the potential needed to serve rural populations much better. In recent years, foreign investment has been allowed to improve load capacity and speed. Work is currently underway to double and triple some railway lines. The entire network is to become electrified by the year 2024—and an intention is expressed of involving solar energy by 2030. This initiative also aims to improve both services and train punctuality (India Brand Equity Foundation, 2020).

The UNESCAP (United Nations Economic and Social Commission for Asia and The Pacific) in conjunction with the Trans-Asian Railway initiated a railway project that connected Singapore to Istanbul—that is, approximately 14,000 km in length and crossing 28 countries. However, there is as yet no such continuous link. Currently, companies continue to promote sustainable rail transport (Florento & Corpuz, 2014).

It is worth remembering that in Southeast Asia the economy grew significantly in the last decade—which represented an increase in investment, a factor that should be maintained. One of the investments taking place in the region involves high-speed rail lines—such as the line between Kunming, Laos and Thailand, which will be built by a group of Chinese companies. The construction of high-speed rail lines benefits the economy, as it promotes trade and tourism. However, above all, high-speed lines facilitate access to rural areas, which may lead to their development.

3 Objectives of the Study and Methodology

The objectives of this chapter are mainly to proceed with the social economic characterization of the rural areas of the region under study in order to understand the main problems involved. We also aim to recognize some successful examples that can prove how the development of rail transport can enhance the quality of life among rural populations.

The methodology is based on bibliographic research, oriented towards the preparation of a conceptual framework on the subject under study. So, we proceed with a synthesis of the most important concepts for the development of this chapter. Likewise, through a bibliographic search, a characterization of the study area was carried out. After identifying the problems in the rural areas of the region under study, we found countless success stories exemplifying how rail transport can improve the quality of life of rural populations.

4 Results and Discussion

4.1 Advantages of Railways

Rail transport has numerous advantages, particularly in terms of the environment, since the train is a sustainable means of transport. At the economic-social level, the railroad brings territories closer to each other and expands opportunities. It also often reduces asymmetries. The train presents itself as a safe means of transport, especially when compared to the road.

Across the world there is a growing phenomenon, the urban sprawl, which often causes the exhaustion of large centres. A direct consequence of this situation is the expansion of cities to the periphery and nearby rural areas, where houses are cheaper. Also, in tourist cities, in addition to the pressure caused by the number of buildings and high rents, there are investments in large hotels or apartments rented to tourists, which forces the population to live farther and farther from the city centre. In the last decades, as car ownership became trivial, many railway lines were closed, isolating rural areas even more. The rehabilitation of railways allows not only to relieve road traffic but also to restore accessibility to rural areas, especially those most remote.

– a revealing number. In developed countries, there are both modern and extremely efficient railway lines. Links to peripheries and rural areas, especially if they are influenced by more complex geographic factors, are crucial (Patarchanov, 2019). Connectivity can provide several benefits for rural areas, as these are more vulnerable to isolation and poverty. Connectivity could allow better access to specialized services (such as health and education), greater political participation, increased social action and the extension of market chains (UNCRD, 2017).

A study on quality of life in South Asia's rural areas has shown that several people believe that improving the transport system often facilitates access to non-agricultural jobs outside the area of residence, especially if there is a direct and speedy form of transportation. The same study concluded that in the early 2000s, the quality of life was lower in rural areas than in urban areas, largely due to difficulties in access to education (Bloom et al., 2001).

4.2 Examples of Success Stories

In recent years we have seen good examples of the use of rail transport in the economic conversion of impoverished areas. Efficient railways can attract passengers to live in areas not previously attractive (especially due to the time and/or costs involved in reaching the nearest city). Likewise, railways can help attracting tourists and allow them to visit the area without in a fast and efficient fashion. Even in the UK and Canada new rail-based initiatives are developed so as to boost sustainable rural areas. It is worth remembering that when the first railway lines were opened, many rural areas benefited, improved their economic situation and increased their population even before the lines involves were opened—development began as soon as the planned routes became known.

Some countries such as the UK have also invested in community partnerships so that private investors would be able to rehabilitate lines and guarantee their maintenance. This allows keeping ticket prices as low as possible and improving the services provided. This action has also allowed for the recovery of stations and rural lines.

As mentioned above, one example that demonstrates the benefits of connectivity between rural and urban areas is the case of "Bangladesh: Rural Infrastructure Improvement II" (Asia Development Bank, 2013). Another already cited example occurred in India (Sharma, 2015).

5 Conclusion

In addition to being an affordable and reliable means of transport, rail transport is increasingly faster and more convenient. It is important that lines are rehabilitated and that efforts to improve services continue, so that ever more citizens are served. The lines that integrate rural areas facilitate access to education and health services, allow easier access to employment and provide quick access for tourists. The train's large load capacity also allows industries to settle in rural areas, generating employment and attracting new residents.

References

Asia Development Bank. (2013). *Bangladesh: Rural Infrastructure Improvement II.*

Asim, M., & Qanita, I. N. (2014). Pakistan Railways at the verge of collapse: A case study. *International Review of Management and Business Research, 3*(3), 1728–1739.

Bloom, E., Craig, P., & Malaney, P. (2001). *The quality of life in rural Asia.* Oxford University Press.

Business Standard. (2013). 70% Indians live in rural areas: Census. https://www.business-standard.com/article/economy-policy/70-indians-live-in-rural-areas-census-111071500171_1.html.

Covas, A. (2006). A Ruralidade do nosso tempo: decálogo para uma 2ª Modernidade [Rurality In Our Times: decalogue for a second modernity]. In M. L. Fonseca, *Desenvolvimento e território: Espaços rurais pós-agrícolas e novos lugares de turismo e lazer – homenagem à Professora Dra Carminda Cavaco* [Development and territory: Post-agricultural rural spaces and new places of tourism and lesiure—A homage to Dr. Carminda Cavaco]. M2 - Artes Gráficas, Lda.

Elbers, C., & Lanjouw, P. (2019). *Inequality in rural India—Are non-farm jobs the driver or a brake?* WIDER Research Brief, 2019(4). UNU-WIDER. https://www.wider.unu.edu/publication/inequality-rural-india.

Ferreira, A. (2016). *A multifuncionalidade e a relação tradição inovação em áreas rurais: O caso de estudo do concelho de Cinfães* [Multifunctionality and the relation between tradition and innovation in rural áreas: The case study of Cinfães district] (Master's thesis). Faculdade de Letras, University of Porto.

Florento, H., & Corpuz, M. I. (2014). *Myanmar: The key link between South Asia and Southeast Asia.* Asian Development Bank Institute. http://hdl.handle.net/11540/1094. License: CC BY-NC-ND 3.0 IGO.

Hoggart, K., Bulles, H., & Black, R. (1995). *Rural Europe—Identity and change* (322 pp.). Indian Railways Industry. https://www.ibef.org/industry/indian-railways.aspx.

Nikkei Asia. (2020). Green light for Southeast Asia's old railways. https://asia.nikkei.com/Life-Arts/Life/Green-light-for-Southeast-Asia-s-old-railways.

Patarchanov, P. (2019). *Railway transport in regional and local development of the rural areas—Challenges and opportunities.* In 19th International Multidisciplinary Scientific GeoConference SGEM 2019 Proceedings, pp. 533–540. https://doi.org/10.5593/sgem2019/5.4/S23.070.

Pinto Correia, T. (2007) .Multifuncionalidade da paisagem rural: Novos desafios à sua análise [Multifunctionality of rural landscapes: New analytic challenges]. *Inforgeo. Associação Portuguesa de Geógrafos* (20–21), 67–71.

Sharma, P. (2015). *Effects of railway development on India's transport system—British rule in India.* HistoryDiscussion.net https://www.historydiscussion.net/british-india/effects-of-railway-development-on-indias-transport-system-british-rule-in-india/5976. Accessed on October 15, 2020.

Singh, R. K., Singh, K. M., & Kumar, A. (2015). Socio-economic characterization of rural households: A village level analysis in Bihar, India. *SSRN Electronic Journal*. https://doi.org/10.2139/ssrn.2568037.

Thanh, V. T., & Duong, N. A. (2016). *Promoting rural development, employment, and inclusive growth in ASEAN*. ERIA Discussion Paper Series, ERIA-DP-2016-03.

Thapa, G. (2005). Rural poverty reduction strategy for South Asia. In R. Jha (Ed.), *Economic growth, economic performance and welfare in South Asia* (pp. 384–401). Palgrave Macmillan. https://doi.org/10.1057/9780230520318_19.

Trading Economics. (2020). *South Asia—Rural population*. https://tradingeconomics.com/south-asia/rural-population-percent-of-total-population-wb-data.html.

United Nations. (2015). *The millennium development goals report*.

United Nations. (2018). *68% of the world population projected to live in urban areas by 2050, says UN*. Department of Economic and Social Affairs, United Nations. https://www.un.org/development/desa/en/news/population/2018-revision-of-world-urbanization-prospects.html.

United Nations Centre for Regional Development. (2017). *Bangladesh: Rural Infrastructure Improvement II*.

New or Renewed Kolkata, the Outcome of the Metro, and Railway Network Designs

Sanghamitra Sarkar and Tomaz Ponce Dentinho

1 Introduction

Accessibility and infrastructure have traditionally played a significant role in transforming human habitats. Interestingly, along other determinants, these factors still dominate changing human landscapes—especially when these become urban, more modern, and further developed through a continuous process of rearrangement of different population groups placed in different categories of habitable pace. Because of the natural tendency by man and living beings in general to find comfortable dwellings (in our case, as concerns the availability of goods and daily services), the competition remains undying. This triggers continuous change—regular improvements through new amenities and varied modern assets. As rightly put by Deakin (1991), 'transportation investments are sought to revitalize the central city, spur rural development, support economic growth and competitiveness, reduce environmental degradation, and improve social equity'. Deakin further stated that 'transportation investments are seen by many as instruments for the shaping of metropolitan structure and, indeed, for the transformation of metropolitan living'. Thus, competition for urban land and/or competition among the various allocations of land (particularly in and around the city core and further towards similarly promising areas) emerges so as to reveal the best use to be made by for different groups. Conversely, individuals compete for the best land and upgrade their economic standards over time. Thus, there is a gradual change in both the form and built of an area. As witnessed across the globe, transport links are perhaps the primary factor initiating significant changes

S. Sarkar
Centre for Excellence in Studies On Tribes and Marginalized Communities, Utkal University, Bhubaneswar, India

T. P. Dentinho (✉)
Atlantic Applied Economics Studies Centre (CEEAplA), University of the Azores, Ponta Delgada, Portugal
e-mail: tomas.lc.dentinho@uac.pt

© The Author(s), under exclusive license to Springer Nature Switzerland AG 2021
S. Mitra et al. (eds.), *Railway Transportation in South Asia*, Contemporary South Asian Studies, https://doi.org/10.1007/978-3-030-76878-2_12

in land allocation and administration over different regions. The change involved is often termed as spatial change. It is effected through alterations in infrastructures, including social structures and/or organizations, and socio-economic changes, as population composition evolves either through either the replacement of local populations or through their enrichment. If we focus on either of these aspects, i.e. developed transport links and human settlement types—more precisely population profiles over a particular region—we immediately discern a direct relation between the two.

The aim of this chapter is to make a theoretically sound prospective analysis of different designs of both the metro and rail network in Kolkata. For that, we revise the literature; we also discuss the impacts of the planned rail and metro networks and delineate a different design that might avoid urban sprawl and improve the city's competitiveness.

In Sect. 2 we take a look at relevant literature that analyses the relation between urban networks and urban development. Section 3 looks into the evolution of Kolkata and its transport networks. Section 4 presents the plans for future adapted spaces and transport networks of Kolkata. Alternative solutions are presented in Sect. 5, which is focused on the impact of a stronger Express Network Rail connecting Kolkata with Howrah, on the other side of the Hooghly River, and with the Region at large. Section 6 presents the discussion of the chapter's perspectives regarding the exposed literature and. We conclude and present a few recommendations for urban sustainable development in Sect. 7.

2 A Model to Analyse the Impact of Railway and Metro Networks on Urban Areas: Linking the Reciprocity Between Transportation and Spatial Organization

In discussing the impacts and/or the relationship between transport networks and land use, one must never lose sight of the vital reciprocity between the two aspects, which always keeps them vibrant with transformation and development. In this context, various models and theories appear from time to time that aim to show the interactive nature of the transport systems and land use patterns in different locations (Chang, 2007).

For instance, if we go back two centuries, two prominent Economists—J. H. Von Thunen and David Ricardo—put forth the relation between land, labour, and capital. The latter two were seen as the primary factors of the productive process, while land was seen as intrinsic to location. Well-connected transportation networks within particular locations much add to the land's competitiveness because transport costs are lower at the centre and because the competition for better locations makes land values higher in more attractive places. Rents decrease with distance from the centre, compensating for greater transportation costs (Deakin, 1991). Land prices, influenced by accessibility, also define land uses.

On the other hand, there is this Business Location Theory, which advocates that the prosperity of a business is linked to the means of curtailing transport costs, together with access to both goods and the market. Thus, urban centres grow as more such business units flourish and transport links enhance an improved accessibility of the markets at large. The spatial distribution of various forms of land use also determines land valuation or land rent, thus influencing the urbanization process as the bid-rent theory emphasizes different economic factors influencing the spatial distribution of land use types. Again, the theories put forth by Burgess, Hurd, and Hyot all stressed the social and historical factors that influence the cycles of growth and decline in an area (Kumar, 2014).

There are further, different models developed by Lowry, Putman, Ducca and Putman, Herbert and Stevens, Prastacos, and others (Deakin, 1991) which suggest that jobs and housing facilities over different locations act as functions of accessibility, land availability, and socio-economic characteristics. However, transportation facilities carry the greatest weight in preferences for land locations among various socio-economic groups.

Now, urban areas are often designated as ‘activity systems’, units of multifarious social, economic, and cultural activities. Different land uses are classified as ‘formal land use’ and ‘functional land use’, the former being the qualitative aspect and the latter typically dealing with the economic activities, which accordingly set the preferences and/or interests by different consumers.

Interestingly, almost all land uses and functions have a reciprocal relation with transportation systems, since it is the accessibility that decides the region’s patterns of development. Different land use units are supported by different transport links; for instance, different public transport modes are developed in heavily residential areas so as to ease the movement. Thus, going back to the different models of land use, viz. Von Thunen’s regional land use, the Burgess Concentric model, the land rent theory, etc. (Rodrigue, 2020b), we see that it is transportation that plays the linking role among all land uses and functions—and that which most influences changes and/ or developments in each land use (Fig. 1).

Hence, both near and far from the CBD (Central Business District), the land value is increasing in many places due to local development processes (development of new urban centres, sub-centres, etc.). Still, it may be mentioned what Wrigley and Wyatt stated—after the argument by So et al. (1997) in the context of the influence of transport on property values—that development depends on a valuation based on four factors: ‘the availability of transport, transport costs, travel time and the convenience of transport modes’ (So et al., 1997 as cited in Wrigley & Wyatt, 2001).

This is one reason why land rents equalize or increase so drastically at the city’s periphery. Classical theories of residential land rents are based on the complementarities between land rents and costs of transportation. This theory is written as: $TC + LR = K$ (Fig. 2)—wherein, with the distance increasing from the CBD, transport costs increase and land rent decrease (and vice versa; Soot, 1974). However, there are contrasting theories and opinions regarding the increase and decrease of land valuation from the CBD, including some factors other than the transport system cited as inducing land value fixation.

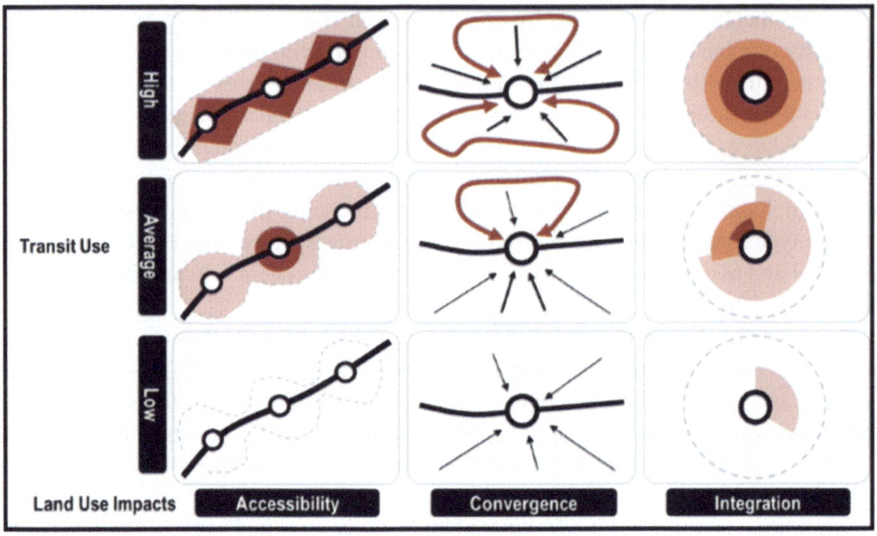

Fig. 1 Transportation and land use (*Source* Rodrigue [2020b])

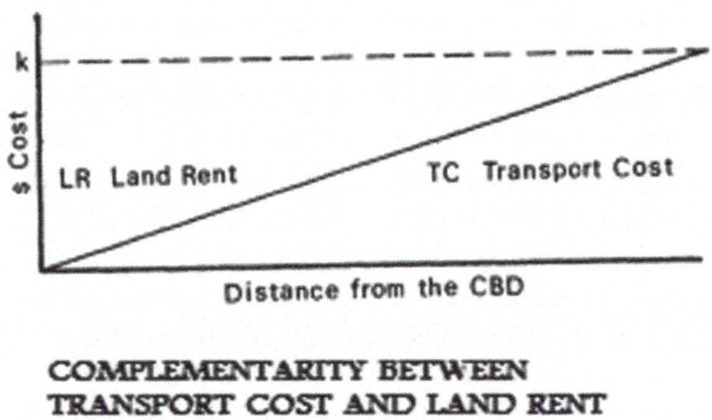

Fig. 2 Complementarity and transport costs (*Source* Soot [1974])

In order to link transportation and land value, we may put forward spatial structures of the urban space as important aspects of studies on modern land rent theory. Wheeler (1970) mentions the contradictory aspect of western economies, wherein 'the poor living on expensive land near the centre of the city and the rich living on cheaper land near the periphery' is common. This is essentially a condition set by the tastes and income levels of individuals, as most people preferring to live in less dense areas, with less pollution and less competition. The construction of a bid-rent curve describing the situation would find those richer having a flatter curve than that of individuals with lower incomes. This means that rich people have the possibility to choose between the size of their house and accessibility to central areas, whereas poor people with lower mobility options have to choose to be closer to their place of employment and occupation in denser, central areas. The development of specific transport links has already started creating such contrasts in Kolkata as well—and we find expensive locations for the rich both at the centre and at the peripheries.

3 Development of the Kolkata Railway and Metro Networks: Though the Lens of Kolkata's Evolution and Urban Development

Kolkata presents a very interesting picture and a good narrative as regards the history of development of a transport system, which is in this case a blend of both old and contemporary modes of transportation. This is perhaps much in concordance with the city's mixed traditional and modern structures, as with its socio-cultural dimensions still present in every aspect of both its tangible and intangible existence. Most of the traffic routing to North East India originates from this city. Kolkata connects well to the rest of the country through extensive National Highways, which ferry passengers and freights through different surface transports, as well as the Indian Railways, which reach different corners of India and further connect to the neighbouring country of Bangladesh. In these modern transport links the airport of Kolkatta connecting to the whole country and to the world plays also an important role (Wikipedia, 2020f).

In considering traditional transport links, waterways come first. The city is still recognized for its ports—the routes through which the British arrived over 300 years ago so as to trade and eventually settle their colonies. Thus, this city developed on the banks of the river Hooghly, stretching from north to south, and also gradually towards east and southwest. Kolkata still carries forward its historical legacy, both through the traditional, hand-drawn rickshaws and through what is nowadays the only city tram service in India (which started as early as 1873, then drawn by horses). Later, in 1902, the electric service started; it is today maintained with pride and connects different parts of the city (Wikipedia, 2020e).

Among other public transport modes connecting every nook and corner of the city as well as some suburbs are buses, autos, taxis, paddle rickshaws, launch/boats, trawlers, suburban railways, the metro rail, etc. Each one is adapted to the path distances and urban tissues, exhibiting a curious combination between modern, rapid means and conventional transport modalities. The city grew rapidly due to these multifarious transport links, inviting people to commute to the urban area in search of employment, education, and better living conditions. This quickly made the city grossly overpopulated, which in turn acted as an impetus for increasing and developing transport routes further.

Here one finds the concern for further renewal of the city—for the benefit of both its citizens and sustainability. With only 2% of the city's area under road cover, there is a rising problem of traffic congestion and environmental problems. As per the records, there were around 1.6 million vehicles and 23.50 million transit trips a day in 2011—numbers expected to reach 13 million and 32 million respectively by 2025. It is already foreseen that the road transport system alone would not suffice to meet the city's growing problems and ever-increasing demand. Thus, we see both the need and feasibility of developing the Mass Rapid Transportation Systems (MRTS)—the railways, which are planned to relieve stress, fulfil the demand, support the growth of urbanization, and sustain the environment by reducing pollution (Kolkata Metro Rail Corporation Ltd., n.d.)

Focusing exclusively on the MRTS—it is, again, foreseen as the most significant means of urban mobility. It would cater to the entire economic region, comprising both the city as the growth hub of economic activities, and the periphery as the zone of influence (Ranjan, 2018). The MRTS, which already acts as Kolkata's lifeline, is composed by the metro rail and suburban railways, which not only provide transportation facilities within both the urban area and the suburbs, but have constituted the primary drivers for a comprehensive growth of the urban agglomeration, influencing both the overall land use and the economy. Railways have been one of the best-fit mediums to carry both large number of passengers and bulk freights over long distances in a short time (*India Times*, 2014). The metro rail, a name with an urban connotation, is the modified, high-speed, smaller in size railway service, which became one of the most favoured means of transportation in Kolkata. The metro rail falls under the category of rapid transit, which began its first operation in London in 1863 (Wikipedia, 2020a). Metro rails have not only gained popularity due to their convenience, but are also predicted to develop in capacity over four times in the future (Mishra, 2018).

Thus, more than cutting down travel times and providing a comfortable ride for commuters, rail will serve as one of the most important means for both local and intercity travel—together with etching out employment opportunities for thousands (MoHUA, 2019). This would accelerate the process of urban restructuring in the near future to a greater degree by creating new scopes for business and work opportunities as well as good living standards. Railway stations will also attract new populations, bringing about socio-economic and infrastructural transformation.

Observation teaches us that, in the context of Kolkata, there is a direct relation between places with metro connectivity and socio-economic arrangement over space. Metro and railway are both means of bulk transport; nevertheless, people from affluent economic background tend not to depend much on such transport links for day-to-day commuting—since they often have their own vehicles. Hence, in parts of Greater Kolkata (the places where the city is recently spreading across its periphery, particularly in the northeast, east, south, and southwest of Kolkata) planned townships have invited the posh class to settle in expensive modern housings equipped with all amenities, in comfortable lifestyle. However, in order to cater to the daily needs of growing populations, i.e. in order to meet the various everyday needs of different socio-economic groups (both low, middle, and high earning), these new areas need different services and hence become new opportunities to work and make a living. Various manufacturing and service units, shops, offices, markets, etc. are opened. Another interesting feature noticed is that the large residential complexes mostly have the name 'city' and 'township' in their suffix, perhaps so as to testify their self-sufficiency in regard to amenities and facilities and thus meet the daily needs of residents. These areas are equipped with grocery shops, schools, gymnasiums, sports facilities (both indoor and outdoor, swimming pools, tennis courts, etc.), medical units, banquets or community halls, open spaces for joggers, and so on. Examples of such units include the Diamond city in both north and south Kolkata; the North City in north Kolkata; the Gems City in South Kolkata; the Larica Township, Emami City, and Fortune Township in the north, the Lake Life township in the south, and many more. Among those mentioned, some are already occupied; some are under construction and some are ready for possession. It is again interesting to note that although the housings mention Kolkata, they are in reality beyond the city, i.e. within the KMC (Kolkata Municipal Corporation area) jurisdiction. Nevertheless, improved transports bridge the distance and ultimately expand both the city area and the urban sprawl. Thus, we need more transport links, as construction of the metro rail service is in full swing to connect new areas in many new parts of the extended city. It is often noticed that real estate agents advertise their business with reference to the metro links, connecting the place with the important areas of the city and/or most importantly the city core—along with the other facilities ownership would provide clients due to its proximity to the main city. Hence, metro links are used as the primary catch for sale, as well as a way to add tangible value to properties (Gupta, 2017). Again, some areas have remained an abode of middle-income class societies for a long time. However, as mentioned, with the introduction of metro facilities, the valuation of the land shot up, inviting economically well-off people. This land thus became highly competitive in relation to other areas. This is bringing a complete transformation in the spatial organization of people, economic establishments, and land values. However, one aspect needs to be mentioned in the context of the metro and normal surface rail in Kolkata: the surface rail covering large areas, connecting the different parts of the city with the suburbs and faraway places, is preferred by people for commuting over long distances, and in many cases it is favoured by the middle and lower economic groups. (The economically well-off class commutes from longer distances in order to combine travel time and comfort.) However, the metro rail has so far remained a transport within the old city and thus invites commuters mainly from within this small urban space (for travel through shorter distances).

Now, it is not only that settlement patterns change as those with more purchasing power quickly become the possessors of the land locations well connected in relation to the metro. The other economic establishments also undergo rapid transformation to meet high-end needs and tastes. Shopping malls, multi-cuisine restaurants, multiplexes, designer boutiques, etc. make an entry, transforming the area and making the land even more expensive, as in the case of newly extended parts of Kolkata alongside the city core.

The places along the metro rail routes witness structural changes, with a multiplying vertical expansion due to the lack of horizontal space to accommodate the growing population (Das, 2016). The original proprietor of the land, if staying in the same location, builds similar modern structures—having earned enough by selling portions of the land at higher prices or completely selling off the property, he can now enjoy a flat in the high-rise. At times also, proprietors migrate to a different location, typically backed by their purchasing power, taste, or any other socio-economic suitability.

All these dynamics are quite similar as concerns surface rails as well. Even in suburbs, the lands nearest to stations are always attractive to those who can afford them; this is largely because other services such as different shops, markets tend to grow near railway stations. This proximity of course ensures greater crowds and hence more customers, along with other suitabilities, making the vicinity of the stations populous and congested with settlement structures. Thus, the pictures above are a definite testimony to the fact that railways and metro rails have a reciprocal relationship with the urban makeover of Kolkata and with the city's further urban expansion (Chang, 2007).

Figure 3 shows the evolution of the city of Kolkata—its spatial spread, along with the transport links that initiated the rapid expansion of the city since 1690. That year was documented in the pages of history when the chief of the British East India Company, Job Charnock, established this important centre for trade and commerce in the Sutanuti village, on the eastern banks of the Hooghly river. Thus, he is often referred to as the founder of Kolkata. After several processes over the years of renaming from Kalikata to Calcutta, the city is now known as Kolkata. (Kalikata

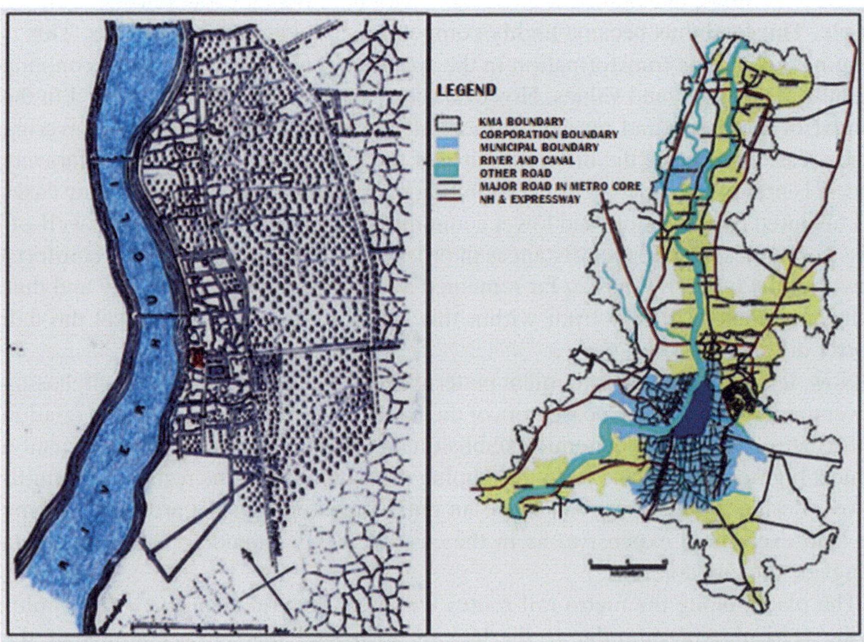

Fig. 3 Evolution of the city of Kolkata (*Source* IDFC and Superior Global [2008])

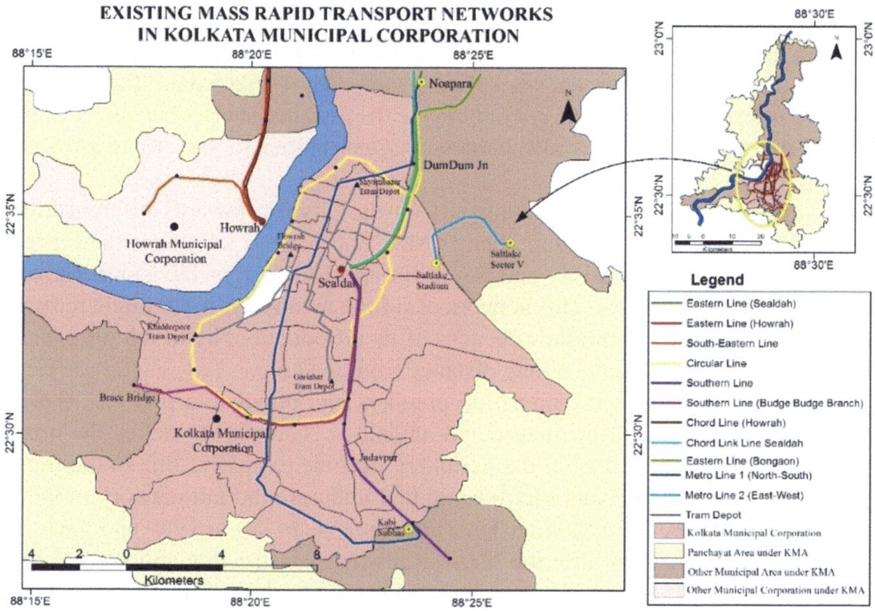

Fig. 4 Tram, metro, and rail lines in Kolkata (*Source* Compiled by the authors from Wikipedia [2020f])

was one of three villages, which merged to form the city of Calcutta—the other two being Sutanuti and Gobindapur.) Thus, this river port evolved as the most important urban centre in eastern India (Archinomy, 2011).

Figure 3 also shows a clear picture of spatial spread of the city area with further agglomeration of the planned townships of Salt Lake, Newtown, and other areas. These new urban areas developed in line with the plan of the suburban railways and metro railways, which are expecting implementation.

From the image of the transport networks in Kolkata (Fig. 4), it is clear that apart from the three lines presenting the metro rail (2) and tram (1) service in Kolkata, the other lines in different colours present the railway networks in the city, which connects the peripheries with the metropolitan area.

The Kolkata Suburban Railway service referred as part of the passenger service, started around the nineteenth century in British India. However, the major revolution came with the electrification of the railways in 1958. The time distance from the central city to peripherical areas situated around 50 km away shrunk with a travel time of only an hour (Mondal & Samanta, 2017). It is the largest suburban railway network in India by both track length and number of stations—it has 393 stations, with a track length of 1,332 km. There are six main lines, with nineteen branch lines. Recent records show that the suburban railway operates over 1,500 services and carries approximately 3,500,000 passengers daily. Another advantage the railways have in the city concerns its connecting stations with the Kolkata Metro at various locations—which makes both networks complement each other in providing a smooth and efficient transport service to the thousands of commuters travelling within and outside the city and its suburbs every day (Wikipedia, 2020c). Thus, these two transport modes have not only made the distance shorter between the city core and the suburbs, but have also harnessed opportunities for the city to grow and prosper.

The suburban railway lines radiating out from the City core towards different directions in the Kolkata Metropolitan Area (KMA) and beyond fall under the Eastern (Howrah and Sealdah Division) and South-Eastern (Kharagpur Division) Railway links of the Indian Railways. This network of suburban railways has brought remarkable transformations within a few decades of its introduction, by developing many peripheral cities located quite a long distance from the core.

We witness alterations in population numbers and population profiles, as well as the development of infrastructure and amenities. This slowly defies the urban rent theory, for lands situated at the periphery become expensive due to the transport connectivity. The Kolkata Metro is the first underground rail network in India, constructed under the Metropolitan Transport Project (MTP) in 1969 and renamed as Metro Railway in 1979 after the Metro Railway (Construction and Works) Act, 1978. In the year 1971, 5 rapid transit lines were proposed for Kolkata by the MTP. Soviet specialists (Lenmetroproekt) and East German engineers prepared a master plan to provide Kolkata with metro lines (Wikipedia, 2020d).

4 The Mobility Plan for the Kolkata Metropolitan Area

The comprehensive mobility plan for the Kolkata Metropolitan Area (IDFC and Superior Global, 2008) is an initiative by the Government of West Bengal to ease traffic congestion in Kolkata. The relevant goals are: (1) Congestion Mitigation, (2) Safety and Security; (3) Improved Air Quality; (4) Improved Quality of Life; and (5) Improved Opportunities for Economic Development.

In tracing the evolution of urban Kolkata, the first thing that comes to notice is the linear growth of the city along the banks of Hooghly River in a north–south pattern, an evolution that began some 300 years ago as the city became an important trade centre.

However, the city's further growth was propelled with the establishment of the Howrah Railway terminus in 1854. The area surrounding Howrah grew into a rich industrial hub. Its area of influence was the entire stretch on both banks of the river—Kolkata and Howrah, which quickly sprawled with huge settlement process and became greatly populated.

The city's sudden spurt of population was further linked to the huge refugee settlements that took place after India's independence in 1947. Kolkata attracted masses because, having once been the capital of the British Indian Empire, it offered many employment opportunities, ease in communication, a fast-growing urbanity, etc. Attracting migrants from different corners of this and neighbouring States, the core of the city become congested over time.

The present developmental plan for a transport system in Kolkata constitutes an attempt to reduce the bottlenecks in the city core and increase accessibility between the peripheries and said core—which in turn was deemed to channel crowds towards the transit nodes and nurture new, transport-induced spatial developments. The major transport networks planned through this approach were the Metro Rails, Suburban Railways, Circular Rails, and Trams (all of course in strategic areas, thus emphasizing environmentally friendly modes), along with buses and ferry services.

The circular line is especially noteworthy and constitutes a tourist attraction. It runs under the Howrah Bridge, through Vidyasagar Setu and parallels to the Hooghly River, thus connecting multiple tourist places and Ghats, as well as presenting scenic views to both daily commuters and visitors. It thus constitutes a popular means of joyride (Wikipedia, 2020c). Now, the North–South and East–West metro corridors have developed and/or were extended, while the suburban railway service too received an impetus, integrating with several metro rail linkages so as to produce smooth traffic across the city and its periphery—as well as making urban expansion and development sustainable dynamics (Fig. 5).

The proposed extension of metro rail connectivity as surveyed by the Rail India Technical and Economic Service (RITES), established by the Government of India in 2012, proposes 16 new routes to connect suburban areas to the city. At present, the Kolkata Metro Rail covers around 300-train trips every day and carries more than 700,000 passengers. The metro rail plans to extend its service to many of the above-mentioned routes by 2021.

In this context, the practicability of extending and advancing the metro and suburban railway networks for the city's transportation becomes significant. The metro, with its underground and elevated channels, is essentially a space-saving medium of transport network. Both the metro and suburban railway networks have special significance insofar as they constitute rapid and bulk transport links. Thus, the metro's new extensions may undoubtedly be referred to as a catalyst for initiating the urban development process of Kolkata and its surroundings—by altering the location and design of spatial development through the process of integrated transport links and changes in land use (Das, 2016).

The impact of transport development in new Kolkata developments is predicted to act through the growth of new urban centres and sub-centres, together with the expansion of new industrial hubs within the Kolkata Metropolitan Development Authority (KMDA) region. Likewise, the completed construction of the East–West metro corridor would integrate the two urban hubs, Kolkata and Howrah, bringing about manifold transformations in Kolkata and the greater urban agglomeration, while renewing the existing city area further.

This has also been a constant catch for realtors and investors willing to develop property at the newly constructed city spreads of New Town, Rajarhat, Barasat, Joka, and Howrah for the purpose of good business and regular rent earnings (Piplani, 2019). The construction of metro rails is thus expected to hike real estate prices all along the passage and buffer zone, i.e. the 'influence zone'. The development of this

Fig. 5 Existing and planned metro lines of Kolkata with planned centralities and industrial parks (*Source* Wikipedia [2020b])

MRTS is perceived as a catalyst to improve the standard of living of a huge section of the urban population, simultaneously ensuring the process of sustainable urban development (Shah, 2015).

5 Express Regional Networks Through the Old Centre

It is already clear that, with the means and opportunities available, Kolkata has been experiencing a rapid urbanization process, which is further projected to increase manifold. However, there are concerns among academics, researchers, and stakeholders regarding the future of the modern transport links not just in Kolkata, but also in most Indian cities.

For instance, Sharma and Newman (2017) have dwelled on the legal aspects of the construction process. There are separate agencies supervising urban transport plans and land use plans—while metro rail affairs are in turn taken care by a separate body. This creates a problem of integration. Again, there is an improper property mapping and valuation—with outdated land use maps acting as hindrances to rapid development. Adding to these factors are the heavy costs of construction as mentioned in a report by the *Traffic Infra Tech Magazine* (2013). Primary costs affect specific areas of metro rail projects, for example the 'cost of metro rail coaches, signalling and telecommunications, automatic fare collection system, air conditioning, etc.'. Thus some means to cut down the expenses are needed from the country's economic standpoint.

The idea is to promote connectivity and have a Renewed Sustainable Town close to the urban centre (rather than a New Sprawled Town). The main goal is to modernize existing railways into an Express Network Rail that crosses the city centre and, like the planned metro, crosses the river to Howrah and beyond as well, using existing railways. This would trains to be stationed further on and liberate the Howrah station for new buildings—of a modern city centre that is close to the old one and which enjoys rapid access to both the airport and port through the Express Regional Railway. As highlighted by Das (2016), the metro rail connecting Kolkata and Howrah is envisaged to develop both cities through the proposed development of commercial and residential structures along the Hugli riverbank (in the location of some closed jute mills). This would not only cater to current needs but would constitute a process of urban renewal, referred as a 'Transit Centric Development Zone'. Again, maybe the Line 6 of the Metro Plan could expand the city into the wet areas in the East, which could be in turn reassessed—and a few and more connected centralities would thus appear. This was again referred as a 'Transit Oriented Development Zone'. The network of Metro Rail and Suburban/Circular Railway, if developed further and capable of connecting the city centre with other relevant areas, would sustain and renew Kolkata's old centre—along with the city's overall development. In future works, we will try to model the urban effects of these two planning alternatives (Fig. 6).

Fig. 6 Express network rail for a renewed sustainable town (*Source* Wikipedia [2020b])

6 Discussion

As is the case in most urban and suburban centres within the KMDA, we find that socio-economic occupancy varies. However, transport links have been designated as a factor that reduces land rent differentials as propounded by Mills (Haring et al., 1976). Alternate transportation networks or well-developed transport means play a vital role in determining land use and transit patterns. The extension of the metro rail, along with already existing suburban railways, has played a similar role in cutting rent differentials—if not completely at least to some extent. The core continues to be expensive, with the range of land rent narrowing over time—not only due to saturation, but also due to the better links with peripheries and the development of self-sufficient sub-centres. Metro rides in Kolkata are bringing comfort and ease in travel over long distances, together with time saving and cost-effectiveness (Verma, 2018).

Land value over Kolkata is slowly equalizing due to transport efficiency. If land valuation in monetary terms is to be estimated, it must be evaluated according to the location and rent of amenities in the place for a given time. While rent location depends on accessibility and demand–supply factors, rents of amenities are mostly site-specific (Kumar, 2014). Notwithstanding this, the alternative to improve the centrality of the old centre will increase land values there—and will allow expected trade-offs between density, accessibility and social interaction.

Let us now bring some light over the scenario of Kolkata's spatial organization and the way it evolved over time. Quoting Rodrigue (2020a), spatial organization relies on the aspects of 'spatial differentiation' and 'spatial interaction'. Places differ in terms of size, population density, and overall built up (as urban areas have different structures, etc.). Interactions/flows between two areas (origin and destination) depend on the inequalities involved. Transportation plays the link between different regions and foster economic development by inducing the movement of labour, changing structures, etc. Thus, in establishing a reciprocal relationship between transport and land, the statement as put by Rodrigue that 'Space shapes transport as much as transport shapes space' is most adequate. The expression is further conceptualized over the aspects of reciprocity between transportation and location, as well as demand. Transportation shapes a region by influencing and developing structures originating from and/or through the nodes and flow links involved. Again, each economic activity being dependent on transportation, whether local or global, the demand of transportation thus influences the built and structures of an area over time. The role of transportation is consequently observed in changing and developing economies through its constant service between the 'core-periphery' and 'gateway-hubs'.

While we are acquainted with the core-periphery relation in Kolkata, it is vital to talk about the ways in which Kolkata acts as a gateway to the peripheries and neighbouring States—North-eastern States in particular. Kolkata serves as the transit location for most passenger and freight movement to diverse locations. Again, this city acts as an important hub for both collecting and distributing major economic goods and services—not only for local markets but also for the peripheries and much beyond. Thus, Kolkata has been the 'growth pole', a place of industrial growth and thriving economic activities, a 'transport corridor' allowing movement of people and goods, an 'employment zone' creating ample job opportunities. Further, the city constitutes an 'attraction zone' providing jobs, education, markets, and recreation made possible with the help of multimodal transport links—thus accelerating urbanization, suburbanization, and also 'exurbanization', the process of extending the progression into places distant from the main city without losing the interaction potential through mobility and integration.

Kolkata's urban expansion is heavily triggered by the developing transportation system, particularly the MRTS, which has brought rapid development of suburban sprawls away from the city core and further transformation of the old city. If we take Kolkata for the purpose of estimating these theories and/or land values, we would find it constitutes a blend of several factors. Most notably, we find that developments in the transport network, particularly railways, with the different land use functions, basically dictate land values. Quoting Kolkata's situation in this context, it was observed that once the industrial belts were located near riverbanks for easy water transportation and utilization of water for industrial activities, settlements promptly grew on these banks. These settlements were mainly comprised of the working class, which contributed to rapid urbanization. The wealthy business class opted for the city core as residence; however, today we witness that as houses became old many have opted for different residential locations—a process wherein transport networks have facilitated an easy accessibility between the city core. Thus, transport facilities, along with other amenities and services, played a significant role in extending Kolkata's urbanization process. Today it is undoubtedly the transport network which structures the competitiveness and sustainability of the city.

The concept of sustainable transport for urban areas as highlighted by Sarkar and Tagore (2011) considers certain aspects as sustainable. To mention a few: (i) the transport should bring down pollution and noise, (ii) it should increase traffic security and safety, while increasing commuters' space, (iii) it should offer a suitable environment to encourage short-distance travel by foot or cycling—and bring down road use, (iv) it should contribute less or zero pollution, (v) it should provide safe routes towards schools, hospitals, workplaces, (vi) it should cut down energy consumption, (vii) it should have ample space for green areas in the city, and (viii) it should bring comfort in travel for all categories of commuters.

Considering all the above aspects, creating a sustainable transport system for Kolkata may seem like a quite difficult endeavour, especially for the old, heavily settled, densely populated, and congested areas of Kolkata. However, some transformation is always possible by altering a few of those transport means that cause pollution. The metro rail is here mentioned again, for it has greatly benefitted the process and is expected to promote safety, convenience, and comfort, of mass passenger movement, while reducing urban pollution.

7 Conclusions

The rapid development of metro rail networks all across India's major cities is often perceived as the ultimate solution to the problems such as mass urban traffic and congestion. However, some academics and critics hold, regarding the metro, that it constitutes a 'capital intensive' transport mode—which is therefore only sustainable in densely populated urban areas and not in less dense suburbs (Chatterjee, 2013).

Kolkata does not fall in any way under the category of less populated urban areas. Nevertheless, there is definitely a blend between different income groups residing all across the city. This may to some extent mean that not all sections of society can benefit of this rapid transport mode due to its high cost. The metro rail's fare for short distances has remained quite nominal in Kolkata, making it a preferred means of communication over bus, auto etc. In fact, metros have become overly packed with passengers during rush hour. This is also the case of suburban railways or Express Regional Railways, which ferry thousands of passengers from the suburbs to the city proper every day. This cheap and fast transportation means has remained popular to the section of population travelling daily between the city and the suburbs.

Moving away from the city core towards the periphery, the picture of this exploding crowd to some extent smoothens. This is possibly due to two reasons: (i) the thinning out of the settlement away from the core and (ii) the fact that planned townships at the outskirts of Kolkata have to some extent done away with unplanned, irrational settlement pattern and heavy population growth.

However, the metro rail has still maintained its popularity throughout urban areas, old lines further being extended to the peripheries and new lines being constructed that connect ever more distant places to the urban centre. The reasons for this are probably the following: (i) a great investment in developing and strengthening the transport network; (ii) a preoccupation in easing out the problem daily commute from the peripheries to the city core, thus saving time and distance; (iii) a preoccupation in spreading out the crowd from the city core towards the outer realms (for the satellite towns to flourish); and (iv) a care to create sustainable land use patterns for Kolkata as the city grows over both space and demographics. Alongside the metro link, suburban rail links have been running with poise, serving bulks of passengers and freight over a large spatial extent and in commune with metro links within the urban space. Thus, these two rapid transport means are no doubt an apt fit for Kolkata's huge population and road congestion.

India's urban development programmes fall under the aegis of three planning bodies, which supervise the urban and spatial development process: (i) the Ministry of Urban Development (MoUD), the Ministry of Housing and Urban Poverty Alleviation (MoHPA), and the Town and Country Planning Organization (TCPO). These together look after the urban and regional development planning. Another organism is (ii) the Planning Commission of India, which was replaced by the National Institution for Transforming India (Hindi: NITI Aayog, Policy Commission). This is a new planning system, which envisages working in cooperation with the economic policy-making process led by the State governments of India (Alliance Experts, 2016). Transport networks being part of a region's development process, stakeholders must work in unison, i.e. with the Ministry of road transport and highways, the national body for the overall planning, implementing, monitoring, and managing the country's transport system. Thus, the concept of a new Kolkata and a renewed Kolkata will certainly materialize—for the foundation stone has already been laid with the further development of the suburban and metro rail networks, bringing transformation in the economy, society, environment, and the overall organization of the spatial built. The challenge is now to combine the New Kolkata with the renewed Kolkata in the old city. In this quest, the design and connectivity of rail, metro and tram are essential to promote sustainability and competitiveness of this city—within the Bengal Area, the country and the World at large.

References

Alliance Experts. (2016). *Urbanisation in India: Trends, opportunities and smart cities*. https://www.allianceexperts.com/en/knowledge/countries/asia/opportunities-in-urbanisation-in-india/.

Archinomy. (2011). *Settlement and transportation of Calcutta*. https://www.archinomy.com/case-studies/settlement-and-transportation-of-calcutta/.

Chang, J. S. (2007). Models of the relationship between transport and land-use: A review, Taylor and Francis online. *Transport Reviews, 26*(3), 325–350. https://doi.org/10.1080/01441640500468432.

Chatterjee, V. (2013, January 6). Making urban travel easier. *Business Today*. https://www.businesstoday.in/magazine/columns/vinayak-chatterjee-on-indian-transport-system/story/190689.html.

Das, P. (2016). *Impact of metro on the development of the twin city: Kolkata and Howrah* (Master's thesis). 9th Urban Mobility Conference and Expo 2016. http://www.urbanmobilityindia.in/Upload/Conference/14172705-499d-4d7e-8a69-179364d73327.pdf.

Deakin, E. A. (1991). *Jobs, housing, and transportation: Theory and evidence on interactions between land use and transportation*. In: Transportation, urban form, and the environment: proceedings of a conference, Beckman Center, Irvine, California, December 9–12, 1990. http://onlinepubs.trb.org/Onlinepubs/sr/sr231/231-003.pdf.

Gupta, S. (2017, August 16). *What Makes Kolkata's Tollygunge a perfect investment destination*. Proptiger. https://www.proptiger.com/guide/post/what-makes-kolkatas-tollygunge-a-perfect-investment-destination.

Haring, J. E., Slobko, T., & Chapman, J. (1976). The impact of alternative transportation systems on urban structure. *Journal of Urban Economics, 3*(1), 14–30. https://doi.org/10.1016/0094-1190(76)90055-3.

India Times. (2014, December 15). Metro rail: Facts & history. https://www.indiatimes.com/culture/travel/metro-rail-facts-history-313969.html#:~:text=Railway%20system%20occupies%20significant%20place%20in%20the%20realm%20of%20transportation.&text=World's%20first%20urban%20underground%20railway,now%20part%20of%20London%20underground, on 30 June 2020.

Infrastructure Development Finance Company Ltd. (IDFC), & Superior Global Infrastructure Consulting Pvt. Ltd. (Superior Global). (2008). *Comprehensive mobility plan: Back to basics—Kolkata metropolitan area*. https://wricitieshub.org/sites/default/files/Comprehensive%20Mobility%20Plan%20for%20Kolkata%20Metropolitan%20Area.pdf.

Kumar, P. (2014, June 5). *Urban rent and land use theory*. Technology, Economy & Finance. https://www.slideshare.net/yashwanthhero/lecture7-guide.

Mishra, S. (2018). *8 Facts about India's metro system*. Proptiger. https://www.proptiger.com/guide/post/8-facts-about-indias-metro-system.

MoHUA. (2019). *Metro projects in India- Rapid strides in urban transport & mobility*. Press Information Bureau, Government of India, Ministry of Housing and Urban Affairs (MoHUA). https://pib.gov.in/Pressreleaseshare.aspx?PRID=1564876.

Mondal, B., & Samanta, G. (2017). Commuting and metropolitan development of Kolkata. *Hill Geographer, XXXIII*(2), 61–79. https://www.researchgate.net/publication/328319153_Commuting_and_Metropolitan_Development_of_Kolkata.

Piplani, G. (2019, June 3). *Top 5 Kolkata localities that are worth your money*. Proptiger. https://www.proptiger.com/guide/post/top-5-kolkata-localities-that-are-worth-your-money.

Ranjan, R. (2018, March 13). *Background paper for the lead-up event on 'Mass Rapid Transport Systems for Urban Areas: Opportunities and Challenges'*. Kolkata. http://www.ris.org.in/sites/default/files/AIIB%20MRTS%20Meeting%20Background%20Note.pdf.

Rodrigue, J. P. (2020a). Transport and spatial organization. In *The geography of transport systems*. Routledge. ISBN 978-0-367-36463-2.

Rodrigue, J. P. (2020b). Urban land use and transportation. In *The geography of transport systems*. Routledge. ISBN 978-0-367-36463-2.

Sarkar, P. K., & Tagore, P. (2011). An approach to the development of sustainable urban transport system in Kolkata. *Current Science, 100*(9), 1349–1361. www.jstor.org/stable/24076600.

Shah, N. (2015, July 16). *In seven Indian cities, expect the metro rail to drive urban development this year*. https://www.metrorailnews.in/in-seven-indian-cities-expect-the-metro-rail-to-drive-urban-development-this-year/.

Sharma, R., & Newman, P. (2017). Urban rail and sustainable development key lessons from Hong Kong, New York, London and India for emerging cities. *Transportation Research Procedia, 26*, 92–105. https://doi.org/10.1016/j.trpro.2017.07.011.

So, H., Tse, R., & Ganesan, S. (1997). Estimating the influence of transport on house prices: Evidence from Hong Kong. *Journal of Property Valuation and Investment, 15*(1), 40–47.

Soot, S. (1974). Transportation costs and urban land rent theory: The Milwaukee example: 1949–1969. *Land Economics, 50*(2), 193–196. https://doi.org/10.2307/3145372.

Traffic Infra Tech Magazine. (2013, April 1). Urban transport and the role of metro rail. https://www.trafficinfratech.com/urban-transport-and-the-role-of-metro-rail/.

Verma, R. (2018, February 20). *History of metro rail in India*. Newsgram. https://www.newsgram.com/metro-history-india/.

Wheeler, O. J. (1970). Transport inputs and residential rent theory—An empirical analysis. *Geographical Analysis, 2*(1), 43–54. https://doi.org/10.1111/j.1538-4632.1970.tb00143.x.

Wikipedia. (2020a). *History of rapid transit*. https://en.wikipedia.org/wiki/History_of_rapid_transit.

Wikipedia. (2020b). *Kolkata Metro*. https://en.wikipedia.org/wiki/Kolkata_Metro.

Wikipedia. (2020c). *Kolkata Suburban Railway*. https://en.wikipedia.org/wiki/Kolkata_Suburban_Railway.

Wikipedia. (2020d). *Metro Railway, Kolkata*. Last modified 18 June 2020. https://en.wikipedia.org/wiki/Metro_Railway,_Kolkata.

Wikipedia. (2020e). *CESC Limited*. https://en.wikipedia.org/wiki/CESC_Limited 11 January 2021.

Wikipedia. (2020f). *Transport in Kolkata*. https://en.wikipedia.org/wiki/Transport_in_Kolkata modified 28 July 2020. https://en.wikipedia.org/wiki/Trams_in_Kolkata.

Wrigley, M., & Wyatt, P. (2001, June 27–29). *Transport policy and property values*. Paper presented at ERES 2001, Alicante. https://eres.architexturez.net/system/files/pdf/eres2001_300.content.pdf, on 6 July 2020.

Bangladesh-India Rail Connectivity: Foreseen Opportunities for Tourism

Pinaki Bhattacharya and Shuchita Sharmin

1 Introduction

Bangladesh and India share a boundary of 4156 km—which is 94% of Bangladesh's total land boundary and the 5th largest boundary between two countries in the world. As 80% of Bangladesh consists of floodplains and wetlands, historically the country's transport systems have relied heavily on waterways (namely river transport); this is particularly true south of Dhaka. However, in order to be connected to the outside world, railway transport has always been of extreme importance. The history of railways in Bangladesh dates back to 1862, when the first railway track was laid. Since then and until till 1947 there were multiple railway tracks with India—from both west and east of present-day Bangladesh, connecting this part of the land-mass with the rest of the world. During the partition of 1947, the total railway track in Bangladesh was 2604 km. This went up to 2855 km in 2006—an addition of only 451 km in 60 years. The only major landmark happening in this period was the Bangabandhu Setu on the Jamuna River, which became the highest toll earner in Bangladesh ever, according to information mentioned in the Dhaka Tribune (Numan, 2018). Mr. Khaled Saifullah had mentioned in his work on floods in Bangladesh that 22–30% of the country is affected by floods every year—which obviously affects the economy and thereby the development of the country in major ways (Saifullah, 2009). Realizing the need of the hour, in 2010, the Government of Bangladesh approved 10 railway projects—and in 2011, the Prime Minister initiated the process of connecting Cox's Bazar with Dhaka. As of date, four railway corridors with India have been restored from Bangladesh's west side, i.e. through West Bengal. Among these, two have already started transporting passengers. The Maitree Express started operating between Kolkata and Dhaka in 2008; it initially ran only three days a week,

P. Bhattacharya (✉)
Begum Rokeya University, Rangpur, Bangladesh

S. Sharmin
Department of Development Studies, University of Dhaka, Dhaka, Bangladesh

© The Author(s), under exclusive license to Springer Nature Switzerland AG 2021

S. Mitra et al. (eds.), *Railway Transportation in South Asia*, Contemporary South Asian Studies, https://doi.org/10.1007/978-3-030-76878-2_13

a frequency increased to four days in 2017 and to five days in 2019 because of high demand. Keeping this heavy demand in mind, the Bandhan Express has been operating between Kolkata and Khulna since 2017—two days every week. As of date, the corridor between Radhikapur in Uttar Dinajpur District (West Bengal), connecting to the Rangpur Division (Bangladesh), to Singhabad in the Malda District (West Bengal) and to the Rajshahi Division (Bangladesh) is limited to freight movement. They are expected to open up for passengers in the near future. The railway corridor between Haldibari in the Cooch Behar District and Chilahati in the Rangpur Division is set to be restored—as mentioned in a detailed research work by Ahmed Murshed on railways of Bangladesh. Four more railway corridors connecting with India are expected to open up on Bangladesh's eastern boundary, of which three were in existence in the pre-1947 days (Murshed & Firozi, 2018). Once all these corridors open up, the benefits would not be restricted to freighting alone—they would be manifold. Bangladeshi citizens would be able to cover longer distances in a more affordable manner. Medical assistance would become more easily available. As concerns India, an additional destination for tourism would open up in Bangladesh, which would again be beneficial in terms of both cost savings and time savings.

2 Literature Review

As David Geary in his text "India's Buddhist Circuit" describes, while the western part of India has conventionally constituted the focal point for tourists seeking to experience heritage and history, the eastern part has always been the area of attraction for Buddhist followers all around the world (Geary, 2018). In order to cover this circuit, travellers used to come to New Delhi from all across the world. From there tourists travelled to Sarnath near Varanasi, Rajgir, Bodhgaya, Kushinagar—and from there further to Nepal so as to visit Lumbini. Kathmandu used to be the exit point for travellers visiting this circuit. However, after the earthquakes in Nepal in the last decade, tourists travelling to that country for this segment are now much less in number—and instead a new Buddhist triangle has emerged through Lalitgiri-Ratnagiri-Udaigiri. The International Finance Corporation has observed that travellers generally transit back to New Delhi these days—thus making the city both an entry and an exit point (International Finance Corporation, 2017).

In addition to the Buddhist circuit itself, there is a high level of interest on the Buddhist way of education—as it is being replicated across various countries over the years, constituting a topic of enquiry for researchers all across the world. Taxila's remains being almost non-existent, Nalanda has been the focal point for those visiting the land of Buddha to gain knowledge on this subject. However, Nalanda has not been the only university of repute imparting education to monks. During the Pala dynasty, out of the five Mahaviharas rising to prominence, Somapura and Jaggadala are in located Bangladesh—and have been recognized by UNESCO as such (Centre, 1985). Somapura was supposedly the largest of the Mahaviharas. It carries the legacy of the Chinese scholar Xuanzang, who stayed there so as to learn to master the texts of Buddhism during his visit to India in seventh century AD. Also Atish Dipankar, the scholar who travelled to Tibet to establish Buddhism, was Principal of this noted education institute for Buddhist monks. Incidentally, Atish Dipankar was born in Bangladesh—at Munshiganj, near Dhaka; the Bangladeshi government has accepted China's help to restore the relics of this place—this has been officially confirmed by Xinhua, the government news agency of the Republic of China (Chuntao & Karim, 2018). There are five ancient Buddhist Universities of Viharas in the Rajshahi Division and two more in the adjoining Rangpur Division. The Viharas Rajshahi Division is already connected by railways through India through the Singhabad corridor. If this corridor, which is restricted to goods trains only, opens for passenger movement, and if a few of the mail/express train connectivity from Patna to Malda Town can be extended to Singhabad (which is located only 25 km from Malda Town) this would provide a major thrust to Bangladesh's tourism through Buddhist circuit travellers. As Dhaka enjoys an international airport (which connects to almost all the countries across the globe), international travellers can use this city as exit route from the circuit—thus replacing Kathmandu.

In addition to Buddhist relics, Bangladesh also holds five of Hinduism's 52 Shakti Peethas. Hindu mythology holds that pieces of Sati's body spread across the Indian subcontinent. Of this, 17 are located in West Bengal, 1 in Tripura and 2 in Assam and Meghalaya. These peethas draw huge numbers of devotees all through the year both in India and in Bangladesh. The Dhakeswari is located in Dhaka, which is already well connected to India through road, rail and air. Except Sugandha in Barishal, all the others—Jashoreshwari in the Satkhira District, Bhabanipur in the Bogra District and Chandranath in Chittagong—are in close proximity to Indian border. Existing railway tracks, once open for commuting passengers, would witness a sizeable increase in the numbers of pilgrims who try to visit as many Peethas as possible—as that is considered an extremely pious job for a Hindu believer. All these places are not only important for Hinduism but also for Bajrajana Buddhists, as these followers of Buddhism are worshippers of Shakti as well. Bangladesh's Buddhist universities were once destinations for students of Bajrajana Buddhism.

3 Results and Discussion

Indians have a liking for the sea and to enjoy sea beaches. For this end, Indians travel to Goa and even to international destinations such as Bangkok and Fuket. Cox's Bazar beach in Chittagong, in spite of being the longest sea beach in the world (running for a stretch of 150 km), and despite being famous for its serene beauty, has not yet become a destination for Indians and does not feature in an Indian's travel map. The local revenue is primarily obtained from domestic tourism. An Indian spends approximately INR 20,000–INR 25,000 for one person to travel to Goa or Thailand (from Kolkata). The time taken to travel from Dhaka to Chittagong by train was approximately nine hours in the metre-gauge railway track. This was reduced to five and half hours in 2017, when the 2nd Bhairav Bridge became operational with broad-gauge rail tracks over the river Meghna—as reported by UNB NEWS (United News of Bangladesh, 2017). The Dhaka-Chittagong high-speed railways which are proposed to become operational in 2022 would reduce travel time from Dhaka to Chittagong to two hours, thereby reducing the travelling time from Kolkata to Chittagong to 10 hours by train. The travelling cost to Cox's Bazar would be nearly 50% that of what one spends on average so as to reach Thailand. Furthermore, young people comprise over 55% of leisure travellers. With all railway corridors from West Bengal opening up for passenger communication, the revenue from tourism would increase manifold for Bangladesh. Chittagong would become a weekend destination for Indians staying in West Bengal (and neighbouring states); this would increase Bangladesh's tourism-related revenue manifold. This revenue will certainly further increase when the railway line (a 15.054 km-long stretch) is completed between Agartala, capital of Tripura, India and Akhaura, Bangladesh. This railway corridor will be endeavoured, as mentioned in details by Anusua Basu Ray Choudhury in her works (Chatterji et al., 2015). The Financial Express confirms this would not only reduce travelling costs for those headed to the beaches from India's north-eastern states but would also reduce travelling time from Tripura to Kolkata from 38 hours to only 10 hours by train (as well as reduce the distance involved from 1551 km to mere 550 km through Bangladesh; Financial Express, 2018).

The benefit is not restricted to Bangladesh alone, as once the railway corridors to India opens up, Bangladeshis themselves will have an umpteen number of newly available tourist destinations. When the Singhabad railway corridor opens up for passengers and the railway service between Haldibari and Chilahati resumes, the Himalayas will come within the reach of Bangladeshis—and won't be restricted to richer citizens alone. Ajmer Sharif, a destination for almost all pious Bangladeshi Muslims, would also become accessible to all—it at present is beyond the means of many due to logistic reasons.

In 2017, a survey was undertaken by the Indian Institute of Tourism & Travel Management (IITTM), Bhubaneswar—under the Ministry of Tourism, Government of India. The project involved sample size of 3000 respondents were interviewed at the Haridaspur border, i.e. the Benepole check post (primarily a road corridor between India and Bangladesh). 1500 respondents were interviewed at the Kolkata station (the Indian hub for the railway corridor) and 500 respondents were interviewed at the NSC Bose Airport, Kolkata, on Bangladeshi flights (IITTM, 2018). The survey revealed that less than 10% of the travellers were of the age group of 60 + —which reveals that the number of travellers visiting the ancestral home is very low. Taking into account all the modes of transport, it was revealed that nearly 60% of the travellers were of the age group 25 to 50 years—travelling mostly for leisure, religious and business purposes. Within this 60%, nearly 60% are leisure travellers, 25% are religious travellers and only 15% travel for business purposes. The number of travellers from Bangladesh to India was 1.38 million in 2016, 2.15 million in 2017 and 2.25 million in 2018—thereby substantially contributing to India's 9.2% GDP obtained from tourism sector. Only 2 trains currently carry passengers between the two countries. It is evident from the findings of the survey that the number of people travelling between the two countries for leisure purposes and in religious trips will increase many times when all operating railway corridors from West Bengal to Bangladesh engage in passenger transportation. When the proposed Sitaldah-Chilabari railway corridor is inaugurated, along with the three railway corridors running from north-eastern states to Bangladesh, Bangladesh's tourism earnings in terms of GDP will increase substantially. The number is at present 4.4% only.

Specialized treatments are almost not available in Bangladesh—and factors such as overcrowding, high costs, demographic patterns, inadequate infrastructure and medical technology, lack of qualified professionals and lack of overall quality in healthcare services lead many patients to travel across border to various cities in India for medical treatment (Dasgupta, 2011). In this regard, Ali (2012) feels that there is an increasing evidence of outbound medical tourism—wherein Bangladeshi patients travel to neighbouring countries for treatment due to "inefficient human resources in healthcare management industry". Medical tourism is an example of bilateral trade in healthcare services between India and Bangladesh (Rahman, 2000). Almost all Indian state capitals are popular destinations for medical tourism, given the infrastructure, tourist attractions as well as the corporate, privately managed hospitals with state-of-the-art medical facilities and qualified as well as accredited professionals. While India has patients travelling for medical treatments from most Central Asian countries, the Sunday Guardian holds that more than 50% of these patients are from Bangladesh (Kumar, 2018). However, the treatment is availed mostly in Kolkata due to the logistics of connectivity—and only a selected few can avail the facilities also available in other parts of India. Once the railway corridors open, there would not be any need to route the patients through Dhaka alone—but one would be able to avail the medical benefits of other states which also lie near Bangladesh.

4 Conclusions and Recommendations

Bangladesh has a strong and preserved heritage. Improved railway connectivity with India would allow more Indian tourists and medical tourists to come to Bangladesh and more Bangladeshis to travel to India for both medical purposes and leisure. Lowered travel costs will much improve access to both leisure and health services. The investments involved in connecting corridors are very high; however, the benefits would be much higher. This is even more true if those endeavours were complemented by a proper railway network within Bangladesh and multiple railway corridors in India.

References

Ali, M. M. (2012). Outbound medical tourism: The case of Bangladesh. *World Review of Business Research, 2*(4), 50–70.

Centre, U. (1985). Ruins of the Buddhist Vihara at Paharpur. *Whc.unesco.org*. Retrieved from https://whc.unesco.org/en/list/322/.

Chatterji, R., Chaudhury, A. B. R., Basu, P., & Sarkar, G. (2015). *India-Bangladesh connectivity: Possibilities and challenges* (Part 1, pp. 1–88). Observer Research Foundation, Kolkata. Retrieved from https://www.orfonline.org/wp-content/uploads/2015/06/IndiaBangladesh.pdf.

Chuntao, L., & Karim, N. (2018). *Feature: Bangladesh, China unite to unearth ancient Buddhist heritage site*. Xinhua. http://www.xinhuanet.com/english/2018-01/08/c_136880174.htm.

Dasgupta, D. (2011). *Medical tourism*. Pearson.

Financial Express. (2018). *Agartala to Kolkata in just 10 hours by train? This new Indian Railways link via Bangladesh to make it possible*. https://www.financialexpress.com/infrastructure/railways/agartala-to-kolkata-in-just-10-hours-by-train-this-new-indian-railways-link-via-bangladesh-to-make-it-possible/1165574/.

Geary, D. (2018). India's Buddhist circuit(s): A growing investment market for a "rising" Asia. *International Journal of Religious Tourism and Pilgrimage, 6*, 47–57. https://doi.org/10.21427/D7PT46.

Indian Institute of Tourism and Travel Management. (2018). *Study on visit of nationals of Bangladesh to India*. http://tourism.gov.in/sites/default/files/2020-04/Final%20Report%20on%20Visit%20of%20Nationals%20of%20Bangladesh%20to%20India.pdf.

International Finance Corporation. (2017). *Investing in the Buddhist circuit: Enhancing the spiritual, environmental, social, and economic value of the places visited by the Buddha in Bihar and Uttar Pradesh, India*. World Bank. https://openknowledge.worldbank.org/handle/10986/26096.

Kumar, N. (2018). Over 50 per cent medical tourists to India are from Bangladesh. *The Sunday Guardian*. https://www.sundayguardianlive.com/news/50-medical-tourists-india-bangladesh.

Murshed, A., & Firozi, M. (2018). *Position paper on Bangladesh Railway*. Presentation, Bangladesh.

Numan, A. (2018). *Record-setting traffic on Bangabandhu Bridge*. Dhaka Tribune. https://www.dhakatribune.com/bangladesh/nation/2018/08/21/record-setting-traffic-on-bangabandhu-bridge.

Rahman, M. (2000). *Bangladesh-India bilateral trade: An investigation into trade in services*. South Asia Network of Economic Research Institutes (SANEI). https://www.eldis.org/document/A28871.

Saifullah, K. (2009). *Causes of floods in Bangladesh*. Feature written in http://en.wordpress.com/tag/floods-inbangladesh/.

United News of Bangladesh. (2017). *2nd Bhairab Railway Bridge to open Nov 9 Hasina, Modi to open it through video conferencing*. http://old.unb.com.bd/bangladesh-news/2nd-Bhairab-Railway-Bridge-to-open-Nov-9percentC2percentA0/55061.

Slums on Railway Land in Guwahati City, Assam: A Sociological Review

Trinity Borgohain

1 Introduction

The twenty-first-century world has witnessed a massive urban growth all around. Cities are now the abode of a significant section of the global population—almost half of the world's population is urban, and three-quarters of the world's population growth now occurs in the urban areas of the developing world. Lemanski and Marx (2015) stated that the world today is not only urban but also south centred, for Asia and Africa are most dominant. This trend of urbanization is also reflected in Indian cities. While 52.8% of the world population lived in cities in 2010, only 31.16% of India's total population lived in urban areas in 2011 (Census of India, 2011). India has witnessed the second largest urban growth in the world, with an increase of 10.57%—next to China (19.54%). Sociologists have emphasized that the factors most contributing to urban growth are rural to urban migration and natural population increase. Cities and towns provide a variety of choices and economic opportunities, which attract millions of migrants from rural hinterlands. Push and pull factors operate together to increase rural to urban migration, which in turn increases pressure on the existing urban infrastructure. With an increasing pace in world urbanization, the growth of slums is seen practically everywhere—and most especially in cities of the developing world.

South Asian megacities include a high percentage of the world's slums. Misra (2013) has stated that according to some estimates, at least 40% of South Asian city dwellers live in slums. Significantly, India has witnessed a rapid growth in the slum population's proportion of total urban populations in the last few decades. In 2001, India's slum populations constituted 15% of the total urban population. This has increased to 17.4% in 2011 (e.g. an increase in roughly 1.37 crore households; Census of India, 2011). About 25.09 million citizens live in slums within the largest, million-plus cities of India—that is, about 38.32% of the total slum population in

T. Borgohain (✉)
Government Model College, Kaziranga, Golaghat, Assam, India

© The Author(s), under exclusive license to Springer Nature Switzerland AG 2021
S. Mitra et al. (eds.), *Railway Transportation in South Asia*, Contemporary South Asian Studies, https://doi.org/10.1007/978-3-030-76878-2_14

Table 1 Slum populations in million-plus cities in India, 2011

Sl. no.	Name of major slum reporting million plus city	Slum population	Percentage of city slum population to all India slum population (in million)	Percentage of city slum population to all million-plus cities slum population
1	Chennai	1,342,337	2.05	5.35
2	Delhi	1,617,239	2.47	6.44
3	Greater Mumbai	5,206,473	7.95	20.74
4	Kolkata	1,409,721	2.15	5.62
5	Hyderabad	2,287,014	3.49	9.11
6	Nagpur	859,487	1.31	3.42
7	All India slum population	65,494,604	–*	–*
8	Slum population of million plus cities	25,099,576	38.32	–*

Note *Not available
Source Census of India (2011)

the country as reported at the 2011 Census. In absolute numbers, Greater Mumbai has the highest slum population with around 5.2 million, followed by Hyderabad (2.2 million), Delhi (1.6 million) and Kolkata (1.4 million). Thus, six municipal corporations—namely, Greater Mumbai, Kolkata, Delhi, Nagpur, Hyderabad and Chennai—together account for around 20% of the total slum populations in the country and around 50.7% of the total slum population in million-plus cities (see Table 1).

The rapid growth of slums in India's million-plus cities was partially caused by the expansion of trade and commerce in cities. Urban economic sectors create a wide range of economic opportunities for migrants, especially when compared to rural areas. Cities have pulled cheap labour forces, especially from rural hinterlands. However, rural migrants are mainly from low-income groups—often illiterate, unskilled, landless people, thus socially and culturally vulnerable. They come to urban areas in search of adequate subsistence. After arriving in the city, and because of expensive housing markets, they mostly find themselves in cheap accommodations, i.e. squatter settlements to be identified as slums. Slum settlements develop mainly in vacant plots, public and private lands, along riverbanks, low lands, water bodies or Railway lines. According to the 2011 Census of India, about one in six Indian city residents lives in an urban slum—with unsanitary conditions that are unfit for human habitation.

2 Statement of the Problem

In the state of Assam in India, the urban slum phenomenon has become an emerging and serious problem in the past few decades. The major reason for the growth of slums in Assam is found to be the influx of migrants from rural hinterlands. The large number of slum settlements in Assam was witnessed in its capital city 'Guwahati'. Slums in Guwahati are overcrowded spaces emerging in different types of land—public, private and Railway lands. We aim to study the growth of slums in Guwahati city in general and on Indian Railway lands in particular. Slum dwellers live with multiple deprivations at diverse levels. They lack access to basic services—and of course lack of education, skills, resources and tenure security.

A large number of slums have grown in Indian Railway lands across the country. The Railway administration has termed slum dwellers as '*encroachers*' for occupying these vacant lands. The same situation arises in the slums of Guwahati city settled on Railway lands. Citizens neither receive any housing benefits from the government of India nor by the state government. The Railway slums' residents even face evictions without any prior notice or information from Railway authorities. They are always at the mercy of bulldozers and basic fight for survival. The Railway administration refuses to provide any sort of accommodation after eviction.

Guwahati city has been included in the list of 20 selected cities under the Smart City Mission led by the Indian government. In collaboration with the Indian government's scheme, Assam's state government aims to transform its capital city—the gateway to Northeast India—into an environment friendly, clean and a smart 'world-class city'. Of course, one may wonder whether vulnerable communities such as slum dwellers have any place in this lofty plan. The chapter thus seeks to understand not only the growth of slum settlements in Guwahati but also the diverse areas of vulnerability involved.

3 Context of the Study Area

'Guwahati' (formerly 'Pragjyotishpur'), meaning the 'areca nut marketplace' in Assamese, is a growing city in Northeast India (Gait, 1981). It is located in the south-east of the Kamrup district in Assam. It is surrounded by Assam's Nalbari district in the north, the Darang and Morigaon districts in the east, and the Indian state of Meghalaya in the south. To the west, it neighbours with the Assam districts of Goalpara and Barpeta (Fig. 1).

Though Guwahati city had grown and developed through different historical stages, it met greatest success with the shifting of Assam's capital from Shillong (capital of the Meghalaya state) to Dispur (in Guwahati) in the year 1972. Since then, Guwahati grew in terms of both political and administrative importance. It also emerged as the hub of several industrial, trade and commercial activities—thus generating employment. The city is undergoing rapid change in terms of population

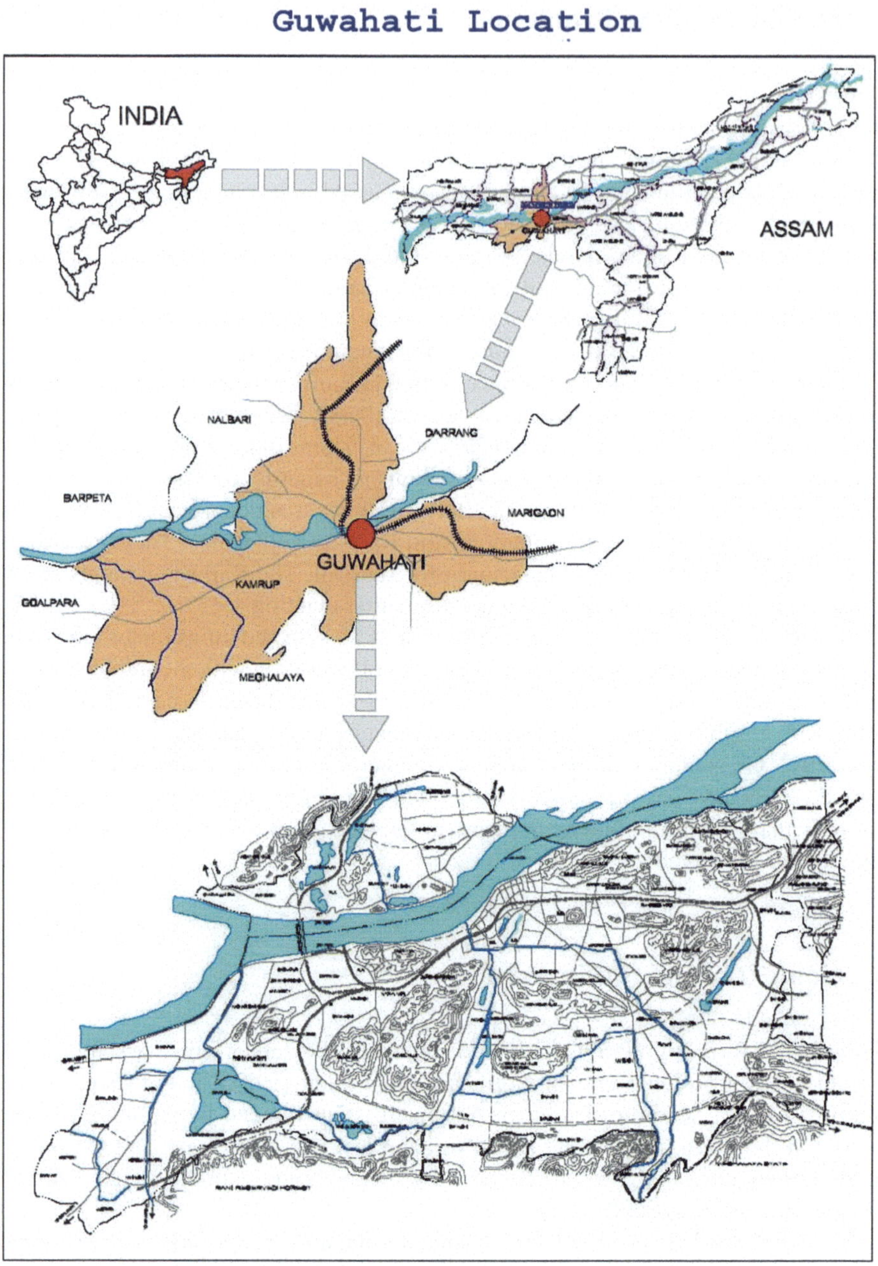

Fig. 1 Locational map of Assam and Guwahati City (*Source* Guwahati Development Plan, July 2006, Guwahati Municipal Corporation)

growth due to urbanization and growth of slum settlements. These have increased both numerically and spatially. Slum growth has become an integral part of Guwahati. As per the Census of India (2011), the city's population is 963,429. Guwahati is governed by the Guwahati Municipal Corporation, which was formed in 1974. Since then, Guwahati became a Municipal Corporation.

4 Study's Objectives and Methodology

The focus of this chapter regards the growth of slums in Guwahati city in general and that of slums in the Railway lands as owned by the Central Government of India in particular. The chapter attempts to assess the socio-economic status of slum dwellers in the present set-up of Guwahati city. The chapter also seeks to understand the livelihood patterns and challenges faced by slum dwellers.

The database used in the present study includes both secondary and primary data. Secondary data was collected from various reports as published by public institutions and government agencies such as the Guwahati Municipal Corporation, the Census of India and the Guwahati Metropolitan Development Authority. Other sources such as books, journals, periodicals, monographs, published articles, magazines and newspapers were also used. As regards the number of slum areas, populations and locations within Guwahati, the data used in the research was based on surveys carried out by the GMC (Guwahati Municipal Corporation) on slum growth in Guwahati during 2006, 2009 and 2011.

Primary data was collected through questionnaires, and field observations were carried out in the year 2017. The research chapter is based on a case study employed in a research site known as the Kumarpara slum (a newly emerging slum) settled on Railway land in Guwahati. With the objective to collect data from the study area, a field survey was administered to 75 households out of which 35 sample households were selected on a random basis. Quantitative primary data emphasizes the socio-demographic characteristics of the Kumarpara slum's inhabitants. On the other hand, qualitative data focused on the information regarding the livelihood issues of slum dwellers—mostly related to apprehensions of land insecurity or evictions. Table 2 gives an account of the Kumarpara slum and its sample size for the study.

Table 2 Kumarpara slum and sample selection

Duration of settlement	Total no. of households	Sample households	Total population	Sample population	Community composition	Land ownership
30 years approx.	75	35	356	173	SC—Bihari Hindu/Bengali Muslims	Private/Railway vacant land

Source Field Survey, 2017

5 Growth of Slums in Guwahati (2006–2012)

Guwahati city has witnessed the fastest growth of slums in the Northeast region of India in the past few decades. Several surveys were carried out by the public institutions such as the GMC so as to estimate growth trends in slum settlements in Guwahati during the last few decades. The GMC conducted three slum surveys in Guwahati in the years 2006, 2009 and 2012. Over the years, the GMC has officially identified some of the informal settlements as 'slums' so as to extend governmental assistance and welfare policies to these areas (Desai et al., 2014). However, these studies have failed to affect much change on the ground.

The 2006 slum survey identified 26 informal settlements as slums, with a total population of 1.6 lakh people (Table 3). Later, in 2009, the GMC undertook relevant observations through the survey on the rapid growth of slums in Guwahati. It achieved this under a scheme by the Jawaharlal Nehru National Urban Renewal Mission (JNNURM). In this survey, GMC notified 90 slum pockets under four different categories (Table 4).

The first category of slums in Guwahati is identified by the GMC as per 2001 Census definition: a compact area of at least 300 inhabitants or about 60–70 households of poorly-built, congested tenements in unhygienic environments, usually with inadequate infrastructure and lacking proper sanitary and drinking water facilities.

Table 3 Growth in the number of slum settlements, households and population in Guwahati, 2006–2012

Year	Total population of Guwahati	Total population of slums	% of slum population to total population	No. of slum settlements	No. of households
2006	809,895	1,60,000	19.7	26	–*
2009	809,895	1,67,796	20.71	90	27,966
2012	963,429	1,39,000	14.42	217	26,069

Note *Not available

Source Guwahati Municipal Corporation (2006, 2009 and 2012) Guwahati City Slum Policy, Phase 1, 2009 (GMC)

Table 4 Categories of slums as per the 2009 slum survey

Categories	Number of slum pockets	No. of approximate households	Approximate population
1	52	17,056	102,336
2	24	5380	32,280
3	6	2850	17,100
4	8	2680	16,080
Total	90	27,966	167,796

Source Guwahati City Slum Policy, Phase 12,009, GMC

52 slum pockets fall under this category. Said category thus involves the highest number of slum pockets within a total of 90 notified slums.

The second category of slums is distinguished by not congested but scattered housing, mostly settled on hillsides—where people are fighting for survival due to lack of infrastructure and live without any proper road or circulation networks. Here, inhabitants often experience natural disasters such as landslides. This category includes over 24 slums (with 5,380 households), all 'scattered hillside housing' where inhabitants are mostly poor and marginalized migrants. These slum settlements have been growing in Guwahati since the 1970s when Meghalaya was formed as a separate state and Assam's capital was shifted from Shillong to Dispur. These rural migrants came to the city in search of a better livelihood and progressively settled in the hills (as they were displaced from the plains).

The third category of slums of Guwahati is settled on the Railway lands, including the small Railway quarters. The size of the quarters is less than 40 sq.mt. The growth of these slum settlements is related to the growing families in the Railway quarters; residents of these quarters need space for new couples following the construction of huts as an extension of their congested rooms with temporary materials (until, eventually, the entire area looks like *zupadpatti*).[1] These slums are further analysed in three sub-categories discussed in the next section.

The fourth category of slums includes rented slums. Here, over 70% of slums are rented because of non-availability of '*miyadi land*' in Guwahati city, and because slum dwellers cannot afford to stay in the expensive rented houses in non-slum areas. The inhabitants of these slums are mostly landless people who informally inhabit the state government land, often called '*dakhal land*'. Sometimes, private landowners also construct huts rented to migrant labourers. These huts are extremely congested and located in very dirty locations. This is why they are rented at a very low price.

The 2009 slum survey by the GMC was updated in 2012. 217 slum pockets were identified, with a population of 1.39 lakh. Out of 217 slums, 99 are notified (87,457 persons & 15,701 HH); the others (51,771 persons & 10,368 HH) are non-notified slums. The surveys in 2012 were undertaken by several NGOs such as Scorpions, the Rural Women Upliftment Association of Assam (RUWAA), the Assam Centre for Rural Development (ACRD), the Strategic Alliance Management Services (SAMS), Eight Brothers, the Society for Research, Development and Communication (SRDC) and Usgravika. The drastic difference in the number of slums settled in Guwahati was due to changes in the definition of slums. When slum pockets were first identified in 2009, any pocket with 25–30 households and lacking basic amenities was considered a slum. By contrast, for the 2012 survey, any pocket with 10–15 households without basic amenities was considered a slum. Hence, we find a curious situation wherein, whereas the number of officially recognized slums increased, the population data actually decreased (Desai et al., 2014).

[1] Zupadpatti is a ghetto, a part of a town in which many poor people of a particular race, religion or nationality live in a congested manner without basic amenities along a railway track.

6 Slums on Railway Land: A Profile of Guwahati City

As per 2001 Census, most slums in Guwahati have settled on Railway sides because of a huge landfall under the Indian Railways of NFR (Desai, 2013). GMC's 2009 slum survey reported that 23% of slums are situated on Railway land (GMC, 2009). This comprises three types of slums. Type 1 includes slums settled on vacant land adjacent to the 25 Railway tracks or on narrow strips of land (e.g. from Lakhtokia gate number one to gate number five, respectively). Here, one finds absolutely no basic services provided for the hutments. Indeed, evictions by Railway authorities are a great threat to slum inhabitants. These evictions take place every year along these Railway tracks—shelters get almost entirely wiped out. After the eviction, slum inhabitants recollect their commodities and immediately rebuild their shelters. Where land parcels along the tracks are somewhat larger (e.g., in Bhootnath and Santipur), evictions used to take place earlier but have ceased for some years now. During one of the evictions, citizens occupied thc tracks in protest. Another time, they approached their respective Member of the Legislative Assembly (MLA), who called the General Manager (G.M.) of the Railways and pressured him to stop the eviction (Desai et al., 2014). Evictions have now ceased because residents have left a buffer space between the tracks and their houses.

Type 2 includes those slums which are the hutments once created either in Railway colonies or on the colonies' vacant lands (e.g. in the Railway colonies in Bamunimaidan and Gotanagar). Demolition is a common issue among slum dwellers in these settlements.

Type 3, however, comprises slums settled on large parcels of Railway land which are not located near any Railway track. In these slums, evictions have ceased many years ago—although some residents fear that evictions might again take place someday because of informal landownership (e.g. the Shakuntala Colony and Kailash Nagar in the Pandu area; Desai, 2012).

Slum dwellers are poor rural migrants who have occupied vacant lands of the Indian Railways due to an expensive housing formal sector—including formal rental, which is unaffordable. State government agencies have failed to do much about fulfilling the housing needs of low-income groups in Guwahati. The Assam State Housing Board (ASHB), established in 1974, has built a total of 1,824 rental housing units at several locations in Guwahati (Desai et al., 2014). This involves the Economically Weaker Section (EWS), the Low-Income Groups (LIG) and the Middle-Income Group (MIG) units for Class III and IV government employees—including retired employees. However, many of ASHB's EWS and LIG rental units are taken up by better-off families in the city. As a result, low-income groups are forced to turn to informal land housing and informal rental, including across the Railway and forested hills. An important issue as regards informal settlements or slums in Guwahati concerns the sheer insecurity of many Railway settlements.

Demolitions by Railway authorities in the first two types of Railway land-based slums are often brutal. The Railway police assault people and damage their meagre properties. Sometimes, evictions also cause conflict and violence between poor squatters and the Railway administration. In November 2013, local citizens injured four Railway Police Force representatives trying to remove unauthorized structures in the Railway Colony at Central Gotanagar (*The Assam Tribune*, 2013).

7 The Kumarpara Slum: A Case Study

The Kumarpara slum is one of the recently emerging slums of Guwahati city. It formed about 30 years ago and was notified by the GMC in 2009. It is situated along a Railway track crossing the north end of the Kumarpara locality as well as the A.T. Road (Assam Trunk Road) that serves as the northernmost boundary (Fig. 2). The slum consists of 200 households as per the GMC Slum Report, 2009. During the field survey in 2017, only a total of 75 households were found in the Kumarpara slum area. The change in the number of households is due to displacements and evictions faced by inhabitants in previous years. The present research has focused on slums informally settled on vacant Railway land in the Kumarpara area, especially along the Railway track and on those settlements adjacent to the Railway tracks that are controlled by private landowners. Assam has the longest Railway track and maximum number of Railway stations among the north-eastern states (Roy & Mitra, 2016).

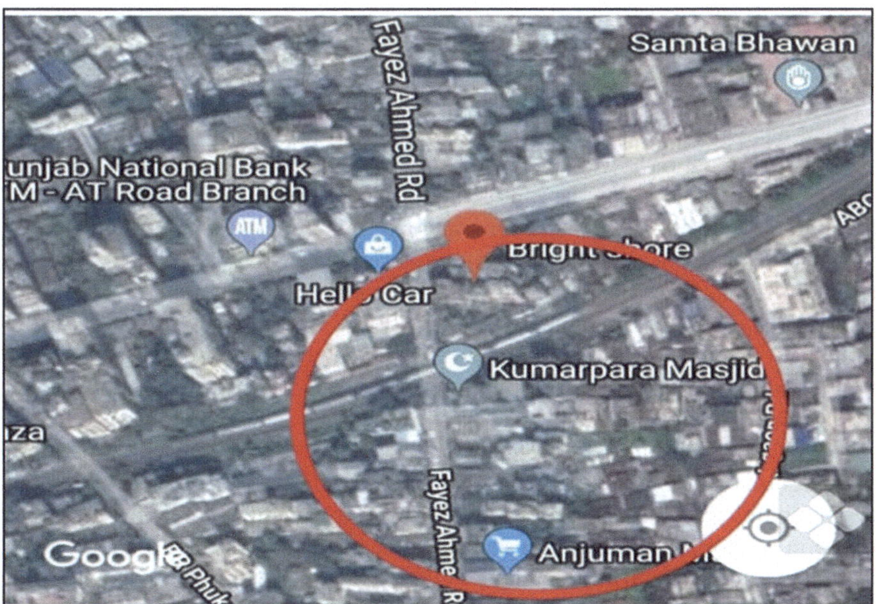

Fig. 2 Google map showing the location of Kumarpara slum (*Source* Google Map, 2017)

The entire Kumarpara slum is inhabited by citizens from two different communities—Bihari Hindus and Bengali-speaking Muslims. Slum dwellers are from Scheduled Caste groups found to constitute politically marginalized communities in Guwahati. These community dwellers come from Scheduled Caste groups in Bihar, India. These groups are economically vulnerable groups. They comprise different caste groups, such as *Yadav*, *Telly*, *Kurmi* and *Kandu*. These caste groups play more significant roles in their native villages than they do in the city. Their caste identification in the city can be understood through their traditional occupations. The *Yadav* caste group was engaged in the traditional occupation of cattle raising and harvesting, whereas the *Telly* caste group was engaged in cultivation. The other caste groups are the *Kurmis,* who were traditionally engaged in cultivation in agricultural fields. The last caste groups, the *Kanu* or *Kandu*, were traditionally engaged in grain parching.

These citizens have considered Guwahati as their destination and hope; today, they are engaged amid all the intricacies concomitant to slum setting. Slum dwellers even lack basic amenities such as adequate sanitation, proper drinking water, electricity, ventilation in huts or indented rooms. Slum inhabitants use the vacant land of the Indian Railway not only for settlement but also for the economic purposes—just to earn extra income while raising their petty business, for example, pan shops, food stalls, etc. This is how livelihood challenges in the urban structure are dealt with.

7.1 Socio-Demographic Analysis

The socio-demographic characteristics of our slum population were grouped into age and sex composition, migration patterns, economic adaptation. The charts are based on field data collected from 35 sample households comprising 173 inhabitants (out of a total 75 number of households with 356 inhabitants; see Table 5).

Our sample comprises 46% of the total households and 48% of the slum's universe population. The selection of the sample households is made while using the purposive sampling method, based on the criterion of religious distribution of slum populations. Thereafter, from each religious community, the sample was collected through a lottery method. It was found that there is a total of 23 Hindu households comprising a total of 88 persons; the remaining 52 include 268 persons belonging to the Muslim community. The slum is mainly inhabited by citizens of these two communities. The study found that among the two different religious communities, Muslims account for 71.43% of sampled households (75.14% of the total sample population).

They all speak variants of the Bengali dialect. These citizens seem to be those most deprived, disadvantaged and culturally marginalized in the city. They are marginalized on multiple grounds and prevented from participating in larger fields of city life such as in economic, political and social structures. The influence of another religious community (Hindu community dwellers) constitutes another undeniable fact regarding the Kumarpara slum. 28.57% households belong to the Hindu community

Table 5 Distribution of sample households and populations from the total no. of households in Kumarpara slum

Sl. no.	Religious community	Total no. of Households	Percentage of households	Total no. of population	Percentage of populations	Sample size of households from the total no. of households	Population of sample households
1	Hindu (Bihar origin)	23	30.67	88	24.38	10	43
2	Muslim (Bengali origin from Assam)	52	69.33	268	75.62	25	130
3	Total	75	100	356	100	35	173

Source Field Data, 2017

(24.86% of the sample population). These Hindu community dwellers were originally from the neighbouring state of Bihar. Our study focused on the analysis of the socio-economic status of these slum populations.

7.2 Age and Gender Composition

In the Kumarpara slum, it has been observed that in every age group the share of male population is higher than the female (Table 6) (Graph 1).

The male population constitutes 60.11% of the total slum population; females constitute 39.89%. In other words, almost 2/3 of slum dwellers are male. This suggests non-female-led households and varied family patterns; households comprising several non-related male members who left their womenfolk back at their place of origin are probably recurrent.

113 persons of the sample population lie within the age category between 18 and 50 years; this accounts for 65.31% of the sampled population. 29.47% of this population, i.e. individuals, are young people yet to reach adulthood (7–17 years old). Only 0.58% of the sample population is senior. The highest number of individuals sampled in the Kumarpara slum are aged 31–40; this comprises 24.28% of the sampled population. The next highest percentage is from the age group of 18–30 years, with a total percentage of 23.12%. The remaining population belongs to the age groups 0–6, 7–17, 41–50, 51–60 and over 61. In these age groups, the shares of slum population are 9.83, 19.65, 17.92, 4.62 and 0.58%, respectively. In each age, males outnumber females.

Table 6 Distribution of slum population by their age-gender composition

Sl. no.	Age group	No. of males	Per cent	No. of females	Per cent	Total population	Per cent of total population
1	0–6	11	6.36	6	3.47	17	9.83
2	7–17	20	11.56	14	8.09	34	19.65
3	18–30	23	13.29	17	9.83	40	23.12
4	31–40	24	13.87	18	10.40	42	24.28
5	41–50	19	10.98	12	6.94	31	17.92
6	51–60	6	3.47	2	1.16	8	4.62
7	61 above	1	0.58	0	0	1	0.58
	All	104	60.11	69	39.89	173	100

Note The percentages of male, female and total populations in each age group are calculated out of the total 173 sampled inhabitants
Source Field Data, 2017

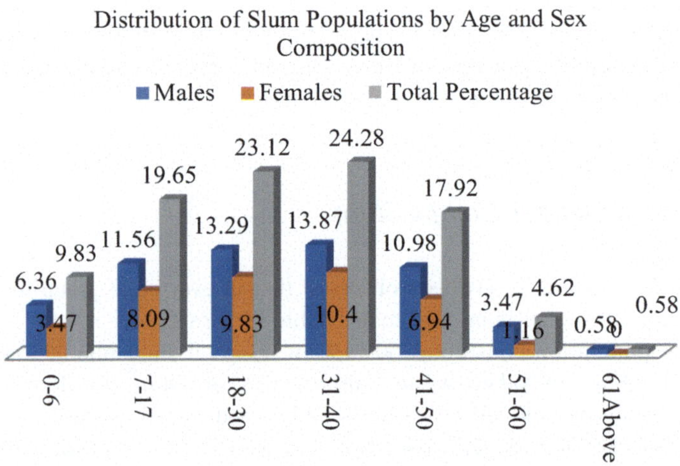

Graph 1 Distribution of slum populations by age and sex composition

7.3 *Migration and Social Origin*

Slum dwellers in Kumarpara migrated to Guwahati city from different districts in Assam as well as from other Indian states, especially Bihar. The study found that the slum population is heterogeneous in terms of its social origin.

The dominant patterns of migration in this slum are found to be inter-state and inter-district migration. Inter-state migration occurs from Bihar to Guwahati city. Citizens from Bihar migrated to the city for a series of economic and social reasons. However, most of these people are de facto landless people and were somehow coerced to migrate towards this city, fundamentally so as to eke out a minimum living. On the other hand, inter-district migration has been experienced by Bengali Muslim populations in the slum, who have migrated from the nearby areas of Guwahati city. 23 slum households out of total 75 migrated from Bihar; this accounts for 30.67% of the sampled population. On the other hand, rural migrants from the western districts of Assam are distributed among 69.33% of households—with a population share of 75.62%. Table 7 shows that in the sample 28.57% of the households (with a population representing 24.85% of the total) are of Bihar origin. 71.42% of households (with a population of 75.14% of the total) are from most of the western districts of Assam.

Slum populations in the sample households migrated from various districts in Assam—mainly from Dhubri (34.29%), Barpeta (22.86%), Goalpara (2.86%), Kokrajhar (2.86%), Bongaigaon (2.86%), Nagaon (2.86%) and Kamrup (2.86%). The slum population originating from the Bihar state accounts for 20% of households and 19.07% of the population; these citizens come from the district of Samastipur. 2.86% of the households have migrated from the Bihar state districts of Darbhanga, Begusarai and Khagaria. A very small percentage of slum households (2.86%) have migrated from Mandakata, a rural village near Guwahati, located in the Kamrup(rural) district of Assam at a distance of 44.8 kms at the north of the city.

7.4 Economic Adaptation

Guwahati city is a place of hope for low-income groups from different regions of India because it provides economic and employment opportunities in different sectors of its economy—especially in the informal sector. Table 8 shows how slum dwellers in Kumarpara are engaged in a series of occupations in the urban informal sector.

Table 8 shows that 35.39% of the total slum population in Kumarpara are engaged in regular, salaried jobs in the informal sector—as shopkeepers (0.88%), tailors (1.77%) and domestic workers (32.74%). The employers of these citizens are mostly private employers located in richer areas of the city.

Daily wage earners and casual labourers in Kumarpara account for a total 29.2% from the total employed sampled population. They are locally called as 'majdoors' and work for about 24 h in day. They hardly even get one to two hours to rest. This type of occupation encompasses a wide range of skilled, unskilled and manual labour. These slum dwellers are mainly masons and bricklayers. For this type of occupation, slum dwellers become informally trained by relatives or friends already settled in the city.

The other type of primary occupations among slum dwellers in Kumarpara was to be found among the self-employed. These slum dwellers comprise 35.39% of the total sample populations. They are mainly vegetable vendors, petty traders, rickshaw pullers, thela pullers and washermen. Slum dwellers who are vegetable vendors by occupation often sell vegetables in markets near the slum. These slum dwellers often pay around €2.26–€5.56 to the police for occupying the vacant prohibited area of the locality for selling vegetables. The study finds that slum dwellers in Kumarpara

Table 7 Place of origin of slum population

State	District	Village	Household (village wise)		Population		District-wise distribution		State-wise distribution	
			No	%	No	%	HH	Population	HH & %	Population & %
Bihar	Samastipur	Garhi Bishanpur	3	8.57	13	7.51	7 (20.0%)	330 (19.07%)	10 (28.57%)	43 (24.85%)
		Sabhaipura	2	5.71	10	4.62				
		Bhirar	1	2.86	6	3.47				
		Jakhar Dharampur	1	2.86	4	2.31				
	Darbhanga	Kusheshwar Asthan	1	2.86	5	2.89	1	5		
	Begusarai	Bindpur	1	2.86	2	2.31	1	2		
	Khagaria	Durgapur	1	2.86	3	1.73	1	3		
Assam	Dhubri	Bangalipara	4	11.43	21	12.14	12 (34.29%)	62 (35.83%)	25 (71.42)	130 (75.14)
		Bilasipara	3	8.57	18	8.67				
		Rowarpara	1	2.86	5	2.89				
		FokiRanijhar	1	2.86	2	2.89				
		Birsanagar Pt III	1	2.86	5	2.89				
		Silghagri	1	2.86	6	3.47				
		Gauripur	1	2.86	5	2.89				
	Barpeta	Kadamtola	2	5.71	11	6.36	8 (22.86%)	41		
		Garemarihabi	2	5.71	10	5.78				
		Alupati	1	2.86	5	2.89				

(continued)

Table 7 (continued)

State	District	Village	Household (village wise)		Population		District-wise distribution		State-wise distribution	
			No	%	No	%	HH	Population	HH & %	Population & %
		Chanpur	1	2.86	5	2.89				
		Palhaji	1	2.86	6	3.47				
		Bayasha	1	2.86	4	2.31				
	Goalpara	Baniapara	1	2.86	6	3.47	1	6		
	Kokrajhar	Kursakati	1	2.86	6	3.47	1	6		
	Bongaigaon	Baghmara	1	2.86	4	2.31	1	4		
	Nagaon	Haoraghat	1	2.86	5	2.89	1	5		
	Kamrup	Mandakata	1	2.86	6	3.47	1	6		

Source Field Data, 2017

Table 8 Occupational (primary) structure of slum populations

Employment status	Occupational categories	Male	Female	Total population	Total percentage	Total percentage of employed persons/dependents
Regular salaried jobs	Shopkeeper	1	0	1	0.88*	35.39
	Tailoring	2	0	2	1.77	
	Domestic worker	0	37	37	32.74	
Casual/daily wage earning	Mason	32	0	32	28.32	29.20
	Cook	1	0	1	0.88	
Self- employed	Vegetable vendor	5	0	5	4.42	35.39
	Petty trader	2	1	3	2.65	
	Rickshaw puller	8	0	8	7.08	
	Thela puller	22	0	22	19.47	
	Washer man	2	0	2	1.77	
Total no. of employed persons		75	38	113	100	100
Dependents/non-working population	Home-based unpaid work	0	19	19	31.67**	31.67
	Unemployed but still seeking work	10	0	10	16.67	16.67
	Attending Education	7	3	10	16.67	16.67
	Other Non-working population (children of 0-6 years and others)	12	9	21	35.00	35.00
Total no. of dependents/non-working persons		29	31	60	100	100
Grand total population		104	69	173	100	100

Note *Calculated from the total sample employed population, i.e. the total of 113 employed persons. **Calculated from the total dependent sample population, i.e. 60 persons
Source Field Data, 2017

slum are largely dependent on the informal sector. They are mostly adapted to menial jobs so as to sustain their family.

7.5 *Livelihood Challenges Faced by Slum Dwellers*

The Kumarpara slum has been in existence for 30 years, yet slum dwellers still struggle to meet livelihood challenges. The settlement is still devoid of any physical infrastructure or social amenity. Slum dwellers are still denied basic services such as water supply, electricity and proper sanitation. In the absence of drinking water facilities, sometimes they raise water pipes by themselves near the Railway tracks. The accessibility to basic services in slums is restricted by the categorization into notified and non-notified slums. Notified slums are those slums recognized as per census data, while non-notified slums are not. Slums located on Railway land are not recognized, and inhabitants are denied basic amenities.

Slum dwellers use the Railway tracks for many reasons, especially for sanitation purposes and economic reasons as some residents raise pan shops, food stalls, etc. These are struggling, low-income groups. Indeed, they were tagged as 'illegal residents'. They are termed as 'encroachers' by the Railway authorities. Their major challenge concerns displacement due to the 'eviction' policy of the Indian Railways. Some slum residents have often faced eviction during the last 30 years—without any prior notice or any provision for alternate accommodation. Most slum families were displaced from Railway lands, and many have lost their lives. The research found that two years ago some slum dwellers have informally occupied the vacant land beside the Railway track. They were forcefully evicted by Railway authorities, and most of these landless and homeless slum dwellers moved to other places of Guwahati or left the city. The other slum dwellers, those residing in huts and other tenements under the private landowners in the Kumarpara locality, have not yet faced any such evictions.

The study found that after every 15–20 days the huts and other assets of slum dwellers along the Railway track get demolished by the Railway authorities. This leads to many economic hazards and livelihood challenges among slum dwellers. After demolition, they collect commodities and resettle. On the other hand, slums are in general known as the city's criminal zones. These are plagued by problems such as gangs, drug peddling, alcoholic abuse and social conflicts. The lack of life opportunities can trap slum inhabitants in a vicious cycle of poverty and violence. Another major problem the slum dwellers face in the Kumarpara settlement is the label of 'criminal zone' as used by both the police and mainstream society. Slum dwellers are also found by the police to indulge in illegal activities, for which most of them become victims of harassment. The incidents of pick-pocketing or stealing are common among slum residents. They also face an identity crisis in the city. The overall picture of the slum and its people is dreary. These people and their lives are the epitomes of the heightened and increasing inequality of our modern times

8 Conclusion

The unprecedented growth in population and related unplanned urban growth in Guwahati have led to the emergence of slums (marginalized spaces) in the last few decades. Slum dwellers, especially those settled on the Railway vacant lands in the city, live with multiple deprivations in association with their livelihood-related challenges such as landlessness, lack of amenities and identity crisis. Government benefits such as housing benefits and other basic amenities are not offered to slums. Slum dwellers even deal with the risk of displacement and harsh role of the Railway administration due to the policy of eviction. The issues and the troubles affecting slum inhabitants are persistently ignored by subsequent governments and civil societies. Poverty perpetuates. Successive governments have treated slums as mere vote banks and paid no heed to their interests and development. The smart city plan has not even proclaimed any commitment regarding the sustainable development of slums.

References

Census of India. (2011). *Primary census abstract for slum, a report by Office of the Registrar General and Census Commissioner of India, New Delhi.*

Desai, R. (2012, October). *Mapping of community structures in Guwahati* (Unpublished Research Report, Center for Urban Equity). CEPT University.

Desai, R., & Mahadevia, D. (2013, August). *Land and housing development processes as determinants of rental housing for the urban poor: The case of Guwahati City* (CUE Working Paper 19). Centre for Urban Equity.

Desai, R., Mahadevia, D., & Mishra, A. (2014, August). *City profile: Guwahati* (CUE Working Paper 24). Centre for Urban Equity.

Gait, E. (1981). *A history of Assam.* LBS Publication.

GMC. (2006). *City Development Plan.* Guwahati Municipal Corporation.

GMC. (2009). *Guwahati city slum policy, Phase-1: Identification of slum/urban poor.* A report prepared with technical support from AAPIL Planning Consultancy, Surat and Associated Builder, Guwahati, Guwahati Municipal Corporation.

GMDA. (2009). *Master Plan for Guwahati Metropolitan Area-2025, Part 1: A Report by Guwahati Metropolitan Development Authority (GMDA), Govt. of Assam.*

Lemanski, C., & Marx, C. (2015). *The city in urban poverty.* Palgrave Macmillan.

Misra, R. P. (2013). *Urbanisation in South Asia.* Cambridge University Press India, Pvt. Ltd.

Roy, S., & Mitra, S. (2016). Infrastructural status of railway transport system in North East India: A geographical analysis. *Asian Journal of Spatial Science, 4*(1), 89–100.

The Assam Tribune. (2013, November 28). 4 Rly employees hurt in mob attack. *The Assam Tribune.*

Railway Transport System: Future Planning and Policy Perspec-tive

Railway Modernisation in India: A South Asian Case Study

Chitresh Shrivastva

1 Introduction

Railways constitute a crucial means of transportation—now perhaps more than ever. As mentioned above, Indian railways suffer from severely misguided investment policies. Revenue generation through core areas, which include freight and passenger train operations, and non-core areas including station modernisation, cater through the PPP model (announced during the 2018 budget) provide a much-needed impetus for FDI.

2 Context of the Study Area

Indian railways are vital to the country's economy. Railway development was much emphasised from the nation's second five-year plan on, as the government focused on strengthening the industrial and infrastructural base of the country and considered railways to be crucial in this endeavour. This led to the establishment of institutions with socialised functions for the systematic execution of projects in the field of track building, signalling and telecommunication, rolling stock and locomotives.

The first institution to be established was the Research, Designs and Standards Organisation (RDSO), created in 1957 with its office at Lucknow (Bhandari, 2006). 1957 was also an important year for Indian railways, as India joined the Global conglomeration of railway networks—International Union of Railways, an organisation of which the country is still a member (UIC, 2017).

Critical infrastructure projects linked to modernisation witnessed an improvement in locomotive hauling power as well as greater train lengths. After the operational crisis of 1980, the sector experienced fast-paced progress through the guidance of

C. Shrivastva (✉)
Department of Political Science, Jain (Deemed to be University), Bengaluru, India

© The Author(s), under exclusive license to Springer Nature Switzerland AG 2021

S. Mitra et al. (eds.), *Railway Transportation in South Asia*, Contemporary South Asian Studies, https://doi.org/10.1007/978-3-030-76878-2_15

M. S. Gujral, the railway board chairman (Kumar & Mehrotra, 2009). The decades between 1960 and 1980 were marked by an investment in high-speed railways, with the introduction of high-speed, long-distance and intercity travel. Objections were raised from various quarters concerning the ambitious Rajdhani express conceived in early 1960 to connect metropolitan cities to the capital (Aklekar, 2019).

During this time operational speeds were much slower than today—115 km/h (71 mph miles per hour) at most. Superfast trains were of course unheard of. From a technical perspective, one of the major concerns expressed concerned the immense speed differential between the Rajdhani and conventional trains—which could result in the underutilisation of several railway lines. Further doubts concerned the substantial infrastructural investments involved, which would be studied by a board of directors.

This committee concluded after two years of discussion that the signalling systems and bridges already in use at the time were capable of handling higher speeds, while improvement in braking (through the introduction of air braking) would further improve the efficiency and safety of high-speed operations. India's first high-speed project—the Rajdhani Express—was finally completed and dedicated to the nation on 3 March 1969, paving the path for future high-speed railway projects such as the Shatabdi Express in 1988—much before India could even think of bullet trains or the Train 18. In this chapter, we will attempt to correlate developments, both domestic and global, which continue to shape Indian railways—as compared to railway networks in other parts of South Asia.

3 Literature Review

3.1 Foreign Direct Investment in Indian Railways

Conceptions regarding FDI have changed according to changes in the broader Indian economic scenario. The idea of FDI is of course not new. Before independence, railways were developed in India solely by the British—through the import of both locomotives and rolling stock from Britain. Since independence, which was followed by partition, the development of railways was riddled with challenges—no less the division of railways between two nations. Post-partition India also needed an alternative means to attain self-sufficiency. The first step to self-sufficiency was the establishment of the Chittaranjan Locomotive Works at Chittaranjan, West Bengal, in 1950. With paradigm shifts in railway technology, a great number of countries have shown interest in partnering with India to help develop its physical infrastructure—while also inducting the latest technology in the field of rolling stock and locomotives.

Bhandari (2006) in his book titled "Indian Railways' Glorious 150 Years" has discussed India's first foreign assistance in 1951, during the establishment of the Integral Coach Factory in Chennai (then Madras). The idea of setting up a coach factory in India took birth during B. Venkatraman's visit to the European Railway

Congress, where he visited the Schlieren Company so as to gather an idea of coach-building techniques. After being approved by the railway board, arrangements were made for the training of apprentices in coach manufacturing. After the submission of a detailed project report in 1951 regarding the establishment of a factory capable of building 300 unfurnished coaches annually, the production unit was inaugurated in 1955—and a technical training school was established in Perambur in 1954, with a capacity to train around 75 personnel annually.

Misra (2009) in his book titled "Indian Railways Turnaround: A Study in Management" discusses India's collaboration with the American Locomotive Company (ALCO) in the establishment of the Diesel Locomotive Works at Varanasi in 1961. The succeeding decades (between 1961 and 1991) were marked by domestic developments, with the commencement of high-speed rail services in the domestic scenario and the introduction of air-braking systems in 1988—which also marked the commencement of high-speed intercity travel. A greater reformation was endeavoured after the 1991 economic liberalisation—1991 was a year of transition in all sectors of the Indian economy, and railways were no exception.

Holt (2003) in his report "Restructuring the Railways" discussed the fact that railways seem free from any competition. This is due to full government participation in all railway operations—the Ministry of Railways is involved everywhere and there are no defined boundaries between infrastructure and operations (unlike countries such as Sweden). Railways were long isolated from private players (especially when compared to aviation). These private players eventually promoted efficiency and improvement of both core sectors—including operations as well as non-core sectors such as passenger facilities and areas such as catering and maintenance, all of which were undertaken by private agents.

In the initial phases of economic reforms focus areas involved the strengthening of both rolling stock and locomotives so as to keep up with the growing demand for traffic and the improvement of both speed and hauling capacity. An agreement was achieved between Indian Railways and ALSTOM LHB Gmbh for the production of modern-age ALSTOM LHB coaches—which are currently emerging as a substitute to conventional coaches manufactured by the Integral Coach Factory and other production units (Rail Coach Factory at Kapurthala and Modern Coach Factory at Rae Bareli). The transition process is unfortunately slow in comparison with the growing demand. This is in part due to high production costs, as each piece costs approximately INR76 lakhs (with a maximum value of INR1.8 crores or 214,334.42 EUR)—in contrast to conventional coaches, which cost INR64 Lakhs (or 76,207.80 EUR). A high level of optimism was built through the previous budget allocations in the railway sector, amounting to INR1.60 lakh crore (20.4 billion EUR) in the year 2019.

The global railway system is highly dynamic. India gradually progresses towards attaining greater technology milestones so as to keep pace with global railway development. This is exactly where the expertise of global powers is put into good use. India has collaborated for the manufacture of electric locomotives with Alstom, Siemens, GE and Bombardier so as to meet its growing demand for freight segment—which is set to increase with the commencement of Dedicated Freight Corridor. At the same

Table 1 Foreign Direct Investment inflows from various countries

Foreign collaborator	Country	Indian company	FDI inflow (in million EUR)
ALSTOM Transport Holdings B.V	Netherland	ALSTOM Transport India Ltd	75.39
Bombardier Transportation Holdings	Singapore	Bombardier Transportation India Pvt. Ltd	34.95
Ansaldo STS Australia Pty Ltd	Australia	Ansaldo STS Transportation Systems in India	19.04
GE Capital International	Mauritius	Titagarh Wagons Ltd	13.03
Inversiones EN Concessions	Spain	CAF India Pvt. Ltd	10.23

Source Government of India (2017)

time, railways are pushing for complete electrification by 2024 so as to help promote greener and cleaner forms of transportation.

Foreign Direct Investment between 2016 and 2017 has included investments worth USD1.2 billion (1.06 million EUR). Bombardier has emerged over the last 20 years as one of the major investors in rolling stock, specifically focused on metro railways. The FDI was summarised in Table 1.

3.2 *The Role of Financial Institutions in Railway Modernisation*

Railway modernisation is not solely characterised by an interest from major players in the manufacturing sector. It must be understood that in a developing region such as South Asia, which is also part of the global south, and perhaps in India itself more than elsewhere, bureaucratic challenges hamper the government's use of internal sources for railway development. The railways have diversified both across long-distance freights and passenger services, suburban railways and metro. Financial institutions have supplemented the government's efforts to expand the railway network. Over the years, and taking into account the increment in traffic since 2000, major routes have become highly saturated (Lok Sabha, 1999).

4 the Study's Objectives and Methodology

The study has been conducted with the following objectives:

- To understand the role of financial institutions in promoting the modernisation of Indian railways;
- To understand the role of European and Southeast Asian countries in the promotion of railway modernisation projects;
- To understand the role of Memorandums and Special Purpose Vehicles in promoting area-specific developments in Indian railways.

The nature of the study is analytical. We explore, through the use of both primary and secondary sources, various policies in the railway sector which help strengthen the modernisation of railway infrastructure in India. Primary sources include Ministry reports, white papers released by the parliament and reports by expert committees. Secondary sources of data include books, studies by financial institutions, industrial reports, newspaper reports, MoU briefs and working papers intended to help us understand the railway modernisation programme in a globalised context.

5 Aspects of Modernisation in Indian Railways

5.1 World Bank

Indian railways have witnessed over the past decade a magnanimous increment in traffic in both passenger and freight segments. Unfortunately, the rapid expansion of railway lines and the doubling and even quadrupling projects in some saturated sections of the Indian railway network have been slow to develop. As India prepares itself for the dedicated freight corridor, the Indian freight scenario in the railway sector does not look very good. We witness a rapid decline in freight transportation through railways, which is being substituted by road transportation and now amounts to a mere 39% of the total freight market. This is largely due to the slow and unpredictable nature of freight movement by rail.

In order to counter the rapidly declining freight market by railway, India embarked on an ambitious endeavour to create dedicated freight corridors—the idea being to help separate passenger traffic from freight traffic—which would help increase freight train speeds from 75 km/h (47 mph) to 100 km/h (62 mph) and allow larger loads being hauled so as to augment train carrying capacity.

Though the project was conceived in 2004, it was shelved over issues regarding the cost-effectiveness and practical prospects of the project. Australia, by contrast, completed its Dedicated Freight Corridor from Port Augusta to Adelaide Springs in the same period—and drastically reduced freight shipment times (the most important freight commodity being iron ore). The foundation stone for the Indian project—with a total length of 1,176 km (730 miles)—was finally laid in 2009, with financial support from the World Bank amounting to US$2.725 million (1.96 million EUR). The Dedicated Freight Corridor will run across various sectors of Indian railways (Table 2).

Table 2 Proposed Dedicated Freight Corridor

Freight Corridor	From	To	Length (in mikm)
North–South	Delhi	Chennai	23431456
East–West	Kolkata	Mumbai	23301448
Eastern	Ludhiana	Dankuni	18561153
Western	Dadri	Jawaharlal Nehru Port	1504934
East Coast	Kharagpur	Vijayawada	1100683
Southern	Chennai	Goa	899559

Source World Bank (2014)

The creation of Dedicated Freight Corridors will also make the railways an ecologically viable mode of freight transportation, with an estimated drop in CO_2 emissions by 67 million tons by 2041–2042.

India's neighbour China as well as other countries across America, Africa and the European Union are already operating Dedicated Freight Corridors—providing India with an opportunity to draw on this expertise. Collaborating with these countries could help improve operating efficiency, but also make train operations ecologically viable and yield higher returns on investments. Freight remains the major source of revenue generation for railways, as it contributes with nearly 75% of total numbers. Currently, the World Bank is providing loans for three projects under the Eastern Dedicated Freight Corridor, amounting to US$2.72 billion (2.72 billion EUR). These are currently at different stages of implementation. The World Bank is also cooperating with the Dedicated Freight Corridor Corporation of India Ltd (DFCCIL) in the domains of Research and Development, long term commercial and marketing plans, approaches to non-discriminatory access, on-track safety, pilot projects on energy optimisation, and freight logistics—to mention but a few.

Pangotra and Shukla (2012) in their study titled "Infrastructure for Low-Carbon Transport in India: A Case Study of the Delhi—Mumbai Dedicated Freight Corridor" have discussed the development of the Delhi-Mumbai Industrial Corridor along the Western Dedicated Freight Corridor, which is handled by authorities from the railway sector in collaboration with the Ministry of Commerce. The project was first conceived in 2008, as part of the Delhi-Mumbai Industrial Corridor Development Corporation under the Public–Private Partnership Model—with 49% equity held by the Government of India and 51% by the World Bank.

The operationalisation of Dedicated Freight Corridors will benefit critical heavy-manufacturing industries. Locally the project will constitute a major catalyst to the economy of Uttar Pradesh because of the logistics parks and industrial corridors involved. Much will be improved as regards not just access to points of origin and consumption, but also speed and punctuality for both consumers and manufacturers—which will now be able to meet their defined delivery schedules. A boost to Make in India initiative is also expected (World Bank, 2014). In recent years, India has started developing its non-core railway segments, for example through the redevelopment

Table 3 Metro projects financed by the Japan International Cooperation Agency

City	Total length (in miles kms)	Project cost (billion yen)	Loan amount (billion yen)	Completion year
Delhi	329204	1274	652	2016
Bengaluru	4226	307	65	2017
Mumbai	3320	347	71	2019
Kolkata	149	140	30	2017
Chennai	4528	331	150	2016

Source Matsumoto (2019)

of stations and facilities such as revamping waiting halls and the Rail Yatri Niwas. Other domains of the non-core sector affected include the Railway University, setting up of a Railway Tariff Authority, Strategic Planning, Digitization and Technological Development. The World Bank will also perform advisory functions and act as a consultant for programme management. These endeavours should take 2–3 years (*Business Standard*, 2017).

5.2 The Japan International Corporation Agency (JICA)

The Japan International Cooperation Agency (JICA) was created in 2003 so as to provide Official Development Assistance, with an annual budget of 1.478 billion yen. It was incorporated as an administrative agency under the Act of the Incorporated Administrative Agency (Act No. 136, 2002). The agency has played an important role in the promotion of international cooperation for socio-economic development, recovery, or economic stability of developing regions. JICA's contribution to the development of the transport sector amounts to 55.3% of total funds. The JICA has played an important role in supporting metro rail construction—the Delhi metro being one example. Currently, Japan has contributed to five metro projects across India (Table 3).

5.3 Diamond Quadrilateral

High-speed railway has an 87-year-old history. The first high-speed railway was globally rolled out in Germany in 1933. It presented an operational speed of 160 km/h (95 mph), later increased to 200 km/h (119 mph). Japan followed in 1964 with the Shinkansen Bullet train making its maiden journey from Tokyo to Osaka at a speed of 300 km/h (178 mph). India saw its first "high-speed" long-distance train service between Howrah and New Delhi in 1969 with an operating speed of 115 km/h (71 miles per hour; Shrivastva, 2018). This was followed by the inauguration of

Table 4 Global high-speed railway networks (in miles)

Country	In operation	Under construction	Total
China	11,806	7456	31,000
Spain	1926	1118	4900
Japan	1665	486	3446
France	1265	470	2793
Turkey	882	936	2926
Germany	828	266	1762
Italy	573	245	1318
Russia	403	478	1419
South Korea	256	349	974
Taiwan	214	0	214
Uzbekistan	213	0	213
Belgium	130	0	130
Netherlands	74	0	74
UK	70	127	70
Total	20,305	11,931	32,236

Source Raghuram and Udaykumar (2016)

an intercity express service in 1988—The Shatabdi between New Delhi and Jhansi, which was later extended to Bhopal. Today, high-speed railways span over 29,792 km (18,511 miles) across countries such as Japan, China and France—carrying 1,600 million passengers across the world. Table 4 highlights the global high-speed railway network.

5.4 High-Speed Railway Networks

India first expressed its interest to construct high-speed rails in 2008 during the United Progressive Alliance regime. In his 2008 railway budget Mr Laloo Prasad Yadav envisaged pre-feasibility studies for trains with operating speeds of 300 to 350 kms per hour (186–217 mph) in the Northern, Western, Southern and Eastern regions of India—thus covering 600 kms (372 miles) in two to three hours (Kumar & Mehrotra, 2009). This ambitious project has gained momentum since the National Democratic Alliance came into power. The government planning to upgrade the existing golden quadrilateral to a diamond quadrilateral, thus helping pace up electrification (*Business Standard*, 2017).

India has collaborated with Japan for Technology Transfer of High-Speed Railways. Other collaborating countries include Russia, France, Korea and Germany.

India has also collaborated with Austria, Switzerland and the USA in the field of high-speed rail, train operations, management and safety and manufacturing of locomotives for Dedicated Freight Corridors. Japan has been playing an active role in India's high-speed dream through both technology transfer and financial assistance. The total cost of the project is estimated at INR98,000 crores (13.3 billion EUR)—with 81% of the funding provided by Japan. The anticipated cost of track laying is between INR100–200 crores (17–27 million EUR) per mile kilometre—in contrast to conventional track construction, which costs INR3—10 crores (approx. 357,225.27 EUR to 1.19 million EUR) a mile kilometre. Thus, each train set would cost INR120 crore (approx. 1.2 billion EUR). The ambitious high-speed rail will also envisage the redevelopment of stations and upgrading of existing Rail Corridors, which is being executed in partnership with France on the Delhi-Chandigarh route to 200 km/h (124 mph; Shrivastva, 2018).

With the proposed bullet train project between Mumbai and Ahmedabad, JICA's role will become more prominent, facilitating the import of Shinkansen technology into India—making it the first country to import this Japanese technology. Greater participation by Japanese companies such as Kawasaki Heavy Industries, Hitachi and the East Japan Railway Company is envisaged. The UK-based private company Virgin is also looking forward to investing in the High-Speed Rail Corridor. Since 2014, the advent of high-speed railways in India has attracted global attention. In the last 17 years, FDI has amounted to $897 million (793.8 million EURO), with $291 million (259.5 million EURO) in the form of equity flows between April 2014 and March 2017.

The development of high-speed railways has further given impetus to Research and Development through both domestic and international collaboration. At the domestic level, railways have collaborated with the Department of Science and Technology, the Ministry of Human Resource Development, and various industrial representatives. Some of the countries collaborating with Indian railways include Sweden, France, Japan, and South Korea (Press Information Bureau, 2014a, b, 2015, 2016, 2018). The Memorandums of Understanding (MoUs) signed envision the following objectives:

- To facilitate technical visits.
- To exchange technical experts, reports and documents.
- To exchange training programmes, feasibility studies and pilot projects.

5.5 Asian Development Bank (ADB)

Growing populations and increasing levels of congestion on Indian roads have greatly impacted general productivity and much increased pollution levels, prompting citizens to seek greener and swifter alternatives for work commutes. This and other factors led to the construction of metro railways in Indian cities.

The first city to initiate metro operations was Kolkata in 1989. Delhi followed suit with an operational metro in 2002. Cities across India are now adopting the metro as the future mode of transportation. The government seeks to support this dynamic and to that end developed the National Metro Policy of 2017, which identifies three significant conditions for developing metro projects. These are as follows:

- The responsibility by state governments to explore and choose various options for urban mobility and implement projects accordingly.
- The presence of a comprehensive mobility plan before executing any metro railway project.
- The creation of the Unified Metropolitan Transport Authority for an integrated approach in planning and management of urban transportation.

The Asian Development Bank—apart from an ambitious railway electrification project—is playing an equally important role in the funding of urban mobility projects, e. g. metro railways. The Asian Development Bank has assisted India in the construction of the Jaipur Metro—spanning over 2.3 kms (1.4 miles) of underground rail and two stations. A large part of participation by the Asian Development Bank was invested in the electrification project, which has been gaining traction since the government first pitched for 100% electrification in 2017. The goal is to save 13,510 crores (9.1 trillion EUR) today spent in mass consumption of diesel—2.8 billion litres of diesel are utilised annually.

A loan agreement was undertaken amounting to $120 million (86 million EUR) for track doubling and electrification on the high-density corridor. This is the third instalment of the $500 million (359.71 million EUR) financing facility for the Railway Sector Investment Program approved by the Asian Development Bank in 2011. The initiative involves the states of Andhra Pradesh, Chhattisgarh, Karnataka, Maharashtra and Odisha, along with the golden quadrilateral connecting Mumbai, Delhi, Kolkata and Chennai. The current status (as of 2017) of the states in terms of electrification is laid out in Table 5.

The electrification scenario in Karnataka as of the 2017 report released by the Ministry of Railways lies at a staggering 19%—only 632 km (392 miles) out of the 3281 km (2,038 miles) of track running through the state.

The Central Organisation for Railway Electrification has currently earmarked 1,728 km (1,073 miles) of track for electrification in Karnataka, which will increase train speeds by 10–15% by 2023, helping reduce the carbon footprint by 24% by 2027–2028 (Philip, 2020).

Table 5 State-wise railway line electrification

State	Network (miles)	Electrified (miles)	Percentage
Odisha	1598	1259	79
Andhra Pradesh	2300	1764	77
Chhattisgarh	753	544	72
Karnataka	2038	393	19

Source Ministry of Railways (2017)

As per reports available up to the year 2016, 23,555 kms (14,636 miles) of the total network has been electrified—only 35.32% of the total route, carrying 64.8% of freight and 51.3% of the passenger traffic. Passenger traffic accounts for 38.7% of total traction costs of Indian Railways. In the last 4 years over 16,815 kms (10,448 miles) of electrification projects were proposed by the ministry, with a projected cost of INR17,165 crores (13 trillion EUR). 40% of the electrified route now carries 55% of the total passenger traffic and 65% of the total freight traffic (Thakur, 2018). Railways have embarked on an ambitious plan to electrify 21,000 kms (13,048 miles) of the track—with 10,500 kms (6,524 miles) targeted over the next two years (2020–2022).

5.6 *Public–Private Partnerships and Corporate Practices*

Over the years the global transport scenario has been subjected to continuous changes. Railways, especially in developing countries, have not been kept up with the dynamism, much because of undue political interference and bureaucracy, which hinder development.

Railways have now become a sunset industry, and traffic is shifted to roadways and airlines, which have greatly expanded greatly. The first time railways experienced Public–Private Partnerships was in the construction of the Konkan Railways under the Build-Operate-Transfer principle (Ranade, 2009). The need for the corporatisation of Indian railways, though first proposed by the Rakesh Mohan Committee, only gained momentum after the Bibek Debroy Committee's recommendations in 2014.

From the Indian perspective, 2004 was a turning point for the aviation industry, with the pioneer advent of the Low-Cost Carrier Air Deccan (now discontinued). This was followed by other low-cost carriers such as IndiGo, SpiceJet, and Air Asia, which are now the top players in the aviation sector. Railways suffered a further decline in traffic after the introduction of the Flexi-fare policy on premium trains in air-conditioned classes in 2016. This brought train fares at par with airlines for the same distance, resulting in low occupancy on prominent trains. Indian railways ownership was strongly impacted by governmental supervision and specifically highly subsidised fares for the passenger segment.

Freight fares were cross-subsidised so as to compensate for the losses incurred in the passenger segment. Railways reported their worst operating ratio[1] in the year 2018. At this point, the government decided to involve private players in train operations. This was carried out on a Public–Private Partnership model. The first was the Indian Railway Catering and Tourism Corporation, which operated two trains, between New Delhi and Lucknow and between Mumbai and Ahmedabad, offering compensation to passengers in case of train delays or house theft while passengers are onboard the train. Under the arrangement, private players were held responsible for train maintenance, luggage handling, ticketing and onboard facilities for

[1] Operating ratio is the ratio of total working expenses excluding gross traffic earnings.

passengers, while railway services were responsible for providing operating crews and infrastructure.

This has attracted attention from the airline sector as well. Players such as IndiGo and SpiceJet are now showing an interest in train operations so as to compensate for losses incurred in certain aviation sectors. The government is therefore planning to privatise 150 more trains so as to help supplement its efforts in railway modernisation.

6 Strengthening Modernisation Through International Collaborations

The railway modernisation programme has acquired a new dimension, as India shows a new interest in revamping its aged railway network through contemporary, global-scale projects such as High-Speed Rail Corridors and Dedicated Freight Corridors, which carry both modernisation and Eco value—valuable assets in today's environmental conditions.

At this nascent stage, as India strives to match the global railway system, it is unviable to think of rapid progress without interaction between nations. As concerns railways, India must cooperate with countries experienced in rail modernisation. India has entered into agreements with Sweden, South Korea, France, the Czech Republic and Slovakia so as to seek expertise and knowledge transfer in the field of policy and regulation, as well as development through MoUs of high-speed and suburban railways, station remodelling and semi high-speed railways.

The ongoing interactions with countries such as France and South Korea are crucial because of the importance of suburban railway expansion and construction in the golden quadrilateral. France actively participated in India's quest for high-speed railways. This includes the goal of meeting the requisites of the Dedicated Freight Corridor by setting up a locomotive manufacturing plant at Madhepura for 12,000HP Electric locomotives—capable of hauling at speeds of 100 km/h (62 mph)—and redevelopment of stations. France is highly experienced in both the suburban sector and infrastructure upgrading.

7 Administrative Restructuring

Inspired by corporate organisations, Indian railways has opted for the corporate management model—while adopting measures such as downsizing and retrenchment, so as to make human resource management more efficient and prevent the law of diminishing marginal utility which develops when high numbers of employees are hired to perform the same function.

The government has proposed to reduce the workforce in the apex body—the Railway Board—by 50%. It also plans to merge the three cadres—Traffic,

Accounts, Personnel—into a single organism, the Indian Railway Management Services (*Business Standard*, 2019).

Recruitment for this cadre will be conducted through a separate exam monitored by the Union Public Service Commission, which recruits candidates for different cadres. This will help eliminate departmental conflicts and improve the prospects of railway development.

8 Conclusion

Railways in South Asian countries have witnessed large-scale modernisation programmes taken at both the domestic and global levels. Financial constraints, which are most natural given the capital-intensive nature of the modernisation programmes involving infrastructure overhaul and expansion, created the need to ask for loans from financial institutions with a grace period for repayment. The fragile nature of the global currency market regarding specific currencies has posed threats to ongoing projects as it escalated project costs. The appreciation of the yen against the Indian rupee by 64% between 2017 and 2020 resulted in the escalation of project costs from 8.6 billion EUR $970 billion to 10 billion EUR $1,090 billion. This was further accompanied by delays in land acquisition and overruns in project deadlines. Projects like high-speed rails on the diamond quadrilateral network challenge the notions of railways as mass means of transportation. This is because of high fares and reduced accessibility of services to a selected class of people. The promotion of Public–Private Partnerships in train operations also raises the possibility of services becoming inaccessible to lower-income sectors of society. These trains could even become premium trains being exclusively managed by private players. However, this management model though possesses several advantages, no less the new governmental ability to focus more efficiently on infrastructure overhaul. Decision making, competitiveness in the rail industry and network development were all improved. Specialised institutions, supervise and guide crucial infrastructure aspects.

The growing collaboration between countries in the railway sector as concerns strengthening technical domains and exchanging expertise and management techniques will contribute to the strengthening of core railway assets as well as infrastructure. This in turn will promote faster rail connectivity and expansion of railways across the sub-continent, paving the way for greater economic diplomacy via rail. Countries can now collaborate not only as regards physical infrastructure but also human resources through the establishment of training institutes—crucially instrumental in exposing employees to current practices in the field of railway operation and management, through either specialised institutions or Special Purpose Vehicles.

References

Aklekar, R. B. (2019). *A short history of Indian Railways*. Rupa Publications.

Bhandari, R. R. (2006). *Indian Railways' glorious 150 years*. Ministry of Broadcasting and Information.

Business Standard. (2017, June 4). *Railways to raise Rs 35,000 crore from World Bank to boost infrastructure*. https://www.business-standard.com/article/economy-policy/railways-to-raise-rs-35-000-crore-from-world-bank-to-boost-infrastructure-117060400093_1.html.

Business Standard. (2019, December 24). *Cabinet approves merging of railway cadres, restructuring board: Report*. https://www.business-standard.com/article/indian-railways/cabinet-approves-merging-of-railway-cadres-restructuring-board-report-119122400578_1.html.

Government of India. (2017, October 11). *70 years of India-Russia: Partnership in the railways*. https://www.makeinindia.com/article/-/v/70-years-of-india-russia-partnership-in-the-railways.

Holt, J. (2003). *The restructuring of the railways*. United Nations Economic and Social Commission for the Asia Pacific.

Kumar, S., & Mehrotra, S. (2009). Political economy of reforms. In S. Kumar & S. Mehrotra, *Bankruptcy to billions: How the Indian Railways transformed* (pp. 38–42). Oxford University Press.

Matsumoto, K. (2019). *Metro booklet*. Japanese International Cooperation Agency.

Ministry of Railways. (2017). *Indian Railways year book*. Ministry of Railways. https://indianrailways.gov.in/railwayboard/uploads/directorate/stat_econ/pdf_annual_report/Railway%20Year%20Book_2017_18.pdf.

Misra, R. N. (2009). *Indian Railways turnaround: A study in management*. Jaico Books.

Pangotra, P. P., & Shukla, P. R. (2012). *Promoting low carbon transport in India: A case study of Delhi-Mumbai Dedicated Freight Corridor*. Indian Institute of Management.

Railway Convention Committee. (1999). *Development of alternative routes for decongesting existing routes*. Lok Sabha Secretariat.

Philip, C. M. (2020, January 29). *Green move: 1,728 km rail route in South Western Railway to be electrified*. https://timesofindia.indiatimes.com/city/bengaluru/green-move-1728km-rail-route-in-south-western-railway-to-be-electrified/articleshow/73704551.cms.

Press Information Bureau. (2014a, August 20). *India and Czech Republic sign MoU on Technical Cooperation in the field of railway sector*. Ministry of Railways.

Press Information Bureau. (2014b, November 17). *Indian Railways and Republic of Korea sign MoU on Technical Cooperation in the rail sector*.

Press Information Bureau. (2015, August 12). *India and Slovak Republic sign MoU on Technical Cooperation in railway sector*.

Press Information Bureau. (2016, April 7). *Memorandum of Understanding between India and Sweden on Technical Cooperation in the railway sector*.

Press Information Bureau. (2018, May 16). *Cabinet apprised of MoU between India and France on Technical Cooperation in the field of railways*.

Raghuram, G., & Udaykumar, P. D. (2016). *Dedicated high-speed rail network in India: Issues in development*. Indian Institute of Management.

Railways, International Union. (2017, December 9). *Vademecum* (List of UIC Members). https://vademecum.uic.org/en/.

Ranade, P. S. (2009).*Infrastructure development and its environmental impact*. Concept Books.

Shrivastva, C. (2018, March). *High- speed rail corridor: Role of external actors*. https://doi.org/10.13140/RG.2.2.31075.37929.

Thakur, A. (2018). "Statoistics." Is this why govt wants Indian Railways to keep running its diesel engines? *Times of India*.

World Bank. (2014, December 11). *India: Eastern Dedicated Freight Corridor Project*. https://www.worldbank.org/en/news/feature/2014/12/10/eastern-dedicated-freight-corridor-project.

What Role for Railways in the Eurasian Supply Chains?

Hülya Zeybek

1 Introduction

The Silk Road, the most important land transportation corridor in trade between East and West Eurasia for over 3000 years, was replaced by container shipping and air cargo due to technological developments and a dramatic decrease in transport costs (Li & Schmerer, 2017). Currently, around 80% of world trade is carried by sea, since this is the most cost-effective way (UNCTAD, 2018). Although rail and air transport only transport less than 1% of the total world trade's volume, air transport carries 35% of total traded value (IATA, 2019). However, in recent years, the need to diversify existing routes and open new alternatives between Europe and Asia has provided railway transport with more opportunities for active development. In particular, the use of railways in trade between the two continents has gained momentum through China's Belt and Road Initiative (BRI), also known as the New Silk Road (Zeybek, 2019). Railway container transport carries lower costs and higher load capacity when compared to both air and road freight—and shorter travel times than maritime container freight. Transporting a product from Europe to China takes approximately 5–9 days by air, 15–19 days by rail and 37–50 days by sea (Jakóbowski et al., 2018). Railways also present important advantages over maritime transportation when it comes to reach isolated, landlocked hinterlands. These characteristics enable rail freight to fill a market niche and even compete with other types of intercontinental transport (Rodemann & Templar, 2014).

With regular, long-distance container trains, new connections established between European and Asian cities result in the revival of Historical Silk Road. Approximately 80% of Europe–China cargoes are carried in containers. While the container traffic between Europe and Asia by rail in 2011 was 2,500 TEU, it reached 324,700 TEU in 2018 (Pieriegud, 2019) and is expected to reach 742,000 TEU in 2027 (Berger,

H. Zeybek (✉)
Vocational School of Transportation, Eskisehir Technical University, Eskisehir, Turkey

© The Author(s), under exclusive license to Springer Nature Switzerland AG 2021

S. Mitra et al. (eds.), *Railway Transportation in South Asia*, Contemporary South Asian Studies, https://doi.org/10.1007/978-3-030-76878-2_16

2017). Container use guarantees cargo protection, standard dimensions, low packaging costs, expedited cargo handling, combined shipping documents and streamlined shipping. Container trains are 20–30% more effective and efficient than conventional (classic) trains, mainly due to faster border crossings and simplified transport documents (UNECE, 2018). Intercontinental container trains usually organized in the form of express block trains, especially preferred in the transport of high-value technology and seasonal products such as automotive parts and accessories, electronic goods and clothing that require quick delivery.

The first regular express container train between Europe and China started in 2011 between Chongqing, China's largest laptop computer production centre, and Duisburg (Germany), Europe's largest inland port. Currently connection points increase day by day. New destinations and routes are first tested with pilot trains, then become regular container services (UNECE, 2018). Moreover, the offer of LCL (less container load) services is increasing (UIC, 2017). Intercontinental container trains also started to attract the attention of e-commerce companies. For instance, in 2019, Alibaba's logistics company Cainiao and the Chinese block train company ZIH jointly launched regular container services between Zhengzhou and Liege, Belgium (van Leijen, 2019a).

2 The Silk Road and Eurasian Railway Freight Corridors

The Historical Silk Road, beginning at the Chinese city of Xi'an and stretching all the way to Anatolia and finally Istanbul, has connected land transport routes to the Mediterranean, Europe and the Persian Gulf—thus turning Anatolia into a commercial centre connecting Asia, Europe and the Middle East. The Historical Silk Road constituted the main route of economic trade and cultural exchanges at least from the second century BC and until the seventeenth century, when it lost its significance due to the rise of more effective sea trade routes (Ekinci, 2014; Schramm & Zhang, 2018). In our days, new technologies and a renewed increase in trade between Europe and Asia led to the renaissance of the Rail Silk Road intended to complement maritime transport. This issue has been in the focus of many international organizations such as the European Union (EU), the UN European Economic Commission (UNECE), the UN Asia Pacific Economic and Social Commission (UNESCAP) and the Economic Cooperation Organization (ECO) (Zeybek, 2019). The use of railways in intercontinental trade has gained momentum with China's Belt and Road Initiative (BRI).

Literature contains various maps of Eurasian railway corridors and routes. It is possible to categorise these corridors into two main groups: (1) Northern Corridor and (2) Southern Corridor (Fig. 1).

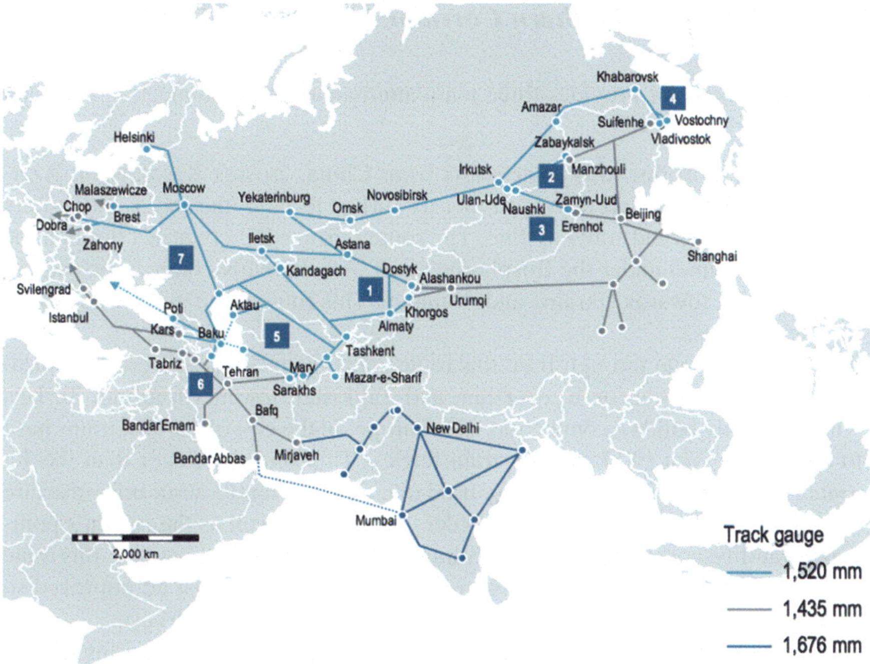

Fig. 1 Eurasian Railway Routes (Different colours show different track gauges) (*Source* Berger [2017])

2.1 The Northern Corridor

The Northern Corridor provides three alternative routes connecting China and Europe through the Russian Trans-Siberian Railway (TSR), the First Eurasian land bridge (Galushko, 2016; Schramm & Zhang, 2018):

- China–Russia through the TSR: via Alashankou/Dostyk; transit through Kazakhstan (Kazakh route—Second or New Eurasian land bridge);
- China–TSR via Erenhot/Zamyn-Uud; transit through Mongolia (Mongolian route);
- China–TSR via Manzhouli/Zabajkalsk (Manchurian route).

Currently, the Northern Corridor is the most popular Eurasian route for containerized freight shipments due to several competitive advantages (Jakóbowski et al., 2018).

2.2 The Eurasian Southern Corridor

The Southern Corridor provides three main alternative routes connecting Asia and Europe via Turkey:

- The Middle Corridor: this route starts from China, crosses Kazakhstan, Azerbaijan and Georgia, and travels further to Turkey through the Baku–Tiflis–Kars (BTK-Iron Silk Road) railway line. The route reaches EU countries through the Marmaray Tunnel under the Bosphorus Strait. An alternative branch runs across the Black Sea towards Ukraine and Romania. This route is the most used among Southern routes.
- The Southern route through Iran: this route starts from China and crosses Kazakhstan, Turkmenistan or Kyrgyzstan, and Tajikistan towards Iran and Turkey, where it finally builds a connection with Europe. In the east of Turkey, trains have to cross Lake Van (90 km) with train ferries. This route is longer than the 1st route (Middle Corridor) and hardly used due to its currently weak infrastructure (UIC, 2017). The high number of border crossings, as well as the region's political instability, constitute main disadvantages (Sahbaz, 2014). However, this route contains a connection to the Pakistan-India route—thus it must be improved so as to support trade from South Asian countries to Europe via Turkey.
- The Istanbul-Tehran-Islamabad (ITI) Corridor: the ITI corridor, covering around 6,500 km, was launched in 2009 within the framework of the Economic Cooperation Organization (ECO). Various test journeys were conducted; however, a regular service is yet to be provided (ECO, 2020). India has put forward a proposal to extend the ITI corridor towards both India and Bangladesh so as to establish a Dhaka (Bangladesh)—Koslkata-Delhi-Amritsar and (India)—Lahore-Islamabad (Pakistan)—Zahedan-Tehran (Iran)—Istanbul (Turkey) route (ITI-DKD). The Dhaka–Istanbul freight route is an international railway network project meant to connect South Asia (from Bangladesh) to Europe by crossing India, Pakistan and Iran towards Turkey and possibly further to the EU. The ITI-DKD-Y route is an integral part of the Trans-Asia Rail (TAR) network. It will open the Indian market, providing much shorter delivery times to Europe (UNESCAP, 2019).

About 95% of Eurasian rail traffic currently uses Northern routes through Russia. The remaining 5% uses the Middle Corridor and other routes (UIC, 2020). Possible congestion in Northern routes and other factors will likely strengthen the role of Southern Eurasian routes.

2.3 The Belt and Road Initiative (BRI)

Infrastructure investments on Eurasian railway corridors have gained momentum with the Chinese Belt and Road Initiative (BRI). Figure 2 shows existing railway

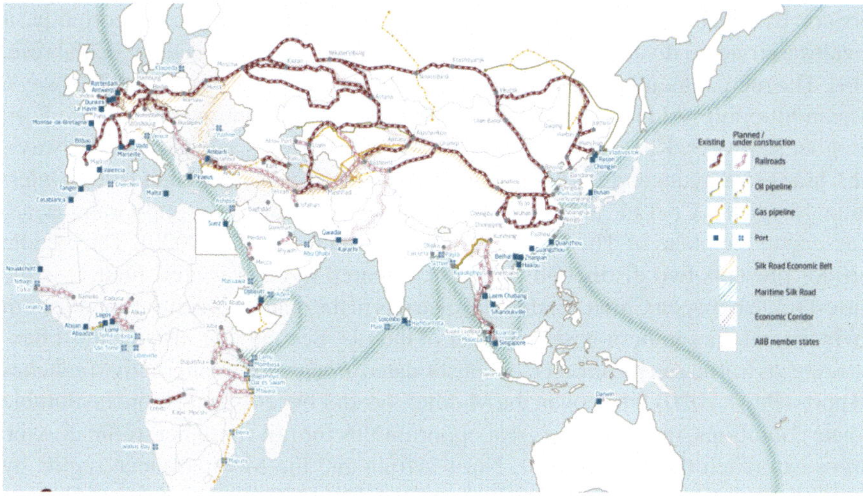

Fig. 2 Location Map: Belt and Road (BRI) Infrastructure Network - Eurasia Railway Lines (*Source* Mercator Institute for China Studies [2018])

lines (red) between Europe and Asia as well as railway lines (yellow) under construction/planned under the BRI.

According to the Belt and Road Portal (2020), currently, 71 countries are taking part in the Initiative—which together represent more than a third of the world's GDP and two-thirds of the world's population. Such grandiose transport infrastructure projects follow essentially historical examples. For instance, the ambitious Baghdad Railway envisaged to run through Turkey and Mesopotamia—thus bypassing the maritime chokepoint of the Suez Canal up to the Indian Ocean—was once intended to accelerate trade and economic growth among Europe, the Ottoman Empire and the Far East (until Britain curtailed the project; Brewster, 2017).

The ambitious Chinese Initiative will create great economic opportunities not only for the BRI region itself and the countries directly involved but also for many other areas (Lu et al., 2018). According to the results of a study by DEIK [Foreign Economic Relations Board of Turkey] (2019), BRI countries increase their trade share both among themselves and in international trade at large by strengthening intra-regional relations.

2.4 Middle Corridor

The Middle Corridor Initiative was launched in late 2013 so as to connect east and west, envisaging a revival of the ancient Silk Road. The Middle Corridor (also called The Trans-Caspian International Transport Route [TITR]) is a multimodal corridor that aims to provide an alternative to the northern routes of the Eurasian Northern

Corridor by creating a synergy with the Belt and Road Initiative. Importantly, it provides a strategic diversification—freer from reliance on Russia alone (Calabrese, 2019). The regular members of the Middle Corridor include the Azerbaijan Railways, Caspian Shipping, Aktau Sea Commercial Port, Baku International Sea Trade Port, Georgian Railways, Kazakhstan Railways, Turkish Railways and Ukraine Railways. The Chinese Lianyungang Port Holdings Group and Xi'an Continental Bridge International Logistics Co. Ltd., along with Romania's private railway group-Grampet, are all associate members of The Middle Corridor Initiative (TITR, 2020). The Middle Corridor presents two distinct advantages compared to other routes. First, it is the shortest route between China's industrial districts in the country's central and Western provinces and the southern-eastern border of the EU. Second, it is also the most flexible route, as traffic can be rerouted or consolidated along various alternative branches and ports (Pepe, 2020). Moreover, the Middle Corridor benefits from more favourable climate conditions, thus offering great opportunities for rail container traffic in Asia. Cargos can reach the Middle East, North Africa and the Mediterranean region by benefiting from port connections in Turkey (Republic of Turkey Ministry of Foreign Affairs, 2020).

The Corridor starts from the Chinese port of Lianyungang and crosses Kazakhstan, the Caspian Sea, Azerbaijan, Georgia and Turkey, further extending into Europe either through said Turkey (the Baku–Tbilisi–Kars [BTK] railway) or through the Georgian Black Sea Ports. There are links to Turkey's major seaports, i.e. the Mersin Port, Turkey's second largest container port (Fig. 3).

The Middle Corridor has gained a significant competitive advantage in terms of transport time and costs with the launch of the 826 km-long Baku–Tbilisi–Kars (BTK) railway line on 30 October 2017 as well as with the opening of the Marmaray Bosphorus Tube Crossing to freight trains on 7 November 2019. The development of Caspian ports has also boosted the Middle Corridor's competitiveness. The Baku International Sea Trade Port (Alat), the largest multi-purpose port in the Caspian

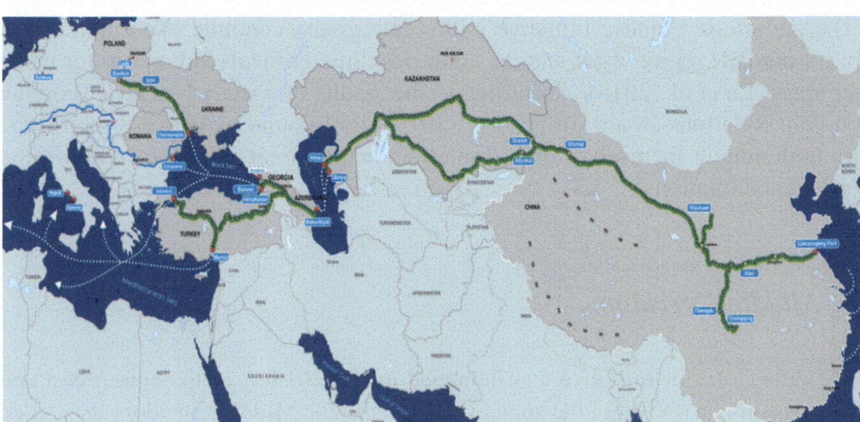

Fig. 3 The Middle Corridor Route (*Source* TITR [2020])

Sea, was inaugurated in May 2018. Kazakhstan's Aktau port has been expanded and outfitted with new loading equipments, rolling cargos and a ferry complex. Georgia is also modernizing its freight train network and expanding the capacity of its ports (Calabrese, 2019).

In 2019, the volume of container traffic in the direction of China—Caucasus/Turkey reached 30 thousand TEU—for reference, it was 0.1 thousand TEU in 2015 (TITR, 2020). The current target is to transport 60 thousand TEU containers by rail through the Middle Corridor (van Leijen, 2019b).

In 2015–2016, several pilot trains were launched: Shihezi (China)—Kishly (Baku, Azerbaijan), Lianyungang (China)—Istanbul (Turkey) and Chornomorsk (former Illichivsk in Ukraine)—Dostyk (Kazakhstan). All of these were part of the "Nomad Express" project (Middle Corridor, 2019). Besides, container freight trains run between Kostenai (Kazakhstan)—Mersin Port (Turkey), Aktobe (Kazakhstan)—Mersin Port (Turkey) and Baku (Azerbaijan)—Mersin Port (Turkey) (UAB, 2020). Container trains serving on the Lianyungang - Istanbul line became a regular service in 2018, departing from the Lianyungang port in China on the 8th, 18th and 28th of each month (van Leijen, 2018). In November 2019, the first container train under the BRI—heading from Xi'an in China to Prague, Czech Republic—ran via the Marmaray Tunnel in Istanbul and completed its 11,483 kms-long route in 18 days. As a result, the Middle Corridor received a new and improved connection. Although currently few regular container trains travel from China to Turkey, the improvements on infrastructure and logistics will contribute to the expansion of containerized services as well as to an increase in the volume of freight traffic on the Middle Corridor. According to Berger (2017), excluding the already well-developed but increasingly overloaded Northern Corridor, both the Middle Corridor and other the Southern Routes could attract up to 8% of all Eurasian rail freight by 2027.

Positioned as an alternative to the northern Eurasian routes, the Middle Corridor has great potential in the transport between Asia and Europe of laptops, hard disks, semi-finished TV panels, textile products, auto parts, food and pharmaceutical products, as well as semi-finished and finished vehicles (Tavsan, 2019).

There are two different track gauges on the route which require bogie replacement or transfer to wagons at border crossings between countries (Table 1). In order to facilitate border crossings, in 2012, Kazakhstan constructed 293 km of a new rail line from Almaty to the Chinese border/Khorgos and established a new border station at Altynkol (UNESCAP, 2016). Moreover, the bogie exchange station at Akhalkalaki was built within the Baku–Tbilisi–Kars (BTK) Railway line project.

In terms of logistics services, except for China and Turkey, Middle Corridor countries demonstrate a poor record in logistics performance, scoring particularly low in customs and infrastructure (World Bank, 2018) (Table 2).

Poor logistics performance among Middle Corridor countries, as well as technical barriers (i.e. unloading and gauge changes), lead to higher costs per container ($6,000–$7,000)—which are therefore still non-competitive in the absence of Chinese subsidies (Pepe, 2020).

Table 1 Technical characteristics of the Middle Corridor

Line	Country	Length (km)	Gauge (mm)	Electrification
Lianyungang-Xi'an	China	1,162	1,435	Yes
Xi'an-Khorgos	China	3,200	1,435	Yes
Altynkol/-Aktau	Kazakhstan	3,157	1,520	No
Aktau-Baku/Alat	Caspian sea ferry	468	-	
Baku-Alat-Boyuk kesik	Azerbaijan	429	1,520	Yes
Gardabani-Akhalkalaki	Georgia	260	1,520	Yes
Akhalkalaki-Kapıkule	Turkey	2,285	1,435	Partly yes
Total	5 country	10,532		

Source Adapted from İNTERNETHABER (2020)

3 Literature Review

As part of the global supply chain, railway container transport is influenced by various external factors such as global trends, macroeconomic developments, technological advances, innovations in infrastructure, international collaborations and legal and environmental regulations. It is necessary to discuss the important variables that may affect the rail container transport market. Rodemann and Templar (2014) aimed to identify the factors that triggered and prevented the development of Eurasian rail freight transport—and propose solutions. In this study, the northern rail routes and the North–South Corridor connecting Iran's Bandar Abbas port to Finland through Tehran are discussed. Li and Schmerer (2017) revealed that although China's Belt and Road Initiative creates great potentials for the countries involved, implementation will not be easy. Sahbaz (2014) focused on rail corridors over the "Modern Silk Road", stressing that all rail corridors should be regarded not as substitutes for but as complements to each other. Nazarko et al. (2016) analysed the New Silk Road from the perspective of potential benefits for countries participating in the BRI, as well as the challenges faced on the road to becoming a well-performing transport corridor. Also, there are many consultancy work or policy studies (UNECE, 2018; UNESCAP, 2016, 2017; OECD, 2018; Galushko, 2016; Jakóbowski et al., 2018; Berger, 2017; Gleave, 2018; Pieriegud, 2019; Lu et al., 2018). Studies on the Eurasian transport corridor tend to focus on the Northern-Trans-Siberian Railway Corridor and related routes.

4 The Study's Objectives and Methodology

This paper aims to summarize recent developments in the improvement and expansion of container traffic on the Eurasian railway routes—with a specific focus on the Middle Corridor—and highlight the main challenges involved. We aim to achieve

Table 2 Aggregated LPI of Middle Corridor Countries 2012–2018

Country	LPI Rank	LPI Score	Customs	Infrastructure	International Shipments	Logistics competence	Tracking&tracing	Timeliness
China	27	3.60	3.28	3.73	3.57	3.58	3.63	3.86
Turkey	37	3.29	2.94	3.36	3.19	3.23	3.37	3.68
Romania	50	3.10	2.73	2.86	3.15	3.01	3.19	3.61
Ukraine	69	2.83	2.46	2.38	2.77	2.76	3.08	3.45
Kazakhstan	77	2.77	2.57	2.59	2.73	2.60	2.81	3.31
Azerbaijan	123	2.45	2.53	2.69	2.56	2.14	2.18	2.62
Georgia	124	2.45	2.38	2.36	2.38	2.27	2.37	2.92

Source World Bank (2018)

this by using the Political, Economic, Social, Technological, Legal and Environmental (PESTLE) analysis. In line with the study's objectives, wide-ranging and varied exploratory research was carried out. Magazine articles, promotional pages, materials and companies' web pages were extensively used so as to take into account current developments, including market movements.

The PESTLE analysis is a strategic planning tool used to analyse the impact of external factors on business, projects or industrial dynamics (Koumparoulis, 2013). PESTLE factors cannot be controlled as such and may pose either opportunities or threats, sometimes both, to industries. In this study, we aimed to determine the external factors affecting the development of the Eurasian Middle Corridor. This route is selected because it is the most used and benefits from a better infrastructure than that present in other Southern routes.

5 PESTLE Analysis

5.1 Political Factors

5.1.1 National or international policies and agreements for the development of the Middle Corridor

The Transport Corridor Europe-Caucasus-Asia (TRACECA) is the oldest initiative for the development of Eurasian railway corridors. It was established in 1993 under the leadership of the European Union so as to connect the Commonwealth of Independent States to Europe through the Black Sea (Zeybek, 2019). Later, in 2013, the TRANS-Caspian International Transport Route (TITR), rebranded as the Middle Corridor, was established with the participation of both railway and maritime organizations of China, Kazakhstan, Azerbaijan, Georgia, Turkey and Ukraine (TRACECA, 2020) as a multimodal route. All these countries are also participants in the BRI. The Middle Corridor constitutes an important issue in Turkey's Multilateral Transportation Policy (Republic of Turkey Ministry of Foreign Affairs, 2020). During a sideline meeting at the G20 Summit in Antalya in 2015, Turkey and China signed an MOU on the harmonization of the BRI with the Middle Corridor Initiative (Talbot, 2018). In mid-May 2017, during the Belt and Road Forum for International Cooperation held in China, emphasis was provided to the opportunity to coordinate the BRI with other networks and Initiatives—including the Middle Corridor (Gleave, 2018). At the corporate level, the State Railways of the Republic of Turkey (TCDD) Transport Inc., which conducts rail transport services in Turkey, has signed various protocols and contracts with companies from other countries on the Middle Corridor so as to facilitate container transport and increase service quality. As a result of these collaborations, rail transport time was reduced and quality increased through operational improvements, particularly the establishment of more developed terminals at borders and the consolidation of customs procedures. All agreements for the development of the Middle Corridor of course positively affect intercontinental rail container shipments. However, in order to improve supply chain performance in the region, cooperation must be intensified so as to reduce costs per container, which are still non-competitive.

5.1.2 Political conflicts between Russia and the Ukraine

Due to the political conflicts between Russia and the Ukraine, in 2016, Russian railways suspended cooperation with Ukrainian railways—and blocked railway transit passages from the Ukraine. Due to this blockage, transit transport was transferred to Belarus. The route via Belarus lengthens transport duration (Gotev, 2019). Again we see that the appeal of the Middle Corridor grows because of its value as a strategic diversification that can reduce reliance on Russia.

5.1.3 China's More Balanced Regional Development Policies

As a result of the policies implemented by China to develop its backward Western provinces, transport distances from ports to the hinterland extended, a dynamic reflected on the logistics of the transport sector in the form of longer transport times and increased costs (OECD, 2018). These logistic developments have made rail freight transport between Europe and the Central Chinese regions more advantageous.

5.1.4 Reduction of Subsidies offered to rail operators by the Chinese government and provinces

Until 2018, China had been applying a standard 50% subsidy to rail transport between China and Europe so as to support rail use. This rate was reduced to 40% in 2019—and is planned to be abolished by 2022 (Rajamanickam, 2019). Usually, these subsidies involve grants to rail operators controlled by the governments of specific provinces and cities, which organize transportation.

According to the International Union of Railways (UIC, 2020) study on Eurasian Corridor development, the reduction of Chinese subsidies for rail transport between China and Europe is likely to have a drastic impact on traffic flows.

5.1.5 Trade Barriers

In 2014 Russia introduced an import ban on a wide range of agricultural products from the European Union. The food import ban was established by the Russian government as an answer to the sanctions put in place by the EU against Russia because of the Ukrainian crisis. The food import ban was extended until the end of

2020 and has been renewed. This ban also applies to transits. Due to this ban, it is not possible to transport food products from the European Union to third countries by train via Russia. Fresh fruits and vegetables, cheese, fish products, red meat and pork are among the main products that Russia prohibits being imported from the European Union. Therefore, these products must be transported to China and other Asian countries either through the Central Corridor or by air and/or sea (van Leijen, 2019b).

5.1.6 Infrastructure Investments

Infrastructure investments by countries on rail, port, terminals and logistic centres increase the capacity and quality of intercontinental rail networks—and shorten transit time. Middle Corridor countries are deploying tremendous efforts in infrastructure upgrades, including setting up logistics centres along the route. Indeed, Turkey is among the best-performing countries within the BRI in terms of infrastructure and customs procedures (DEIK, 2019).

The most strategic developments in this route include the launch of the Baku–Tbilisi–Kars (BTK) Railway line, which enables an uninterrupted rail freight transport from Baku to Europe, and the opening of Marmaray—a railway tube crossing under the Bosphorus Strait dedicated both to passenger transport (since 2013) and freight transportation (since 2019). The construction of the Ankara-Sivas High-Speed Rail is planned to be completed within 2020. The current Ankara-Sivas railway is 603 km-long and the travel time involved is 12 h. The Ankara-Sivas High-Speed Railway line will shorten the existing railway by 198 km. However, as only passenger trains are planned to operate, this line will not contribute to the reduction of freight transport time on the Middle Corridor. There are also projects to upgrade rail line standards to speed line statute (200 km/hour) between Kars and Sivas (UAB, 2018). Investments in Azerbaijan and Kazakhstan so as to develop ports on the Caspian Sea coast also significantly increased the Middle Corridor's competitiveness (Calabrese, 2019). China, too, focused on the improvement of land transport infrastructure so as to strengthen rail links between Western Europe and Turkey (DEIK, 2019) and create opportunities for further developments of the Middle Corridor.

Another important investment involved border terminals and gauge changes, which currently create significant delays. As part of the Baku–Tbilisi–Kars railway, the logistics centre in Kars is under construction jointly by Turkey and Azerbaijan (UAB, 2018).

5.2 *Economic Factors*

5.2.1 Economic growth

According to the OECD, the impact of the COVID-19 outbreak on economic projections for 2020 will be severe. It is estimated that China's economic growth (real GDP growth) in 2020 will decrease to 4.9%, the EU's to 0.8%, Turkey's to 2.7%—while the growth in the global economy at large may decrease to 2.4% (OECD, 2020).

5.2.2 Economic advantages of the railway system

In the age of globalization and internet, connecting continents through intercontinental rail transport constitutes an important breakthrough. A study by Lu et al. (2018) revealed that the existence of a railway connection between commercial partners in the New Silk Road (BRI) region had the greatest impact on global trade development (increasing exports by 2.8%). Moreover, this study showed that infrastructure investments in the BRI region increased the total trade volume not only in the BRI region but also in other areas.

In their study, Egger and Larch (2007) found that the rail network is 50% more effective in developing international trade than other modes. Intercontinental rail transport's costs correspond to one-third of those by air transport, while transport times correspond to a quarter of those by maritime transport. While trains can transport approximately 90 TEUs at a time—far above both the airline and TIR carrying capacity—this figure is still well below ship capacity, which is over 10,000 TEU. However, since the choice of intercontinental rail transport (instead of maritime or road) shortens transport times, the railway is especially suitable for the transport of high-value goods that require a short travel span. Therefore, more high-value and capital-intensive products are currently transported by rail between Europe and Asia. The necessity of special protective packing for electronic products sensitive to humidity and cold increases costs through Northern routes, especially in the wintertime (Rodemann & Templar, 2014). When shipping by rail, the lower interest costs for invested capital partly cover the higher transportation costs involved, making rail freight an attractive alternative to sea freight as concerns high-value goods. Lower speeds basically mean more capital tied up in transit cargo. According to DB Schenker's calculations, **this cost of tied up capital in** in-transit maritime cargo increases the cost of sea freight by two-thirds, whereas the corresponding proportion for rail freight is one-third (Jakóbowski et al., 2018).

5.2.3 Imbalanced Traffic Volume—the Empty Containers Problem

The question of balancing the ratio of train services from China to Europe and from Europe to China remains the key challenge for the development of rail connections.

Until 2017, the imbalance between westbound and eastbound traffic between Europe and China revealed a significant "empty containers problem". To solve this problem, the Chinese Government limited the number of empty containers per train by 10% through new legal regulations—and this problem was significantly reduced in 2018. As a result of this practice, in 2018, the rate of empty containers in trains from the Northern Corridors towards China was reduced from 29% to 18%—and from 6% to 2% in container trains to Europe (Rajamanickam, 2019). However, a lower quantity potential is estimated for transports from the Central Corridor to China. There is a huge trade imbalance between Turkey and China. In 2018, while Turkey's exports to China amounted to 2 billion 913 million dollars, imports from China amounted to 20.7 billion dollars (TUIK, 2019). For this reason, it is necessary to investigate multi-legged transport opportunities from various countries both along and around the corridor, including Pakistan, India, Bangladesh and Iran, as well as identify innovative solutions for balancing trade flows through partnerships and cooperation among Corridor countries.

5.2.4 Organization and Logistics Services in Railway Transportation

In general, the Eurasian rail transport market is controlled by multinational companies. However, operators of Middle Corridor countries are trying to increase their market shares by offering direct forwarding services to customers. Transportation organizations and agency services related to the sale of the train capacity account for about 10% of container transportation costs; forwarding companies earn the highest incomes from the services they usually provide both at the beginning and end of the transportation (Jakóbowski et al., 2018). Therefore, logistics companies are competing for a greater share of the cake. For example, ADY Container, a subsidiary of Azerbaijan Railways, and KTZ Express JSC, a subsidiary of Kazakhstan Railways, both play an active role in rail container transport in the region. The Middle Corridor's future success depends on the ability of service providers to maintain an offer of rail services that are attractive when compared to maritime routes.

5.3 Social Factors

5.3.1 Improvements in social welfare

The World Bank (2019) predicts that BRI transport projects, when completed, could increase world trade between 1.7% and 6.2%—and global income between 0.7% and 2.9%. That would make an important contribution to poverty reduction.

5.3.2 Epidemics

The logistics industry was hard hit by the closure of entrances and exits to both China and many European countries due to the COVID-19 virus epidemic—which started in Wuhan, China, where many global companies have factories spread all over the world. The most affected mode has been air transport due to numerous flight cancellations; there were also cancellations in both sea and railway transportation. It is expected that the demand for rail freight will increase above current capacity after the virus threat decreases, because transport times are shorter than those of sea routes.

5.4 Technological Factors

5.4.1 The development of digital technologies and e-commerce

The development of digital technologies and electronic commerce has increased the importance of carrying e-commerce products. Although e-commerce reduces the size of shipments in the intercontinental freight transportation system, it increases the shipping frequency and thus increases vehicle traffic. Until 2013, only full container transportation by rail (FCL-Full Container Load) was possible. After 2014, as a result of developments in logistics, less than a full container load (LCL-Less than Container Loads) as expressed in m2 could be rented, so shipments not large enough to fill whole containers could be transported on Eurasian rail routes (Rastogi & Arvis, 2014). Both LCL and FCL can be transported on the same train; however, LCL containers require additional handling (UIC, 2017). Reductions in the size of shipments increase the need to consolidate goods and logistics centres. Transport of e-commerce products presents a great potential—and rail could play an important role in this market by ensuring a more sophisticated organization.

5.5 Legal Factors

In contrast to air and maritime transport, there currently lacks a globally unified regime covering rail transport. Lack of consistency in legal regimes for rail operations complicates international rail operations and affects the competitiveness of rail transport because it causes delays and a cost increase in rail freight transport in comparison with sea freight. Two different legal regimes are in force in rail transport between Europe and Asia. The Convention Concerning Internal Carriage by rail (COTIF)'s and the International Convention for Goods Transport by Rail (CIM)'s provisions are applied in the west; the Agreement on International Goods Transport by rail (SMGS)' Convention provisions are applied in the east (Russia, China, etc.). In 2006, in order to solve this problem, the CIM/SMGS common consignment note was introduced (UNESCAP, 2016). So far, border points between China and Kazakhstan, Russia and Mongolia were opened to the possibility of using the CIM/SMGS common consignment note in 2017 (Evtimov, 2017). According to a UIC study (2020), 100% use of the CIM/SMGS consignment note is perceived as a factor with a positive impact on volume development. Besides, the UNECE Inland Transport Committee continues to work on the Unified Railway Law (UNECE, 2019). These studies and practices will provide a significant contribution to the development of intercontinental rail transport—and increase its competitiveness.

5.6 Environmental Factors

5.6.1 Resolution by the International Maritime Organization (IMO) The Marine Environment Protection Committee MEPC 280 (70)

For ships operating outside designated Emission Control Areas, the IMO has set a limit of 0.50% m/m (mass by mass) for sulphur in fuel oil used on board ships from 1 January 2020. Also, the carriage of fuel oil for use on ships will be prohibited from 1 March 2020 if sulphur content exceeds 0.50%. It is expected that this practice will increase the cost of sea transport—and Chinese exporters will search for ways to send their products to Europe by rail (DTO, 2019). Since the price of sea transport from China to the Middle East and Eastern European countries from the south is more competitive than is the case in the north due to short distances and more effective connections (Berger, 2017), this application would constitute an opportunity for rail container transport through the Middle Corridor.

5.6.2 Greenhouse gas emissions

The share of air transport in trade between Europe and Asia is growing. The increase in the use of air in international trade causes greenhouse gas emissions (GHG) to increase by 23%—a rise 42% faster than that of trade (Cristea et al., 2013). As environmental awareness by customers increase, there may be a switch from such polluting air transport towards the environmentally friendly railway.

6 Results

See Table 16.3.

Table 3 Middle Corridor PESTLE Analysis

	Factors	Opportunity	Threat
Political	National or international policies and agreements	Trans-Caspian International Transit Route (TITR), TRACECA, Belt and Road (BRI)—such initiatives and agreements for the development of the Middle Corridor will positively affect intercontinental railway container shipments	
	Political conflicts between Russia and the Ukraine	The problems encountered in the Northern Corridors as a result of the Russia's transit ban to railway transport via Ukraine support the development of the Middle Corridor	
	China's Regional Development Policies	China's policies to develop its Western provinces have made rail transport more advantageous	
	Trade barriers	Due to Russian's transit ban, it is not possible to transport food products from the European Union to third countries via Russia by train. This creates an opportunity for the Middle Corridor	
	Reduction of Subsidies offered to rail operators by China		China is planning to completely remove the subsidy applied to rail transport between China and Europe by 2022. This practice is expected to negatively affect intercontinental rail container transport

(continued)

Table 3 (continued)

	Factors	Opportunity	Threat
	Infrastructure investments	The commissioning of the Baku–Tbilisi–Kars (BTK) Railway line, the opening of the Marmaray Bosphorus Tube Crossing to freight transport and the investments in developing the Caspian ports of Azerbaijan and Kazakhstan increase the competitiveness of the Middle Corridor	The infrastructure quality of Northern Corridor is better than that of the Southern Corridor
Economical	Economic growth		The impact of COVID-19 outbreak on economic projections for 2020 will be severe
	Economic advantages of railway system	Intercontinental rail freight costs correspond to one-third of those of air transport, while transport times correspond to a quarter of those of maritime transport. Approximately 90 TEU can be transported by trains at a time span far above both air and TIR carrying capacity. Climate conditions in Trans-Siberian routes increase the protection and packaging costs in the winter; this creates opportunities for the Middle Corridor	More than 10,000 TEU can be transported by ships Border crossings are less in number both by air and sea transport when compared to railways
	Imbalanced Traffic Volume—the Empty Containers Problem		Lower quantity potential is estimated for transports from west to east via the Middle Corridor

(continued)

Table 3 (continued)

	Factors	Opportunity	Threat
	Organization and Logistics Services in Railway Transportation	The operators of Middle Corridor countries are trying to increase their market share by offering direct forwarding services to customers	Measures required to improve service quality on the Middle Corridor
Social	Improvements in social welfare	It is predicted that Belt and Road transport projects, when completed, would make an important contribution to poverty reduction	
	Epidemics	It is expected that the demand for rail freight will increase above current capacity after the virus threat decreases since transport times are shorter than those of sea routes	The logistics industry was hard hit by the COVID-19 virus epidemic. The most affected mode has been air due to numerous flight cancellations. There were also cancellations in sea and railway transport
Technological	Development of e-commerce	E-commerce reduces shipment size in the intercontinental freight transportation system, it increases the frequency of shipping and thus increases vehicle traffic. Transportation of e-commerce products presents great potentials in the Eurasian intercontinental container transport. Railway will play an important role in this market	The reduction in shipment size increased the importance of consolidating shipments and the need for a more sophisticated organization

(continued)

Table 3 (continued)

	Factors	Opportunity	Threat
Legal	Legal structure of Eurasian rail transport	In 2006, the CIM/SMGS common consignment note was put into practice. Studies were pursued on the Unified Railway Law. These studies and practices will have a significant contribution to the development of intercontinental rail transport and increase its competitiveness	The use of two different transport documents and the existence of two different legal regimes affects the competitiveness of the rail transport as it causes delays and increases costs in rail freight transport in comparison with sea transport
Environmental	Resolution of the International Maritime Organization (IMO) The Marine Environment Protection Committee MEPC 280 (70)	It is expected that the implementation of this Resolution will increase transportation and logistics costs. If so, Chinese exporters will look for ways to send their products to Europe by rail, especially from the western regions. This application creates an opportunity for rail container transport through the Middle Corridor	
	Greenhouse gas emissions	As environmental awareness by customers increases, there may be a switch from such polluting air transport towards the environmentally friendly railway	

Source Compiled by Author from different sources

7 Conclusions and Recommendations

In this study, the PESTLE analysis was conducted so as to reveal opportunities and challenges involved in the improvement of Eurasian Southern rail corridors—with a focus on the Middle Corridor. The analysis shows that opportunities are far greater than the threats. The Middle Corridor has gained a significant competitive advantage in terms of both transport time and transport cost after the commissioning of the Baku–Tbilisi–Kars (BTK) railway line in 2017 and the opening of the Marmaray Bosphorus Tube Crossing to freight trains in 2019. Furthermore, investments in ports of both Azerbaijan and Kazakhstan on the Caspian Sea coast increased the Middle Corrdior's competitiveness. The countries involved in the Middle Corridor are also BRI participants. With the BRI's momentum, Eurasian container trains are also increasing along the Southern routes and complementing the Northern ones. There is an opportunity in the transport of high-tech electronic products such as laptops, hard disks, semi-finished TV panels, textile products, auto parts, food and pharmaceutical products, semi-finished and finished vehicles. Russian's transit ban creates an opportunity to transport food products from the European Union to China via the Middle Corridor. E-commerce transport could be a new market for this route as well. However, all these time-sensitive commodities require special handling, as well as sophisticated, high-quality organization and service—and competitive pricing. These are all dependent on international cooperation among Corridor countries and their rail and logistics companies. A UIC study (2020) revealed that rail transport is more sensitive to prices than speed factors. Therefore, Middle Corridor countries must give priority to harmonizing tariff structures and reducing prices.

One of the major threats is the traffic imbalance between west to east and back due to a severe trade imbalance with China. For this reason, it is necessary to investigate balancing solutions, including the possibility to connect the Middle Corridor with the Iran-Pakistan-India route. The other major threat is China's plan to remove subsidies for the Eurasian rail container transport. Every additional rail connection along the Eurasian Corridors will contribute to a sustainable development of the countries involved—and, ultimately, a sustainable future for their citizens.

References

Belt & Road Portal. (2020). International cooperation. https://eng.yidaiyilu.gov.cn/info/iList.jsp?cat_id=10076&cur_page=1.

Berger, R. (2017). Eurasian rail corridors: What opportunities for freight stakeholders? International Union of Railways. UIC. https://uic.org/IMG/pdf/corridors_exe_sum2017_web.pdf.

Brewster, D. (2017). Silk roads and strings of pearls: The strategic geography of China's new pathways in the Indian Ocean. *Geopolitics, 22*(2), 269–291. https://doi.org/10.1080/14650045.2016.1223631.

Calabrese, J. (2019). Setting the middle corridor on track. https://www.mei.edu/publications/setting-middle-corridor-track.

Cristea, A. D., Hummels, D., Puzzello, L., & Misak, G. A. (2013). Trade and the greenhouse gas emissions from international freight transport. *Journal of Environmental Economics and Management, 65*(1), 153–173.

Economic Cooperation Organization (ECO). (2020). 10th High-Level Working Group Meeting on the Islamabad-Tehran-Istanbul Container Train. http://www.eco.int/news/88199-10th-High-Level-Working-Group-Meeting-on-the-Islamabad-Tehran-Istanbul-Container-Train.html?t=News.

Egger, P., & Larch, M. (2007). The bilateral and multilateral trade effects of road and railway transport infrastructure. Mimeo. http://www.etsg.org/ETSG2007/papers/pegger.pdf.

Ekinci, M. B. (2014). The silk roads from past to the future and their interactions with the foreign trade of Turkey. *Avrasya Etüdleri, 14*(1), 7–42.

Evtimov, E. (2017). Legal interoperability. CIM/SMGS Eurasian corridors stakeholders' group. Paris, 22 November 2017 International Rail Transport Committee CIT.

Foreign Economic Relations Board of Turkey (DEIK). (2019). Türkiye'nin Kuşak ve Yol Girişimi'nde Konumlandırılması [Positioning Turkey on the Belt and Road Initiative]. http://www.deik.org.tr.

Galushko, D. (2016). Study on corridors. OTIF - Intergovernmental Organisation for International Carriage by Rail.

Gleave, S. D. (2018). Research for TRAN Committee: The new Silk Route – opportunities and challenges for EU transport. European Parliament, Policy Department for Structural and Cohesion Policies.

Gotev, G. (2019, February 1). Commission reacts to Russian transit ban via Ukraine. Euractiv. https://www.euractiv.com/section/central-asia/news/commission-reacts-to-russian-transit-ban-via-ukraine/.

International Air Transport Association (IATA). (2019) Air Cargo Matters. https://www.iata.org/whatwedo/cargo/sustainability/Pages/benefits.aspx.

İNTERNETHABER (2020) [INTERNETNEWS, China-Turkey-Prague Train Route] https://www.internethaber.com/cin-turkiye-prag-tren-hatti-guzergahi.

Jakóbowski, J., Popławski, J. K., & Kaczmarski, M. (2018). The Silk Railroad The EU-China Rail Connections: Background, actors, interests. OSW Studies, 72.

Koumparoulis, D. N. (2013). PEST analysis: The case of E-shop. *International Journal of Economy, Management and Social Sciences, 2*(2), 31–36.

Li, Y., & Schmerer, H.-J. (2017). Trade and the New Silk Road: Opportunities, challenges and solutions. *Journal of Chinese Economic and Business Studies, 15*(3), 205–213.

Lu, H., Rohr, C., Hafner, M., Knack, A. (2018). China belt and road initiative: Measuring the impact of improving transport connectivity on international trade. Rand Europe.

Mercator Institute for China Studies. (2018). Mapping the belt and road initiative: This is where we stand. https://www.merics.org/en/bri-tracker/mapping-the-belt-and-road-initiative.

Nazarko, J., Kuźmicz, K. A., & Czerewacz-Filipowicz, K. (2016). The New Silk Road – Analysis of the potential of New Eurasian transport corridors. 9th International Scientific Conference 'Business and Management'.

Organisation for Economic Co-operation and Development (OECD). (2018). China's Belt and Road Initiative in the global trade, investment and finance landscape. *OECD Business and Finance.* https://doi.org/10.1787/bus_fin_out-2018-6-en.

Organisation for Economic Co-operation and Development (OECD). (2020). OECD Interim Economic Outlook, March 2020.

Pepe, J. M. (2020). Value chains transformation and transport reconnection in Eurasia: Geo-economic and geopolitical implications. https://doc-research.org/2020/01/value-chains-transformation/.

Pieriegud, J. (2019). Analysis of the potential of the development of rail container transport market in Poland. European Commission Final Report Contract No 2018CE16BAT079.

Rajamanickam, V. (2019). The freight railway route from China to Europe is not generating business volume. Freightwaves. https://www.freightwaves.com/news/the-freight-railway-route-from-china-to-europe-is-not-generating-business-volume.
Rastogi, C., & Arvis, J-F. (2014). The Eurasian connection supply-chain efficiency along the modern Silk Route through Central Asia. World Bank.
Republic of Turkey Ministry of Foreign Affairs. (2020). Turkey's multilateral transportation policy. http://www.mfa.gov.tr/turkey_s-multilateral-transportation-policy.en.mfa.
Rodemann, H., & Templar, S. (2014). The enablers and inhibitors of intermodal rail freight between Asia and Europe. *Journal of Rail Transport Planning & Management, 4*, 70–86.
Sahbaz, U. (2014). The Modern Silk Road: One way or another? On wider Europe, The German Marshall Fund of the United States. https://www.gmfus.org/file/3277/download.
Schramm, H. J., & Zhang, X. (2018). Eurasian rail freight in the one belt one road era, Conference Paper, June 2018. https://www.researchgate.net/publication/328880505_Eurasian_Rail_Freight_in_the_One_Belt_One_Road_Era.
Talbot, V. (2018). Turkey and China: Towards a stronger partnership. In Talbot, V. (Ed.), Turkey: Towards a Eurasian Shift? Ledizioni.
Tavsan, S. (2019). Chinese freight train skirts Russia on new route into Europe. https://asia.nikkei.com/Spotlight/Belt-and-Road/Chinese-freight-train-skirts-Russia-on-new-route-into-Europe.
Trans-Caspian International Transport Route (TITR). (2020). https://middlecorridor.com.
Transport Corridor Europe Caucasus Asia (TRACECA). (2020). http://www.traceca-org.org.
Turkey Statistical Institute (TUIK). (2019). Foreign Trade Statistics. https://www.data.tuik.gov.tr/Bulten/Index?p=Dis-Ticaret-Istatistikleri.
Turkish Chamber of Shipping (DTO). (2019). https://www.denizticaretodasi.org.tr/tr/sirkuler/gemilerde-dusuk-sulfurlu-yakit-kullanimi-hk-11188.
Ulaştırma ve Altyapı Bakanlığı (UAB). (2018). Ulaşan ve Erişen Türkiye. [Reaching and Accessing Turkey, Ministry of Transport and Infrastructure of Turkey] Kasım.
Ulaştırma ve Altyapı Bakanlığı (UAB). (2020). 2020 Bütçe Sunumu [2020 Ministry Budget Presentation]. https://www.uab.gov.tr/uploads/pages/butce-sunumlari/2020-butce-sunumu.pdf.
International Union of Railways (UIC). (2017). Eurasian Corridors Stakeholder Group Meeting Presentation, Paris, 22 November.
International Union of Railways (UIC). (2020). Eurasian corridors: Development Potential.
United Nations Conference on Trade and Development (UNCTAD). (2018). Review of Maritime Transport 2018. https://unctad.org/webflyer/review-maritime-transport-2018.
United Nations Economic Commission for Europe (UNECE). (2018). Inland Transport Committee Report on Phase III of the Euro-Asian Transport Links Project Informal document ITC, 8. https://www.unece.org/fileadmin/DAM/trans/doc/2018/itc/Informal_document_No_8_EATL_3rd-phase_report.pdf.
United Nations Economic Commission for Europe (UNECE). (2019). *Transport Committee Working Party on Rail Transport Group of Experts towards Unified Railway Law*. Nineteenth session.
United Nations Economic and Social Commission for Asia and the Pacific (UNESCAP). (2016). Report on Documentation and Procedures for the Development of Seamless Rail-Based Intermodal Transport Services in Northeast and Central Asia.
United Nations Economic and Social Commission for Asia and the Pacific (UNESCAP). (2017). Comprehensive Planning of Eurasian Transport Corridors to Strengthen the Intra- and Inter-Regional Transport Connectivity. Study Report 2017.
United Nations Economic and Social Commission for Asia and the Pacific (UNESCAP). (2019). Strengthening Connectivity for Trade and Development: An Assessment of the Southern Asian Container Rail Corridor. South and South-West Asia Development Papers, 1901.
Van Leijen, M. (2018). New railway line Turkey-China now regular service. Railfreight. https://www.railfreight.com/beltandroad/2018/12/11/new-railway-service-turkey-china-now-regular/.

Van Leijen, M. (2019a). Alibaba sends e-commerce goods to Belgian Liege by train. Railfreight. https://www.railfreight.com/beltandroad/2019/10/10/alibaba-sends-e-commerce-goods-to-liege-by-train/.
Van Leijen, M. (2019b). New Silk Road in 2019: more trains, less empties and lots of politics. Railfreight. https://www.railfreight.com/specials/2019/12/30/new-silk-road-in-2019-a-wrap/.
World Bank. (2018). Aggregated logistics performance index. https://lpi.worldbank.org/international/aggregated-ranking.
World Bank. (2019). Belt and road economics: Opportunities and risks of transport corridors.
Zeybek, H. (2019). Uluslararası Ticarette Yeni Avrasya Ulaşım Yolları Arayışı: Demiryolu Yük Taşımacılığına Etkileri. [Search for New Eurasian Transport Routes in International Trade: Their Effects on Railway Freight Transport. Avrasya Etüdleri 56(2): 135–154.

The Role of Railway Transport Systems in Modern Multiscale Spatial Development. Bulgaria in the Orient Express

Plamen Patarchanov, Emilia Patarchanova, and Lyuben Stoyanov

1 Introduction

A system is a set of elements functioning in complex connections and interactions with each other. From this point of view, the most important thing for a transport system is the coordinated work and interaction of individual transport subsystems—which implies the transport system to be considered as a system that includes production activities by different modes of transport in a single organizational and operational unit.

Rail transport occupies today a leading position in the sector, offering a wide range of spatial solutions for individual regions with specific geographical characteristics, as well as answering to social preferences by different segments of the general population. Railway is thus extremely important in day-to-day mobility, for example for work commute, study commute, and every type of day-to-day travel. Achieving an integrated social space is possible by building a competitive transport system making efficient use of both regional and local resources. Environmentally friendly infrastructure and innovative low-carbon technologies play a key role in the development of the Single European Transport Area in open markets. The rail transport system promises to become a pillar of sustainable spatial development in coming decades. The realization of the strategic goals and priorities underlying both European and national transport policies can create conditions for effective spatial models. Other potential achievements by efficient railway transport systems include:

P. Patarchanov (✉) · L. Stoyanov
Geology-Geography Faculty, St. Kliment Ohridski, Sofia, Bulgaria
e-mail: p_patarchanov@abv.bg

L. Stoyanov
e-mail: desiro_man@abv.bg

E. Patarchanova
Faculty of Mathematics and Natural Sciences, SWU Neofit Rilski, Blagoevgrad, Bulgaria
e-mail: emilia_patarchanova@abv.bg

© The Author(s), under exclusive license to Springer Nature Switzerland AG 2021

S. Mitra et al. (eds.), *Railway Transportation in South Asia*, Contemporary South Asian Studies, https://doi.org/10.1007/978-3-030-76878-2_17

reducing the use of fuels and conventional energy sources, organizing efficient and environmentally friendly freight rail corridors, completing the construction and the maintenance of a dense network of high-speed passenger railways, structuring fully functioning multimodal services, etc. The railway transport systems of the future can restore something lost over half a century ago—the role of railways as a major pillar in regional, national, and international space. At the national level, the railway system's leading function is to promote the integrated economic and social development of a country by providing efficient (with maximum benefits and minimum costs) and sustainable (with minimal external influences) transport services to citizens, societies and business. In this way, railways can support a much more active, balanced regional development; they can also contribute to the full integration of a country into international structures such as the European Union. This is extremely relevant for the country of Bulgaria, due to its geographical position "at crossroads" and its consequent transit potential.

2 Context of the Study Area

European Union (EU) transport policies can be systematized into four main areas—design and implementation of integrated and intelligent transport systems; targeted logistics plans; implementation of the large-scale, green transport programmes; implementation of a single railway area (European Parliament, 2018).

The new EU policy guidelines and corridors of the Trans-European Transport Network (TEN-T), based on key transport documents, require the implementation of national and regional policies aimed at developing and enforcing the so-called Green corridors. The main points in the new policies are related to the development of integrated transport infrastructure-ports, intermodal terminals, rail and road networks, and highways. The share of intermodal transports is to be expanded by changing the structure and route of goods through the introduction of intermodal transport units—semi-trailers, interchangeable superstructures, and large containers. These units will require specialized vehicles as well as the provision of secure shuttle transport schemes.

The role of this type of transport for national development is manifold. In peripheral (rural, mountainous, etc.) and other types of functional areas in Bulgarian conditions railway transport does not just constitute an important factor in regional and especially local development, but also a vital tool (sometimes the only tool) for the implementation of basic public transport services. By contrast with both global and European dynamics in the development of railway transport, Bulgarian realities show a steady decline in the sector, in terms of both passengers and freight. This is especially true for rural areas, which have generated the most serious demographic, social and economic challenges in the last three decades (Patarchanov, 2019) (Fig. 1).

Fig. 1 Projects for the modernization of railway infrastructure in Bulgaria (*Source* Republic of Bulgaria [2018])

3 Literature Review

In the literature, a generalized opinion (Bakalova & Nikolova, 2010) considers transport systems as a set of vehicles and transport routes that provide spatial movement for goods and passengers based on the interconnected and coordinated operation of each mode of transport. In this view there must be a certain unity in the technological processes involved in transport, so that (Gerami & Colic, 2014) availability and efficient use of natural resources are ensured. Other concerns include connecting the regions of goods production with those of consumption, expanding the boundaries of product markets, and thus helping to achieve high living standards among the populations involved.

Transport systems (Rodrigue et al., 2006) consist of a complex set of relationships between demand, service points, and the networks that support traffic. These are most dependent on environmental conditions—which affect transport costs, capacity, efficiency, reliability and speed. Such conditions are closely linked to the development of transport networks, in terms of both capacity and territorial scope. At the same time, transport systems are evolving within a complex configuration of relationships between transport supply (determined by the scale of the network's operational capacity) and transport demand—all in the face of a given territory's specific mobility requirements.

Today, in order to increase the role of railway transport systems as an essential part of general transport systems, the concept of "earth bridge" is becoming more and more relevant (David & Stewart, 2008). In this way, the transit time is reduced and routes are made more efficient—which allows for economies of scale and reduces ecological footprints.

4 The Study's Objectives and Methodology

The present study aims to clarify the role of railway transport systems in modern spatial development. We focus on Bulgaria because of its geographical location as a natural intermediary between the West and the East. The country has been part of the "Orient Express" route since its inception.

Various scientific publications were used in this research—analytical research materials, normative, strategic and planning documents, all described in both national and foreign literature. Through these materials, we aimed to analyse opinions and interpretations by different authors and institutions on key issues related to our research interest. Based on various scientific approaches (chorological and chronological) and methods (comparative and spatial analysis, etc.) we articulated concrete views on a number of relevant issues.

5 Preliminary Data Analysis

Spatially speaking our study concentrates on the impact of the railway transport system at the national level. We do highlight some regional features by statistical regions—specifically, the European Union (EU)-based Nomenclature Des Unités Territoriales Statistiques (NUTS)-2.

The railway infrastructure in Bulgaria's South-West Region (SWR) has remained almost unchanged over the last decade. The reason for this is the delay of the projects under OP "Transport" and their expected completion after 2022. The SWR contains the longest railway network in the country (see Table 1).

The length of electrified railways (main and secondary) is higher than both the national average and the EU-27 average—51.9%. The region's territory is unevenly served by railway transport, as the railway network is located mainly in the northern half (districts of Sofia, Pernik, Kyustendil and Sofia-capital) and in the southwest. The railway network's density in the region is the highest in the country. It lags slightly behind the European average—49.1 km/1000 km^2.

Due to the peculiarities of the geographical environment, there is a more serious lag in the region (which also applies to the country as a whole) in terms of the construction of high-speed railways—seen as an analogue of railway connections between the capitals and larger urban centres in long-time EU member states. Another lack concerns high-speed railway networks so as to connect Sofia with capitals of neighbouring countries.

In terms of rail transport accessibility, the SWR again is the most favoured. Access to populations is relatively better due to the high density of railway stations

Table 1 Length and density of the Bulgarian railway network

Zones, regions	Railways in total		Doubled railway lines		Electrified railways		Density of the railway network
	km	% from the country	Km	% from the region	km	% from the region	km/1000 km^2
North and South-East Bulgaria	2,384	59.16	705	29.58	1807	75.80	34.89
North-West	648	16.07	192	26.63	444	68.52	33.98
North Central	627	15.56	89	14.20	436	69.54	41.88
North-East	484	12.01	244	50.42	369	76.24	33.41
South-East	625	15.51	180	28.80	558	89.28	31.57
South-West and South-Central Bulgaria	1,646	40.84	285	17.32	1063	64.59	38.57
South-Central	782	19.41	116	14.84	403	51.53	34.97
South-West	864	21.44	169	19.56	660	76.38	42.54
Bulgaria-total	4,030	100	990	24.57	2870	71.22	36.31

Source NSI (2019) and own calculations

and stops. The main railway lines serving this territory—the "Dragoman-Sofia-Plovdiv" (ETC10 and 4), the "Sofia-Karlovo-Burgas", the "Sofia-G.Oryahovitsa-Varna", the "Sofia-Kyustendil-Gyueshevo" (ETC8) and the "Vidin-Sofia-Kulata" (ETC4)—integrate the region with both the country's overall railway network and those of Southeast Europe as well as Middle and Far-East Asia.

The unsatisfactory condition of railway stations, low train speeds, and the poor condition of the rolling stock much contribute to the outflow of passengers from railway transport in the region (and the country as a whole). Another problem concerns the lack of adequate public access to functioning stations when these are far from the settlements served.

The most significant negative impact on the railway transport system in the region was the new spatial configuration of the highways along the meridional axis of development "north-south", which were built in the period 2005–2020. A huge spatial conflict or contrast exists between these modern highways and the lack of rehabilitation and modernization of existing railway routes in the region. Thus, at present, rail transport has lost the battle in the difficult competition with the dynamically developing road transport, and this concerns both freight and passenger transportation.

The second-longest railway network is that of the South-Central Region (SCR). It has lower values than the national average in terms of density, doubling and electrification. The districts of Plovdiv, Pazardzhik and Haskovo are well served, for here the railway network is best developed—and the density of the functioning railway lines is the highest. These territories are suitable for the development of production activities that require the transport of large quantities of goods. The density of the railway network in the Kardzhali district is low, as there are areas significantly distant from the railways and served only by road. In our country, only the territory of the Smolyan district is not served by rail, which prevents the deployment of new, cost-intensive economic activities.

The main railway lines serving the territory of the SCR—the "Sofia-Plovdiv-Burgas", the "Plovdiv-Svilengrad" and the "Ruse-Gorna Oryahovitsa-Stara Zagora-Podkova" are all elements of the European transport corridors. They spatially integrate the region internationally to the north, west and southeast. Parts of these networks are included in the European agreement for combined transport, through the container terminals in Plovdiv-Filipovo and Dimitrovgrad-North and the newest—at the Todor Kableshkov station, the degree of use of secondary lines is directly related to local economic development. Their load is often much lower than their capacity. Some of them, for example the narrow-gauge line "Septemvri - Dobrinishte", serve as basis for tourist attractions, travel and others—thus as a basis for the development of sustainable forms of public intercity transport.

The total length of built and functioning railway lines on the territory of the North-Western Region (NWR) is lower than the national average. After the commissioning of the Danube Bridge 2, which is a combined purpose, the load on the railway network increased. The main railway lines serving the territory of the NWR are the Sofia-Mezdra-Gorna and the Oryahovitsa-Varna, included in the AGC (European Agreement on Main International Railway Lines), and Vidin-Mezdra-Sofia included in the AGTC (Agreement on Important International Combined Transport). The two, together with the Sofia-Kulata line, form 5/6 of the route along the Priority Axis n°22 of the Trans-European Transport Network on the Bulgarian territory (in the direction of the Trans-European Transport Corridor n°4). The role of the other lines is of intra-regional importance as well.

The railway system of the North Central Region (NCR) has values close to the national average. It is connected to that of neighbouring Romania through the bridge over the Danube in the direction of the Gorna Oryahovitsa-Ruse-Bucharest. A significant spatial problem for the modern development of railway transport concerns the purchased character of the settlements and the disturbed demographic structure of the population, which strongly limits the needs for public transport services.

The railway transport in the North-East Region (NER) is the shortest in the country and includes three main railway lines: the Sofia-Gorna Oryahovitsa-Varna, the Karnobat-Sindel-Varna and the Ruse-Varna. The latter has an important role both for the region and for the functioning of the national transport system, as it provides a link between the international port of Ruse on the Danube and the ports of Varna-West and Varna-East of the Black Sea (respectively ECT 7 and 8). The industrial potential of the Varna-Devnya complex is the basis of both current and future consumption of railway services.

Railway networks are important for the economic development and the development of urban networks, as well as for providing transport access to populations in the Southeast region. After the periodic closure of sections in some inefficient railway lines in the period 2005–2010 in the region (Burgas-Pomorie), the total length decreased. However, there is in the area the highest share of electrified lines in the country—90%, well above the EU average. The network's density is the lowest in the country—although still higher than in some peripherally located EU countries such as Estonia, Greece, Spain, Ireland, Latvia, Lithuania, Portugal and others. In technical terms, a high level of modernization has been reached in the districts of Burgas and Stara Zagora. The territories of the districts of Burgas and Yambol / Sliven are suitable for the development of productive activities requiring the transport of large cargos. The area is peripheral in terms of railway transport services due to the absence of a railway border checkpoint on its territory and the difficult access by the population as a result of the lowest density of service stations and stops.

6 Results and Discussion

The leading condition in the choice of mode of transport concerns its competitive advantages—which include not only the price of transport, but also spatial conditions, environmental friendliness, noise pollution and energetic efficiency. When calculating the price for a transport service, social costs and benefits for its respective region must also be taken into account.

The European Parliament Resolution's on EU logistics and multimodal transport in the new TEN-T (Trans-European Transport Network) corridors emphasizes the importance of ensuring the free movement of people, goods and services—including through an efficient and sustainable freight system (Eur-Lex, 2017). The resolution's main directions concern providing benefits for regional development and growth when imposing a new infrastructure policy in connection with the TEN-T network. Of particular importance for modern sustainable spatial development is the application and implementation of new technologies and innovative solutions in railway transport, which will improve the activity of the sector and accelerate the transition to a secure, reliable and low-carbon transport system. The previous steps will facilitate opportunities to improve connections and develop infrastructure in areas where it is lacking, thus stimulating the European Railway Policy for a profitable perspective.

The adoption of Regulation (EU) nº 1316/2013 of the European Parliament establishing a Connecting Europe Facility provides for the financing of projects of common interest in railway transport and railway infrastructure (Eur-Lex, 2013). The annex to the Regulation also identifies the cross-border routes related to Bulgarian territory: the Sofia-Serbian border railway line and the Sofia-Macedonian border railway line. The horizontal priorities eligible for funding through the Connecting Europe Facility are Single European Sky (SESAR), traffic management systems in the road, rail and inland waterway transport (ITS - Intelligent transportation system, ERTMS - The European Railway Traffic Management System and RIS - River Information Services) and ports, motorways of the sea and airports of the core network (Ministry of Transport, Information Technology and Communications, 2017).

The territory of Bulgaria is crossed by two corridors of the main TEN-T network:

- The Orient-Eastern Mediterranean corridor connects the German ports of Bremen and Hamburg, passing through the Czech Republic and Slovakia, with a junction crossing Austria and continues through Hungary to the Romanian port of Constanta, the Bulgarian port of Bourgas, with a connection to Turkey and Greece. The Piraeus includes railways and highways, airports, ports, rail-road terminals and inland waterways on the Elbe River;
- The Corridor "Rhine River" linking Strasbourg and Mannheim through two parallel axes in southern Germany—and to the Romanian port of Constanta and Galati. It includes railways and highways, airports, ports, rail-road terminals and inland waterways along the Main River, the Main-Danube Canal, the entire Danube valley from Kelheim downstream, as well as the Sava River. Under Regulation (EU) nº 913/2010 by the European Parliament on the development of a European rail network for competitive freight, 10 rail corridors have been set up for priority transport of goods crossing the territories of EU member states. Railway freight corridor 7 (Orient-Eastern Mediterranean) crosses the territory of the Republic of Bulgaria. The corridor's main route is Prague-Vienna/Bratislava-Budapest-Bucharest-Constanta and Arad-Craiova-Vidin-Sofia-Kulata-Thessaloniki-Athens—with alternative routes Videle-Ruse distribution-Sindel, distribution-Nova Zagora-Svilengrad and Sofia-Plovdiv-Svilengrad. Traffic management in the Bulgarian part of the corridor, its maintenance and modernization are the responsibility of the State Enterprise "National Railway Infrastructure Company" (NRIC).

Railways transport constitutes a key element in the Bulgarian transport system. Its development is a prerequisite for the country's economic development. An important component of population mobility concerns the provision of affordable rail transport services.

The main problems for the development of railway transport in Bulgaria are:

- The unsatisfactory state of infrastructure and rolling stock, which leads to low speed and poor level of passenger and freight services;
- The insufficient integration of the national railway network in European-level networks;
- Insufficient connections between the national rail network and both marine and inland ports. Such relationships must be developed so as to increase co-modality;
- Deficiencies in signalling systems;
- The need for implementation of a European railway traffic management system.

The future development of rail transport—passenger and freight—is of great importance for the national transport policy for several reasons:

- Congestion of road infrastructures;
- The national interest in safety and environmental protection;
- High fuel prices.

Insufficient funding for railway infrastructure remains a major problem for the development of the railway sector—and to its ability to compete with other modes of transport. This leads to a reduction in the quality of the service offered. Other negative consequences in this regard are the modernization of "piecemeal" sections. A common situation occurs wherein 80% of a given railway corridor has completed the necessary activities and implemented the relevant projects for modernization of the infrastructure, while the remaining 20% remain as "narrow places"[1] in the railway network, reducing the route's capacity and the quality of the overall transport service. This has a direct impact on the performance of railway undertakings and constitutes a major obstacle to both competing with other modes of transport and attracting new investment.

> The construction of a modern railway network in the Republic of Bulgaria—one compatible with European structures—is of paramount importance and is present in all programmes and planning documents by the MTITC. (Ministry of Transport, Information Technology and Communications of Bulgaria)

The main goal is to build and develop key transport links of both national and European importance, so as to achieve maximum spatial effect and reduce the transport vacuum created in some regions. For Bulgaria, the main point in this direction concerns the fastest possible completion of the main projects for modernization of the railway routes, as well as the connection of Bulgaria's main railway network with that of the neighbouring countries. The implementation of these tasks is of particular importance for overcoming interregional differences.

In line with European requirements for railway transport, the state should play a significant role in reducing the large fund deficit for the maintenance and repair of railway infrastructure by providing cash flow from the state budget according to the annual budget forecast. This will allow the reduction of infrastructure charges and the transformation of railway transport into a highly competitive and attractive sector, and not only from an ecological point of view. The expected introduction of a toll system based on distance travelled for trucks (over 3.5 tons) would contribute to attracting more freight traffic into the Bulgarian railways. It is necessary to much invest in the development and maintenance of railway infrastructure in those main areas capable of absorbing the increase in traffic. The development and maintenance of transport infrastructure are unattainable without the full absorption of funds under the OPTTI, (Operational programme Transport and Transport infrastructure) the Connecting Europe Facility, or without the financial support from the state itself. Taking into account the capabilities of the Medium-Term Budget Forecast for 2019–2021 and the state budget for 2019 (Ministry of Finances of the Republic of Bulgaria, 2019), the management of the National Railway Infrastructure Company expects the state to take adequate measures to correct the net deficit by providing additional financial resources under the subsidy procedure of railway infrastructure. The funds are needed to invest in the current maintenance of the railway infrastructure so as to maintain the quality of the railway service at the current level as well as ensure safety

[1] "narrow places"—sections of railway lines with violated transport and access parameters. For example, a double railway with a section of a certain length that is single and reduces the maximum number of trains that can run on the route per day (and partially violates the speed parameters of the railway).

and security (Ministry of Transport, Information Technology and Communications 2019a).

The Ministry of Transport, Information Technology and Communications will continue to promote all initiatives in support of intermodal transport and strive to provide an environment for the efficient functioning of logistics systems, which are particularly important for the development of settlements and have a strong spatial effect. An example of this is the Plovdiv railway junction, which has further increased its importance as a regional centre for the exchange of passengers and goods, after the completion of the intermodal cargo terminal at Todor Kableshkov station (Zlatitrap village). The construction of this terminal has increased the economic and logistical performance of the South-Central Region. The importance of this intermodal terminal has grown even more since the COVID 19 pandemic transport crisis. The long columns of trucks at the Kapitan Andreevo checkpoint (at the frontier with the Republic of Turkey) reached over 35 km. Under these conditions, most truck containers travelling in the direction of this transport corridor were reloaded at the intermodal terminal and departed as long train compositions running on a clock schedule, which ensured fast and trouble-free delivery of goods to the final destination (Stoianov, 2020). Over the next years, these efforts will be aimed at creating favourable conditions for the modernization and improvement of the network of intermodal terminals, as well as providing high-quality transportation au-logistics services, which will be of direct benefit to capacity development and specialization of space. In this regard, several concession mechanisms will be used, providing flexible financing of the supply infrastructure and new high-capacity intermodal facilities—thus creating preconditions for the growth of the latter into freight settlements: the Zlatitrap-Todor Kableshkov station, in the town of Zlatitrap, the Ruse-Ruse-distribution station, the Svilengrad-Svilengrad station, the Plovdiv-Plovdiv-distribution station, the Burgas-Burgas station-freight.

Reconstruction of key station complexes along the main railway lines will continue during the new programming period. It aims to improve the intramodality of both passengers and freight by connecting these complexes with other modes of transport—as well as by offering communication solutions for their transport and pedestrian connections. Six station complexes are planned for reconstruction: Stara Zagora, Nova Zagora, Karnobat, Poduyane, Iskar and Kazichane.

In terms of spatial development, the railway sector of the Republic of Bulgaria should create the necessary conditions for the country's economic and social development, namely as concerns ensuring effective and resistant transport capable of supporting a balanced regional development. The network of railway lines in our country is entirely structured and configured. Over two-thirds of the routes (71%) are electrified. The share (almost 25%) of the doubled railway lines is not small either. According to these parameters, a much more intensive operation by fixed infrastructure is possible and, accordingly, it is possible to satisfy a future, higher demand for transport services.

The policy pursued so far as concerns railway infrastructure is aimed at the renewal and/or repair of individual sections. Unfortunately, this does not change the overall transport capacity of Bulgarian Railways. The general condition of the railway infrastructure requires urgent measures so as to restore, improve and increase the railway's technical and operational parameters, its contact network, and its communication and security equipment. Another issue concerns the elimination of bottlenecks, especially on the lines that are part of the main Trans-European transport network. It is important from a regional point of view to modernize not only the routes that will generate transit traffic but also the secondary sections with high revenue indicators, mainly from passenger transport, intensively involved in human mobility and facilitating opportunities for daily business trips for relatively short distances. A good example in this respect is the railway line Plovdiv-Krumovo-Asenovgrad, which provides passenger electric trains on a clock schedule, with an easy and fast access to the economic core of the South-Central Region-of Plovdiv. The service has a high level of competitiveness with both road and bus transport. The distance between the two city centres is 20 km. The train takes this distance in 23–25 min, serving a total of 5 passenger stations and stops. One of them is Plovdiv Airport (railway station Mavrudovo), which is of particular importance for dual development of both types of transport (rail and air) at this point—with a strong spatial effect and in line with EU policy of "intermodalizing people".

The practical realization of the aforementioned idea of effective Trans-European Transport Network (TEN-T) is a key element of the Lisbon strategy for competitiveness and UV jobs on the continent. If Europe is to realize its economic and social potential, it must build the still missing communication links, remove the narrow sections of its infrastructure and ensure the sustainability of future transport networks (Ministry of Transport, Information Technology and Communications, 2019b).

Given the increasing traffic between member states—for which expectations (before the COVID 19 crisis) was a doubled traffic in 2020, great investments are needed for the modernization and development of transport infrastructure. This is expected to exceed 550 billion euro in the period 2010–2030 (of which 215 billion euro are to target priority axes and projects). Given the scale of these investments, it is crucial to prioritize projects in close cooperation with national governments, ensuring effective cooperation at the European level.

The Union (Official Journal of the European Union, 2013) plans to structure investments at two levels. A core network comprising the most important international relations should be completed by 2030. An expanded or comprehensive network is to provide full EU coverage and access to all regions; it should be completed by 2050. These two levels cover all modes of transport: road, rail, air, sea and inland waterways, as well as intermodal platforms. The investments from European infrastructure funds should be directed to corridors of the main Trans-European transport network, namely: the direction Vidin-Sofia-Kulata; the Plovdiv-Svilengrad-Turkish border and the Plovdiv-Burgas. By the end of the strategy period, the MTITC's efforts will be focused on the construction of the East–West route through Southern Bulgaria: from the Serbian border through Sofia and Plovdiv to Bourgas and Svilengrad and the Turkish border. Modernization activities will also include sections of the direction "North–South": Vidin-Medkovets and Sofia-Pernik-Radomir. The railway junctions will be built around Sofia (partially), Plovdiv and Burgas—as well as intermodal connections in Ruse. Several key railway stations in the mentioned direction have already been renovated, including Sofia, Burgas, Pazardzhik. The modernization of Stara Zagora, Nova Zagora, Chirpan and other stations is forthcoming. The newly built lines are equipped with modern traffic management systems. For projects beyond the scope of the EU's paramount importance, but of common European interest and included in the comprehensive TEN-T network, which is of great national importance, a "National Investment Program" has been prepared. It includes projects for which no financial sources have been identified at the moment—we must promote the projects to international financial institutions. The scope of this Investment Program includes projects in the direction Sofia-Varna (section Mezdra-Kaspichan), Ruse-Varna and Karnobat-Sindel. Potential sources for the implementation of the programme include national financing and/or state loans from IFIs. At the moment, the projects Ruse-Varna and Karnobat-Sindel have been proposed for implementation through financing from this new external state debt. Taking into account the potential for the development of international and transit railway freight transport, and in order to increase the competitiveness and sustainability of railway transport, feasibility studies should be initiated for the modernization of the railway line Ruse—Gorna Oryahovitsa—Dimitrovgrad, as well as for new railway crossings towards both Turkey and Greece. The EU-funded Sea 2 project aims to create a multimodal transport corridor by creating an efficient railway connection between the Greek ports of Kavala and Alexandroupolis on the Aegean Sea and the Bulgarian ports of Varna and Bourgas on the Black Sea—and the port of Ruse on the Danube. The project identifies the need for a study on feasibility and sustainability for the construction of a railway connection between Komotini and Bulgaria at Nymph. There is a great potential for active development of Europe-Asia relations. The possibility of implementing a railway connection to Tekirdag port on the Sea of Marmara, which is connected to the national railway network of the Republic of Turkey, is currently under study.

Since the opening of the Marmaray tunnel under the Bosphorus for regular use, the importance of the Plovdiv-Svilengrad railway section has increased many times over. Thus, in practice, the stations of Svilengrad, Dimitrovgrad, Plovdiv and Sofia turned out to be located on the same route as the stations of Beijing (PRC), Tehran (Iran), Baghdad (Iraq), Ankara, Istanbul (Turkey), Belgrade (Serbia), Vienna (Austria).

The completion of this transcontinental route from London to Beijing is one of the greatest technical achievements of the twenty-first century. As a significant part of the Orient Express route is a natural "bridge", not only between large geographical areas but also between separate, important periods in the history of two continents—Europe and Asia. It creates a very sustainable infrastructural base for the sustainable spatial development of huge sections of these two regions, as well as that of several geopolitical and economic centres. The construction of a continuous 15,000 km railway system is an achievement that will feature in all textbooks in geography and history—yet more importantly, it is an achievement that will allow to transport millions of tons of freight and thousands of passengers. This endeavour should not just be noted, but deeply studied and upgraded.

That is why the railway project Plovdiv-Svilengrad is so important for our country. In no case should it be seen as a project of national importance alone, but as part of a transcontinental route of great importance. The accelerating development of combined transport and the related construction of intermodal terminals will support the nation's economy and improve the condition of state structures related to the processing of goods and commodities. After the completion of the project, the quality of the service offered by the railway carrier will be improved, as will fast international connections. This will improve communication and relations not only between Balkan countries, but also with those with Middle and Far-East Asia.

7 Conclusions and Recommendations

The implemented policy and guidelines in the spatial development of the railway sector should take into account the trends in the sector's European development and help achieve both short-term and medium-term goals for promoting the use of railway services. The competitiveness of rail transport is a key factor in maintaining and increasing the sector's spatial impact. It is necessary to use the railways' natural advantages.

A serious problem with a negative impact on spatial development is the insufficiently effective Republican transport scheme, which leaves 80% of stations outside settlements, without leading transport. This does not allow to develop the advantages of suburban rail transport, which would reduce congestion in large cities.

The balance between modes of transport is strongly distorted in favour of road transport. This is contrary to EU priorities and has a lasting negative effect on both spatial development and the environment.

The integration of the country's transport system is a necessary condition for achieving full (political, economic and social) cohesion of Bulgaria within enlarged Europe. It is also in full compliance with EU policies for overcoming infrastructural imbalances between the central and peripheral regions. In order to ensure real freedom of movement of people and goods, this priority is extremely important for Bulgaria—which lies at an EU external border (Official Journal of the European Union, 2013).

Based on the empirical research available, some opportunities for sustainable development of the transport sector of Bulgaria have been identified. These are essentially related to improving the quality of railway transport infrastructure in Bulgaria—as well as applying global practices leading to sustainable transport development. In this regard, a great potential related to the construction and operation of quality railways and roads, integrated with the intermodal transport system, has been localized.

References

David, P. A., & Stewart R. D. (2008). *International logistics: The management of international trade operations*. Thomson.

Eur-Lex. (2013, December 11). *Regulation (EU) No 1316/2013 of the European Parliament and of the Council of 11 December 2013 establishing the Connecting Europe Facility, amending Regulation (EU) No 913/2010 and repealing Regulations (EC) No 680/2007 and (EC) No 67/2010 Text with EEA relevance*. https://eur-lex.europa.eu/legal-content/EN/TXT/?uri=celex%3A32013R1316

Eur-Lex. (2017, January 19). *European Parliament resolution of 19 January 2017 on logistics in the EU and multimodal transport in the new TEN-T corridors (2015/2348(INI))*. https://eur-lex.europa.eu/legal-content/EN/TXT/?uri=CELEX%3A52017IP0009

European Parliament. (2018, February 23). *Report on a European strategy on cooperative intelligent transport services*. https://www.europarl.europa.eu/doceo/document/A-8-2018-0036_BG.html

Ministry of Transport, Information Technology and Communications. (2017). *NATIONAL IMPLEMENTING PLAN on the technical specification for interoperability relating to the 'control, command and signaling' subsystems of the rail system in the European Union*. https://ec.europa.eu/transport/sites/transport/files/rail-nip/nip-ccs-tsi-bulgaria-en.pdf

Ministry of Transport, Information Technology and Communications. (2019a). *Draft budget for 2020 and updated budget forecast for 2021 and 2022 in program format*.

Ministry of Transport, Information Technology and Communications. (2019b). *Budget for 2019 and updated budget forecast for 2020 and 2021 in the program format of the Ministry of Transport, Information Technology and Communications*.

National Statistical Institute, Republic of Bulgaria. (2019). *Length of rail network*. https://www.nsi.bg/bg/content/1737/%D0%B4%D1%8A%D0%BB%D0%B6%D0%B8%D0%BD%D0%B0-%D0%BD%D0%B0-%D0%B6%D0%B5%D0%BB%D0%B5%D0%B7%D0%BE%D0%BF%D1%8A%D1%82%D0%BD%D0%B8%D1%82%D0%B5-%D0%BB%D0%B8%D0%BD%D0%B8%D0%B8

Official Journal of the European Union. (2013, December 12). *Regulation (EU) 1315/2013 of the EP and of the Council*.

Patarchanov, P. (2019). *Railway transport in regional and local development of the rural areas—Challenges and opportunities*. 19th International Multidisciplinary Scientific GeoConference SGEM 2019. pp. 533–540. https://doi.org/10.5593/sgem2019/5.4/S23.070

Republic of Bulgaria. (2018). Operational program "transport" 2007–2013. https://www.eufunds.bg/archive2018/archive/documents/1390288027.pdf (in Bulgarian).

Rodrigue, J. P., Comtois C., & Slack B. (2006). *The geography of transport systems*. Routledge.

Бакалова, В., Хр. Николова. (2010). Икономика на транспорта. [Economics of transport] С. УИ УНСС.

Герами, В. Д., А. В. Колик. (2014). Управление транспортными системами. Транспортное обеспечение логистики [Transport systems management].

Стоянов, Л. (2020). Социално-икономически аспекти на пространствено въздействие на скоростните жп трасета [Socio-economic aspects of spatial impact of high-speed railway routes].

Privatization of Indian Railway Services: The Story so Far

Paulose N. Kuriakose and Vallary Gupta

1 Introduction: Deregulation and Privatization

Privatization is recommended by some policymakers as the panacea for structuring and reorganizing ailing industries. Although used interchangeably, deregulation and privatization indicate different meanings. While deregulation refers to the elimination of all types of institutional or legal impediments to the entry and exit of all firms, intending to encourage competition in the market, privatization means the transfer of proprietorship rights from the public to a private entity. Such transfer may involve passing over ownership to a single individual/firm or a group of firms in the private sector.

The theoretical support to deregulation and privatization is attained from the work of the American economists on 'contestable markets' (Baumol, 1982). A contestable market refers to one where present firms are susceptible to hit-and-run entry and all have access to similar methods of production, as a result, their cost functions are alike. Its essential characteristic is that the entry includes no sunk costs, and a firm can enter without making irreversible expenditures. As firms are obliged by the meagre risk of entry to produce at minimum cost, there is internal efficiency at equilibrium in a contestable market. It is entry threats that promote competition in such markets.

The contestability theory has been at the core of the new American enactments on deregulation, like that of telecommunications, road haulage, and airlines. The Austrian school of thought that stresses on ownership rights and incentives provided motivation for privatization in the UK to allow profit maximization in the development of an optimal industry structure. The theory of property rights offers a legal

P. N. Kuriakose (✉)
Department of Urban and Regional Planning, School of Planning and Architecture, Bhopal, India
e-mail: paulosenk@spabhopal.ac.in

V. Gupta
Research and Knowledge Management, Coalition for Disaster Resilient Infrastructure, New Delhi, India

© The Author(s), under exclusive license to Springer Nature Switzerland AG 2021
S. Mitra et al. (eds.), *Railway Transportation in South Asia*, Contemporary South Asian Studies, https://doi.org/10.1007/978-3-030-76878-2_18

ground for the privatization of those public enterprises where losses in productive efficiency compensate for losses arriving from allocative weaknesses. Handover of rights from the public to the private sector (or vice versa) to remaining profits from operating an enterprise implies a relationship change amongst people involved in making the firm's decisions and the profit beneficiaries. The adjustment in the designation of rights prompts an alternate structure of incentives for executives. The British theory of the privatization has been legalized within the system of the impacts of such incentives. In developed countries, privatization of public ventures involves the sale of equity to several investors, with each having a comparatively small part of the total equity of a certain enterprise. Thus, a stakeholder can transfer his property rights and end his association with the company at any time. As a result, considerable variations in the number of principals and shareholdings distributions can take place quite swiftly (Alchian & Demstetz, 1972). The management's search of its objectives thus becomes restricted by the enterprise's shareholders who seek contractual engagements with management that increase their payoffs, new investors who may acquire the firm's shares before altering current arrangements, and the enterprise's creditors who pursue managerial changes under threat or definite default.

Since the late 1970s, many countries have been coping with the need to restore the financial and market performance of their railways. The need for competitive adjustments is a challenge that both privately owned and state-owned railways have had to face. Indeed, well managed privately owned railways are in a perpetual state of strategic realignment. Continuous reinvention is their response to competition. Publicly owned railways, on the other hand, require higher government authority and explicit public policy redirection before they can respond to the challenge of improving their competitiveness.

2 Problem Statement

Railways define one corner of the envelope of state-owned enterprises that are viable candidates for restructuring or privatization. Railways define the frontier between public utilities and infrastructure and the outer limits for large-scale enterprises suitable for conveyance to the private sector. Not long ago, the textbook wisdom was that railways were a natural monopoly and that large economies of scale in producing rail services could be realized only through centralized management. The role of the state was required not only to assure that the economies of scale were realized through massive infrastructure investment but also to assure that they were fairly distributed within the national economy. This perception no longer applies. The technology base of the railway industry has shifted; with advanced technology, the industry has become less labour and capital intensive and significantly more information and communication intensive. Transportation markets themselves have changed since most state-owned railways were organized.

3 Methodology

This study reviews the case studies policies, current status, barriers and opportunities related to the infusing of private finance mobilization options to the Indian Railway. This study relies on exploratory research methods like secondary data available with various stakeholders, focused group discussions with different stakeholders and the information collected were then analysed for understanding the problems and issues better and coming up with innovative solutions. This chapter is heavily based on the existing published information by various Central and State Governments. During the study, the team interacted with relevant government officials, experts in their affiliated fields of technical and academics along with independent research organizations to gain perspectives of the opportunities and growth of the rail projects in India.

4 Privatization of Railways Lessons Learned

4.1 Experiences from Britain

Railway Reforms in 1990 led to increasing competition in many European countries. Privatization of railways in Britain began in 1992 when the government drew its intentions with a White Paper- 'New Opportunities for the Railways' in 1992 and the Railways Act in 1993. The document proposed radical restructuring of British Rail as a step towards privatization. It was done with intent to end British Rail's monopoly and provide long-term improvements in railways. It was based on the rationale that the introduction of competition and the creation of incentives to all parties would lead to better economic efficiency (Foster, 1995).

The 1993 Railways Act acted as the base by vertically segregating the rail business into operations and infrastructure. It further horizontally separated passenger operations into 25 separate operating businesses franchised to the private sector, and over 60 other separate businesses like infrastructure maintenance, rolling stock maintenance were made and transferred to the private entity. A regulatory body was established to ensure fair charges and control competition amongst operators (Preston & Robins, 2013; Shaw et al., 1998). Research shows that privatization led to major efficiency improvements. Privatizing rail services benefited users with lower regulated fares and the introduction of innovations in services and retail distribution like telesales and web-based sales (Preston & Robins, 2013). Output quality has improved as compared to when under public ownership output quality has improved as compared to when under public ownership (Pollitt & Smith, 2002). It also resulted in an increase in punctuality, safety on railways improved, reduced government subsidy per journey, and reduced operating cost per passenger mile (Subir, 2019).

However, other researchers suggest that privatization of Britain's passenger railway services has not yielded an overwhelming success. The process that started with the aim of encouraging competition and lessening the government's involvement has over time resulted in the re-formation of the monopoly of British Railways in the private sector. It has been observed that privatization was more of an application to prepare regulation rather than being an effort to enhance competition. The withdrawal of a mass service necessitated the need to impose social regulation in the form of subsidies. The threat to complete the process of franchising post-privatization stimulated regulation which hindered the very prime intent of maintaining competition. The re-emergence of monopoly also necessitated regulations which were accompanied by other regulations in the form of quality-of-service specifications (Moore, 1992; Shaw et al., 1998).

The government was burdened with financial threats and infrastructure costs. Much of the passenger railway system operated and continues to operate at a loss. As per the Serpell Report, almost 85% of the passenger railway network would have to be closed to run a commercially viable industry (Department of Transport, 1993). The case suggests that vertical separation emphasizes difficulties as private sector entities will not compete over the right to lose money. The vertical separation created the need for monitoring other's performance, resulted in high transaction costs, and added trouble of creating complex performance schedules (Krueger & Anne, 2004).

The complex organizational structure led to increased costs, both direct and indirect. The under-investments in infrastructure further aggravated the costs. It is argued that the benefits accrued to customers and others have not been adequate to outweigh the increased costs.

The British privatization model has been inspired by a multiplicity of objectives that it becomes hard to frame a strong rationale behind it (Kay & Thompson, 1986). The privatization process in the United Kingdom (UK) is based on the belief that it will cause managers to put more weight on profit maximization and industrial and managerial innovations. This is because managers in private sectors are exposed to incentives unlike those applicable to the public sector. While in UK's scheme for privatization, the importance is given to the need to promote internal efficiency and boost innovations through competitions, the main interest of the treasury has been to raise revenue from privatization (Dalvi, 1995). The case shows that high costs and lesser revenues, uncertainty in policy, absence of a regulator to provide a level playing field, and incentives for investors along with operational problems considerably restrict private sector participation (Singh, 2015).

4.2 Experiences from Japan

Different countries have practised privatization of passenger railways in various ways with railway restructuring in Japan been considerably dissimilar to European nations. The former state-owned Japanese National Railways (JNR) was privatized in 1987 and segregated into six vertically integrated passenger railway establishments. Before that, it was a public corporation that played a key role in providing rail passenger services. However, over time it began to lose its competitive edge to cars and airlines so much so that between 1965 and 1985, its market share in terms of passenger kilometres declined from 45% to 23%. JNR showed an operating loss for the first time in 1964 (Mizutani & Nakamura, 2004). The failure to introduce consumption tax to ease its dependency on revenues from income tax and a lack of freedom to set its budget or fares added to the cost burdens. A nationwide fare not reflective of region-based cost differences was in place. In short, JNR encountered various organizational problems and political interference. All this together paved the way to privatization in 1987 (Krueger & Anne, 2004; Nakamura, 1996).

Since the beginning of the privatization, the results of the reforms have been revealed in increased productivity, quality of service, and diversification. JNR privatization was performed in a step-by-step manner. To prevent the negative effects on stock prices due to the poor reputation of JNR, the stock of newly established Japan Railways was not immediately provided for public use during the initial period. It was done until the newly privatized companies established a good public reputation by increasing efficiency. The distinguishing features of the Japanese approach to rail privatization involve horizontal separation or regional subdivision, separation of freight and passenger services, vertical integration or integration of rail operation and infrastructure, establishing an intermediate institution to pay off debts and find jobs for its terminated employees, a subsidy for low-density Japanese Railways to stabilize their management, and an allowance to private companies to engage in non-rail business (Mizutani & Nakamura, 1997).

Unlike the case of the European rail industry and British Railways, vertical integration in Japan Railways continued after privatization. The government since privatization has interfered less in aspects related to investment and financing, in corporate management, and also in fair regulations. Unlike in Britain, special laws like guarantees of a certain degree of autonomy have been implemented to prevent political intervention. Competition introduced due to privatization benefited JNR as it became a viable alternative to aeroplanes. Companies focused not only on reducing transport time but also on the quality of services and price. The privatization attracted businesses away from other transport organizations reducing their ridership. The performance of Japan Railways improved significantly between 1987 and 1990 showing an average per annum increase in passenger and cargo from an initial 5% to 10% (Nakamura, 1996).

The Japanese approach to rail privatization has been successful in many ways as its improved performance in many areas, like finances, service quality, the productivity of labour, and operating costs. Also, fares for nine years after privatization did not increase and the rate of accidents also remained low. Japan's experience suggests that privatization policy helps to transcend vested interest. While the example has been largely a success, caution must be practised that privatization does not result in a simple transfer of monopolistic power from a public establishment to the private sector. It is competition rather than privatization that leads to efficiency. The key objective of privatization is to introduce competition in the market and also inside the organization. The utmost important lesson is that privatization can result in the creation of a powerful platform both for the company and society as a whole. As argued by (Moyer et al., 1992), the restructuring of railways must focus on elements that promote its ability to meet the requirements of potential users. The privatization of JNR encouraged innovative services and created more value-added interests for the public thus bringing social welfare for the society (Huang & Michael, 2019).

4.3 Experiences from United States (US)

The main focus in the US in the absence of any nationalized sector has been to encourage a competitive industrial arrangement through the deregulation of private industries dominated by monopolies. The US policy was framed upon a clear rationale derived from contestability theory and was anticipated that deregulation would transform managerial practices and bring in innovation, causing productivity improvements. This would result in a drop in prices, improvement in the quality of products, and finally removal of cross-subsidy. The bankruptcy of the Penn Central led to the introduction of the Railroad Revitalisation and Reform Act in 1976 and the Staggers Act (rail) in 1980. The legislation segmented industry into competitive and market-dominated domains (Dalvi, 1995). The railway restructuring in the US and Canada was a bottom-up process wherein the restructuring initiative of separating light density lines and unifying business interests were taken up by the private sector. Regulations were however established and enforced by the government. The lessons from Canadian experience primarily deal with the interfering nature of economic regulations and the consequences of cross-subsidies. It suggests that regulations that support uneconomic activities significantly diminish the incentives for reorganization (Kopicki & Thompson, 2012).

Deregulation of railways in the US began in with Staggers Act that allowed railway establishments to compete with one another and resolve transportation charges (Association of American Railroads, 2006). In contrast with Europe, where infrastructure is government-owned, US railway deregulation is built on the basis that infrastructure is owned by the operators. It is a successful example of deregulation with its total modal being 40%. As per a World Bank report, capital investments like total route length and the total number of locomotives have reduced (World Bank, 2006). Since 2004, the selected US Class I Railroad companies have exhibited increasing profits with steadily rising revenue flows (Thomson Financials, 2006). Deregulation benefited as transport costs reduced with rising productivity. For example, the railway rates dropped by 7% and rail movement between 1981' and 1984 increased by more than half (Dalvi, 1995). Literature suggests that US deregulation benefited both companies and customers and led to in volumes have and considerable productivity rise. The major advantage has been delivered to customers as deregulation led to a reduction in price rates. Along with improved efficiency, innovations were also fuelled. The development however may pose a risk to long-term investments and profitability (Rennicke, 2004).

4.4 Experiences from Across Globe

Unlike the case of Britain, vertical segregation yielded a positive experience in Sweden. It was in the 1950s when the Swedish State Railways (SJ) began to enter a phase of deterioration and financial problems. As a result, several lines were shut down and direct subsidies were introduced to cater to unprofitable. Despite an increase in the number of measures taken, the financial situation further deteriorated during the 1970s and 1980s. In an attempt to restructure the industry, Transport Policy Act 1988 was introduced to make the circumstances for rail more similar to that for the road. The Act is widely indicated as the beginning point in the railway reform process of Sweden. Its key element was vertically separation of track infrastructure from train operations. The implementation of the Act started a step-by-step liberalization process and deregulation that continues to this day (Gunnar & Konstantinos, 2013). Unlike the UK, that privatized railway maintenance, Swedish National Railways was split up into independent companies whose ownership stayed with the Swedish state. Railway infrastructure in Britain is under government control with maintenance operations being outsourced.

Sweden has a well-developed trail system, equipped with technical innovation, and exhibiting average productivity in comparison with other European countries (Woxenius, 1998). Passenger transportation has become considerably popular witnessing an increase of as much as 40% compared to values in 1990 (European Union, 2004). Research suggests that actions accomplished in 1988 have made all desired benefits, and there is a need to promote increased competition in passenger transportation operations (Carlson, 2004).

In countries from Latin America, like Argentina, Brazil, and Mexico the major form of privatization was the transfer of responsibility related to services and infrastructure investment to private parties, based on long-term concessions. The experiences from these countries show that parts of railway networks can make profits, but other parts need public subsidy. Concessions did lead to cost reduction but mainly through job loss, which took happened on a very large scale. World Bank report suggests that direct railway employment reduced by 75% (Martin, 2002). Both Latin America and Japanese railway reform followed the segregation of railways into monopoly franchises. It consisted of tracks and service operations allowing competition during the process of bidding for the franchise. UK, Sweden, and Germany on the other hand involved splitting of infrastructure and operations. The company that owned infrastructure was subjected to regulations in the interest of the public.

In the case of both Argentina, and Japan, though rail traffic characteristics varied, and rail network conditions were in total contrast, railway restructuring was required due to huge losses incurred by the railways with their unionized labour force. Even for Chinese railways, that were closest to the Indian system with no segregation of government and railway management, an effort to restructure railways involves the formation of independent enterprises, and the reorganization of the railway along commercial lines. Restructuring in China has already passed ahead of diversifying ownership, introducing competition through franchises, and developing specialized services, and price regulation relaxations (Wu & Zhao, 1998).

Privatization when compared across countries reveals that it leads to an increase in labour productivity, with a major increase in the UK, Sweden, and Japan. The growth percentage for French National Railways between 1987 and 1995 was 1.179, 1.668 for British Railways, 1.598 for Swedish State Railways, and 1.255 for the Japan Railways. The reform led to an increase in rail demand with an increasing demand for passenger rail (Mizutani & Nakamura, 1997). There major two different factors that determine implications for railways deregulation are segregation of infrastructure from operations and free competition. The case of the UK highlighted the extreme presence of the two, while Sweden and the US employed them either partly or completely avoided it. The decision in the case of the US could be a major reason for its success. It is this felt that the government should remain the main owner of both infrastructure and railway operations, to permit medium-term thinking in day-to-day decisions. A move away from ownership of railway operations should be ensured, with infrastructure staying under strict control, and gradually under public ownership.

American model unlike the British approach encourages more competition and organizational and technological adjustments that improve the production efficiency. The American approach is based on the idea of contestability and industries that fulfil the criteria become entrants for deregulation and, in the case of the nationalized industries, for privatization. The British approach on the other hand has a diversity of objectives, with its source essentially in the theory of property rights and a system of incentives. It relies on the faith that private ownership compared to public ownership helps to achieve efficiency in productivity in a better way. Research by (Cowie, 2002; Pollitt & Smith, 2002) suggest that passenger rail privatization promoted market competition that led to increased efficiency in the case of UK, Sweden, and Germany. Experiences from the UK highlight that segregation and private company involvement in infrastructure leads to sub-optimization of the complete sector, and possible failure and malfunctioning in efficiency absence of government involvement. In the case of the US where infrastructure and operations are not segregated, private ownership is a better working model. Sweden so far has been one of the most encouraging instances from point of decision making for infrastructure policy based on European Union regulations (Hilmola & Szekely, 2006).

5 History of Indian Railways

Indian Railways is the fourth biggest railway network in the world operating 21,000 trains across 115,000 km (km). Managed by the Government of India under the Ministry of Railways for the last 166 years, railway services were proposed in India originally during the 1830s but effectively began its journey only in 1853 when the first passenger train was run between Mumbai and Thane (Ather, 2018; Shrivastva, 2020). The first discussions for a rail system in India inspired from England began in 1832, but remained largely on paper. The thoughtful discussions over the first proposal for railways in India started in Great Britain in 1840. The then Governor-General of India Lord Hardinge permitted private businessmen to establish rail systems. Powerful lobbying by banks, traders, and others who had their vested interests was seen in its support (Indian Railways Fan Club, 2020). The railways in India thus initially started as a private entity where eight railway companies including Madras Railway, Eastern India Railway, Bombay Baroda, Great India Peninsula Company, and Central India Railway were entrusted with the responsibility of managing railways (Shrivastva, 2020). A Guarantee System, privileged by the East India Company, was introduced wherein companies that setup railways systems in India were given a rate of interest or an annual return of 5% on its capital investment. This incited new railway establishments supported by UK-based private financiers to invest in India which subsequently led to the rapid creation of railways. In the following years, various kingdoms built their rail systems. By 1875, as much as £95 million were invested by British businesses for constructing railways in India. It was in 1901, for the first time in its history that the railways started to generate profit after years of investment and construction and by 1907, the government took over all the rail companies which were then again leased back to private owners (Ather, 2018; Indian Railways, 2007). By 1920, about 15% or a total track route of 37,000 miles was privately held. However, the arrival of the First World War in 1914 impacted rail development and left the network in a state of disrepair. Railway funds in 1924 were thus segregated from the general budget, a practice that has been continuing since then.

6 Privatization of Railways in the Contemporary India

It was then during 1946 pre-partition that all rail systems in the county were taken over by the Government of India. It was in 1951 that the railways was publicly owned with the entire system becoming a part of the Government of India. The 1989 Railways Act today governs Indian Railways and authorizes government and non-government railways. Some discrete special-purpose railways are thus present as joint ventures amongst the Ministry of Railways and other private parties like the Kutch Railway Company Ltd (World Bank Group, 2017).

The push towards public-private partnership (PPP) began in the mid-2000s to improve Indian Railways, and for the past six years, there has been a strong thrust to involve private companies in the sector. In 2001, after witnessing operational performance failures and a deteriorating financial condition, the government appointed an expert group in 2001 to scrutinize the situation of Indian Railways and submit their suggestions. The Railways Reform Report or the Mohan report criticized the railway's business model. The committee recommended splitting of Indian Railways into two bodies—an operations body and a regulatory body followed by rationalization of fares, the introduction of a business approach to finances, termination of non-profit lines, reduced manpower, and an objective to privatize railways after

15 years. It was in 2007 that private entities were permitted for container transport operation, which ended the monopoly of Container Corporation of India. The first private container train was consequently run from Cossipore to Loni (IRFCA, 2020).

Indian Railways in recent times has been subjected to various high-level strategic assessments, with each looking into several aspects along with areas for improvement in detail. The railways have been facing enormous financial constraints due to subsidies provided in the passenger segment amounting to approximately INR 38,000 crores. The 2012 Report of the Expert Group for Modernization of Railways in India recommended the use of new revenue models including a public-private partnership for growth. More recently, the Debroy Commission Report of 2015 put an increased focus on the participation of the private sector in the railways. It identified mechanisms and new methods of financing for mobilizing resources for railway projects. It included three major reforms that were commercial accounting practices, an independent regulator to encourage competition and safeguard stakeholders, and business-oriented Human Resource strategies (World Bank Group, 2017). It suggested the creation of a separate establishment for infrastructure; access provisions for willing new operators, separate services that run as joint ventures with state governments, and entry of private players in competition with Indian Railway in both freight and passenger trains (Singh, 2015). It was found as an economically feasible measure towards unburdening the government of its financial and administrative responsibilities.

Government of India has set forth the idea of privatization of Indian Railways as it continues to struggle with financial mayhem. The India Transport Report 2014 highlights the necessity to double the yearly investment (as a proportion of Gross Domestic Product or GDP) in railways from 0.5 per cent to one per cent during the following few years. The government believes that the privatization of railways can provide favourable returns on investment and can compensate for operational losses (Shrivastva, 2020).

7 Privatized Rail Facilities in Country

In an attempt to provide improved facilities and world-class service to passengers, the Ministry of Railways along with NITI Aayog has ignited a discussion for the participation of private entities in the operation of 150 passenger trains on 100 routes. Indian Railways has recently decided to outsource a few commercial and onboard services to private entities to include new technology rolling stock with lesser transit time and lower maintenance, enhance employment opportunities and safety, and deliver world-class travel experience. Ministry of Railways in July 2020 has started seeking Request for Qualifications (RFQ) from private entities for passenger train operations through the introduction of 151 modern Trains. The private companies will be involved in financing, operation, procuring, and maintenance of trains. The 151 trains would supplement the already existing trains and would run on routes where the demand is already higher than the existing capacity. As per Railways, the first twelve private passenger trains will be introduced by 2023, 45 in the next year, and remaining of 151 by Financial Year 2027. 70% of these trains will be manufactured in India (Press Trust of India, 2020). These 109 OD pairs have been made into 12 Clusters like Delhi, Mumbai, Chandigarh, Howrah, Patna, and others across the railway network with firms being allowed to bid for maximum three (Ministry of Railways, 2020).

The Indian Government back in February 2019 recommended privatization of passenger related activities like parking, platform ticketing, advertising, catering, cleaning, and waiting room facilities. The private entity would be given the right to set tickets and parking prices and will have the freedom to make decisions related to stoppages, and types of services on offer in the trains. The train will be operated by railways' driver and guard with their operation and maintenance being governed as per standards and requirements laid down by Indian Railways. In case of any failure to fulfil prescribed performance criteria, penalties will be implemented based on pre-specified performance standards. Private operators will be permitted to run these trains for 35 years in return for a share in the revenues they earn, energy charges as per consumption, and fixed haulage charges for using public infrastructure (Bera, 2020; Sharma, 2020). It is expected that railways will receive total haulage charges of around 3,000 crores per year from private trains operations. The first station to be privatized after Bangalore is proposed to be Anand Vihar. Other stations include Pune, Chandigarh, and Secunderabad (Uni AE, 2019).

7.1 The First Private Train

The Delhi-Lucknow Tejas Express was the first private town of India launched in October 2019 and is being operated by private operator Indian Railway Catering and Tourism Corporation (IRCTC), a subsidiary of Indian Railways. An outcome of Debroy Committee recommendations of 2015, Tejas Express, encourages private sector competition. The provision of physical infrastructure like guards, locomotives, coaches, and security staff lies with Indian Railways, with services like ticketing, repayments, parcels, catering, and housekeeping, being contracted to private entities under the PPP model. As per concession agreements signed between IRCTC and private companies, later is required to share their profits with IRCTC, which will pay haulage charges in turn to the railways (Mufti & Sampal, 2019). Inspired by its success IRCTC has since then launched two more trains—the Mumbai-Ahmedabad Tejas and special train for pilgrims. IRCTC suggests that the success of the train can make PPP be introduced into hard infrastructure in the next phase (Fig. 1).

Fig. 1 Origin Destination pairs identified for private involvement

7.2 Redevelopment of Stations

The Centre in 2017 decided to auction 23 railway stations across the nation under its public-private partnership (PPP) projects. It is proposed that the stations will be allotted through auction to private companies for up-gradation and modernization allowing them to provide commercial services like malls, hotels, and speciality hospitals in the station premises. The proposed financial model does not accrue any expenditure to the government and makes the private entity responsible for renovating and maintaining the railway stations for 15 years. The firm benefits by getting access to leasing rights of the developed commercial properties for 45 years. The firms are also permitted to maintain station facilities like lounges or waiting rooms, power, platform, parking facilities, eateries, retiring quarters, etc.

The redevelopment is proposed to occur in different phases with the responsibility of redevelopment lying with Indian Railway Station Development Corporation (IRSDC), a Special Purpose Vehicle created by a joint venture of Ircon International Limited and Rail Land Development Authority during the first phase. In the second phase, 23 railway stations which include Kanpur Central, Thane, Lokmanya Tilak, Pune, Jammu, Visakhapatnam, Ranchi, Howrah, Kamakhya, Bandra, Secunderabad, Chennai Central, Kozhikode, Allahabad, Vijayawada, Bangalore Cant, Indore, Bhopal, Udaipur, Mumbai Central, and Borivali railway stations shall be auctioned. As per Railways officials, Malaysia's national Construction Industry Development Board (CIDB) along with many Korean and Japanese businesses are interested in participating in the redevelopment project (Shaji, 2017).

For instance, the Ministry of Railways in 2017 privatized Habibganj Railway Station, making it India's first private railway station. The responsibility of the station's operation and maintenance is allotted to Bansal Group, Bhopal for eight years. The firm has been given tracts of land on a lease period of 45 years. The company is to handle facilities like food outlets, retiring rooms, platform upkeep, parking, etc. The core operations like train and parcel movement, signalling, and ticketing will not be under the purview of the private firm. The project overseen by IRSDC was commenced as a part of the plan of Indian Railways to re-develop 400 A1 and A category railway stations. Other projects handled by IRSDC apart from Habibganj include Bijwasan and Anand Vihar stations in Delhi, Chandigarh, Surat, Mohali, and Gandhinagar (Sood, 2017).

7.3 Privatization in Practice-Twin Faces

In an attempt to promote competition, including innovation and modern technology along with an array of other benefits like reduced travel time, job opportunities, and increased safety, Ministry is fast making changes to privatize Indian Railways.

As per the Government's estimate, Indian Railways requires a fund of INR 50 lakh crore for the following 12 years of its operations (Sharma, 2020). A continuous loss in market share along with rising expenditure has pushed Indian Railways to the edge of financial crises. Railways in addition to running trains are involved in non-income-generating services like providing security to customers, catering, educational, and medical facilities to its employees that puts added financial stress. A report by the Ministry of Railways suggests that Indian Railways may become a burden on the National economy and thus necessitate a need to redefine its role. The committee thus suggests commercial accounting, effective human resource changes, and an independent regulator. So far, engaging state governments and other remunerative prospects have also been little (Ministry of Railways, 2015). However, the radical new plan to private railways comes with its share of challenges and benefits.

At a time when real estate is expensive, privatization can help the railway's leverage its properties by allowing private entities to develop commercial facilities like malls, hotels, etc. Since a majority of private players will lie in the premium segment, Indian Railways will be able to utilize resources to provide more services to the common man. The move will help expand railway's revenue, attract external funding, and free up capital to be used for network expansion. Private train operators will help relieve the burden from Indian Railways.

The Bibek Debroy Committee of 2015 favoured privatization of the railway's rolling stock. It suggested splitting Indian Railways into two independent organizations with one being in charge of infrastructure and other for operating trains to simplify cost recovery of the government's expenditure in setting up of infrastructure. It would profit the government which would charge the operator. Experiences from countries like the UK, US, Germany, Japan, and Australia suggest that the engagement of private operators opened up to competition, allowed lowering of prices, and created better services.

Privatization in essence refers to the transfer of at least a portion of the operations of a state-owned enterprise to a private entity (Heald, 1984). However, it is not limited to asset transfer but is essentially characterized by competition rather than by monopoly. A general agreement is observed amongst researchers that competition between firms helps to improve efficiency and not only reduce transfer payments to railway operators but also contributes to improved railway services. Cases like Europe, Japan, and North America show that different restructuring models were used to promote competition in railways. A report commissioned by MehrBahnen estimates that use of competitive mechanisms for securing passenger services resulted in a drop in potential subsidies at about 18–38% (Private Sector Participation Consult, 2004; Gunnar & Konstantinos, 2013).

8 The Anticipated Effects of Privatization on Indian Railways

It is argued that privatization when seen from the lens of equal access to services is likely to get impacted as railways will then become a service that is availed based on the purchasing power of an individual. Services under Indian Railways cover all classes of passengers from low-income to high-income groups, connect remote locations, and cross-subsidize for travellers in low-cost trains through greater freight tariffs (Ananthakrishnan, 2020). The decisions related to fare hike, the introduction of new services, and projects are taken up on basis which is not commercial led.

The intervention of the private sector will however cause price hikes. The consequences of the Flexi-fare introduced in 2016 are still being felt across the railways, the challenge is thus also related to the impact that privatization will have on service demand especially the premium segment and the cost compatibility. It may lead to market monopoly and bring forth issues like reduced accountability, mistreatment of the poor, lack of responsibility, etc. Indian Railways moves as much as 2.3 crore people and 3 million tons of goods each day (Indian Railways, 2019). Railways are the most viable mode of public transport for a majority of the population who cannot afford private transit. The 68,155 km long railway network connects remotest corners of the country. Since private entities are profit-led, privatization is likely to affect the connectivity of the mode to these areas as profitable routes would be preferred over non-profit yielding routes (Shaji, 2017).

It will also impact the pensioners and employee benefits which will then be regulated as per the private player goals. It could make employee's interests being overlooked or employee lay-offs for cost-cutting. Indian Railways that currently emphasize on service of the public will turn into a money-making machine. A major point of contention that will follow is associated with fares. At present, Indian Railways' passenger fares are amongst the cheapest in the world (0.35 per kilometre per passenger) (Mahajan, 2020). This is done to provide subsidies for many lower sections of the society which then makes it troublesome for railways to recover its costs. Under the proposal, private entities have the freedom to determine their fares. It is thus felt privatization will widen the already exiting class division (Aurora, 2020). The introduction of high-end services in a country where railways are mostly used by people from low-income backgrounds seems dubious.

Another criticism that follows includes the role of government in situations where railways may suffer a financial withdrawal and determining the workforce in a private organization. It would also involve crucial decision making related to the responsibility of different sectors of railways and the distinction between the jurisdiction of powers of government and the private players. There is a further lack of clarity related to the operation of private trains in the absence of an independent regulator. It is inevitably necessary that the government involves an independent regulator for the private train project to succeed. In the absence of a regulator, a lot of friction is likely to arise between the two parties and a transparent mechanism to achieve a settlement becomes difficult. It is also argued by private players that privatization should not be limited to operations but allowed in manufacturing as well. Considering the already choked network, accommodating new trains is still in a grey area. It is still to be examined if the dedicated freight corridors can free up sufficient capacity. There are added fears about who gets a priority—train run by a private player or railways in critical situations like arriving station on time (Mahajan, 2020). The current private freight trains operators argue that they are overlooked for a preference towards railways run container trains (Table 1).

As per recent reports by the Comptroller and Auditor General of India, the efficiency of the Indian Railways in 2018 dropped to its lowest in a decade (Table 2). For the Financial Year (FY) 2018, railways spent INR 98.44 on every earned INR 100 indicating an operating ratio of 98.44%, the lowest in 10 years. Its surplus revenue dropped to INR 1,665.6 crore; a six-year low (Government of India, 2019). The organization's financial health has been drastically affected as its earnings from passenger traffic in recent years grew at a pace slower than its expenditure on salaries and pensions. The reserved passenger traffic on Indian Railways between 2013 and 2018 grew at less than 5%. This compares to a 13% growth in air traffic during the

Table 1 Overview of Railways' finances (in INR crore)

Sl.no	Particulars	2012–13	2013–14	2014–15	2015–16	2016–17	2017–18 RE	2018–19 BE
	Receipts							
1	Internal Resources	126,180	143,214	161,017	168,380	165,382	187,425	201,090
2	Budgetary Support	24,132	27,033	30,121	37,609	45,232	40,000	53,060
3	Extra Budgetary Resources	15,142	15,085	11,044	39,066	52,579	69,100	81,940
4	Total	165,454	185,332	202,182	245,055	263,193	296,525	336,090
	Expenditure							
5	Total Revenue Expenditure	117,914	139,473	153,352	157,874	160,469	181,000	188,100
6	Total Capital Expenditure	50,383	53,782	58,718	93,520	109,934	120,000	146,500
7	Total Expenditure (=5 + 6)	168,297	193,255	212,070	251,394	270,404	301,000	334,600
8	Operating Ratio	90.2%	93.6%	91.3%	90.5%	96.5%	96.0%	92.8%

Source PRS, 2018

same period (Bera, 2020). Indian Railways faces stiff competition from the road sector due to its competitive advantage and low sunk costs. Also, the high freight charges to cross-subsidize passenger fares have encouraged the shift of freight and short distance passenger traffic to roadways. The constraints with exiting limited capacity and routes have caused trains to steadily lose its passenger traffic.

As per the CAGI report 2019, the growth in earnings per passenger in 2018 significantly reduced for the third successive quarter. Free tickets/passes and fare concessions contribute to the losses significantly. Earnings from passengers grew at a slow pace as operational losses incurred in classes such as 'sleeper', 'second', and 'ordinary' ballooned in FY16 and FY17. For instance, in FY13, the railways made a loss of INR 40 crore in operating AC first class (Table 2). Free tickets/passes and fare concessions contributed to the losses significantly. Introducing private operations is thus critical as it will allow the provision of higher service quality that may increase passenger traffic on railways. However, since privatization will raise passenger fares in exchange for better services; it remains largely unclear if the initiative will encourage a shift away from cheaper and convenient budget airlines and roadways, the primary modes that the railway is losing its passengers and fright to. Privatization has its positive side in terms of encouraging competition and improving

Table 2 Operational losses of various Classes of Passenger Services (in crore)

Class of Passenger services	2012–2013	2013–2014	2014–2015	2015–2016	2016–2017
AC-1st class	(−) 40.86 (7.48)	(−) 47.39 (7.54)	(−) 127.49 (17.68)	(−) 175.79 (23.05)	(−) 139.39 (17.68)
1st class	(−) 61.36 (6 1.26)	(−) 92.06 (75.82)	(−) 69.50 (74.71)	(−) 58.00 (81.03)	(−) 53.31 (80.27)
AC 2 Tier	(−) 348.09 (12.53)	(−) 497.28 (15.26)	(−) 495.59 (13.32)	(−) 463.1 1 (12.01)	(−) 559.27 (13.6)
AC 3 Tier	494.99 (10.29)	4 10.67 (6.84)	881.52 (12.57)	898.06 (11.69)	1040.52 (12.43)
AC Chair Car	(−) 38. 12 (4)	(−) 148.47 (11.32)	(−) 142.26 (9.9)	(−) 5.58 (0.4)	117.83 (8.13)
Sleeper Class	(−) 6,852.72 (45)	(−) 8,407.85 (44.57)	(−) 8,5 10.06 (41.50)	(−) 8,30 1.15 (38.65)	(−) 9,313.27 (40.8)
Sccond Class	(−) 5,167.53 (38.9)	(−) 7,134.42 (44.75)	(−) 7,642. 13 (43.19)	(−) 8,569.77 (45.37)	(−) 1 0,024.88 (49.17)
Ordinary (All Class)	(−) 9,783.80 (67.78)	(−) 11, 105.24 (67.08)	(−) 11,673.80 (65.58)	(−) 13,237.74 (69.14)	(−) 14)647.64 (70. 19)
EMU Suburban services	(−) 3,365.47 (61.7)	(−) 4,027.14 (62.98)	(−) 4,679.1 1 (63.98)	(−) 5,124.74 (65.19)	(−) 5,323.62 (64.74)

Note (1) Negative figures denote losses and positive figures denote profits on passenger services. (2) Figures in bracket represent percentage loss/profit. *Source* Summary of the E1 1d Results Coaching Services Profitability/U nit Costs

operational and cost efficiencies but the transition needs to be transparent and all concerned stakeholders must be taken into confidence.

9 Conclusion

After its structural reorganization post-independence, Indian Railways has been managed as a departmental undertaking of the Ministry of Railways. Its performance since then has been measured more in terms of achieving social objectives, rather than in terms of commercial success. Indian Railways finances its investment programs mainly by borrowing from the central government and partly from resources generated internally (on which railways exercise a limited control). The borrowed funds are not in the form of grants but are non-refundable loan on which railways pay a fixed rate of dividend (at around 6.5%) since the 1924 Separation Act. The railways are also obligatory to make a certain contribution to the General Revenues of the central government.

As railways face increasing financial burdens from competition from other modes of transport and various policy constraints like the uneconomic tariff policy, their dividend compulsions have collected into enormous indebtedness to General Revenues. The financial constraints have affected the required expansion on the network, track renewal, and technological up-gradation affecting the safety of the whole system (Rail India Technical and Economic Service, 1996). Since, 1950–1951, there has only been an addition of 9,500 km of track length resulting in the addition of less than 0.5% annually (World Bank, 1995). Thus, due to the constraints now and then, the necessity for operating Indian Railways as a commercial undertaking has been highlighted in reports of various Railway Convention Committees and inquiry groups. The same is also suggested by Mckinsey-Sweden rail report (Swede Rail and McKinsey & Company Inc, 1997) that emphasizes private sector investments in critical areas for the fast growth of railways as a whole.

However, none of these makes suggestions about how to run the railways on business principles while maintaining the character of a departmental undertaking and also fulfil the various social obligations and concerns forced by the government. Being a departmental enterprise, the Railways practices no control over their investment and pricing policies and have their investment programs is determined by the Planning Commission with pricing policy subjected to parliamentary control. Considering today's competitive environment, there remains a little hope of financial existence unless Railways is given the freedom to regulate pricing and investment to improve productivity. A radical change towards privatization is hence required in the structural organization of Indian Railways that would involve new managerial practices, operational procedures, and profit-oriented work ethics. Privatization will allocate the pricing and investment decisions related control from precedent-led bureaucrats to an independent entity, based on incentives offered to them would manage Railways in the best interests of the consumers and shareholders, thus benefitting both. The question arises whether Indian Railways can be operated inside the framework of principal-agent relationships while continuing to work as a public industry. The Indian Railways since time immemorial have played a vital role in the country's economic development and sociopolitical integration. They have performed many social obligations to achieve the government's non-commercial goals. The dilemma thus faced is if Indian Railways were to operate within the framework of principal-agent relationships, how it would account, and pay for the benefits enjoyed by the country.

After privatization, it would be solely the interests of taxpayers themselves served by railways since the Railways' equity capital would be entirely held by private investors who would not find it in their welfare to accomplish the government's social objectives. Also, to accrue maximum return on their investment they would aim to maximize the commercial output of the railways. These social burdens require careful review to create a fine financial strategy so that the costs are met. Many Indian planners believe that allocative efficiency and equity distribution are still important objectives for India's socio-economic development. This suggests that unlike a publicly owned railway system a privatized system would not be able to realize the social objectives as efficiently.

However, the cases of railway restructuring from across the world have highlighted that public service obligations of railways need not be opposed by commercial concerns. It is indeed significant to guarantee both, but the same can only be ensured with segregation of railway management from government, with the latter giving direct subsidy to targeted groups and simultaneously retaining adequate control to impact private ventures (Mattoo, 2000). Also, if we were to concentrate only on productive-efficiency aspects and not issues related to allocative efficiency, (Dalvi, 1995) suggests that privatization of the Indian Railway would still be hard to justify because of the non-applicability of the contestability model. This is due to the technical indivisibility of resources call for huge investments in railways, which are generally beyond the private investors' scope. Almost all investments in railway assets (tracks, locomotives, and rolling stock) are sunk costs that act as a deterrent to private investors as disposal of such investments become tough in case of traffic reduction. The very rationale of the contestability theory works against the privatization of the railway system. This is not because the funds required for fixed investments are huge but because investments have the probability to be 'sunk' due to the innate uncertainties of railways whose market has been increasingly worn by other modes of transport.

Research suggests that there was some averseness for railway reform even in countries that had restructured other sectors mainly due to the issue of high sunk costs. The reluctance was lastly overcome by a political decision at the highest level, which in some cases took place due to rising financial losses to the government, and in others, due to a strong public sentiment about the environmental effect of road transport (Mattoo, 2000). Further, the operations of Indian Railways unlike many across the world are more vertically integrated like the majority of the equipment that operates the line-haul system are manufactured by Indian Railways in their workshops. Research by (Helm & Thompson, 1991) suggests that privatization may cause underinvestment in the transport sector, mainly in the formation of facilities with long life spans, if private entities face inadequate incentives to invest proficiently. Also, considering the size of railways' network, the co-ordination of the timetabling process through market mechanisms would be a difficult process. As (Adamson et al., 1991) point out, if, for example, the rights to train routes were bought by one company, it would affect access to the platform and rolling stock opening at a station shared with other operators. The platform space available at a given time would, in turn, affect the availability of paths on other routes and so on. This would thus call for scarce capacity being commercially traded and consequently functioning in a co-ordinated way making train timetabling a tough process under a privatized administration unless the complete system was handed over to an operator as a single entity.

If Indian Railways were to be privatized, despite the doubts, it would be useful to speculate the most appropriate models to be adopted. The privatization models considered by British Railways included suggestions like privatizing BR as a single unit, segregating and privatizing BR as regional companies, privatizing BR along with business sectors, and fragmenting BR into track and train operations followed by contracting of train operations independently through competitive tendering with government selected the last option for privatization. The Japanese Government had split JNR's passenger service network into six regional enterprises. The reforms as suggested by (Swede Rail and McKinsey & Company Inc, 1997) include limited privatization or leveraging private participation in designated areas like subcontracting of services such as catering.

Since, Indian Railways is already structured on a zonal basis, with each of the nine zonal railways running as a separate operating and costing unit, (Dalvi, 1995) finds Japanese model as the best fit for Indian Railways. The nine zonal railways are closely combined in terms of ticketing and timetabling operations and also commercial business and marketing operations. Thus, for their successful privatization running them as independent business units is suggested to test their financial viability. The selected model would help eradicate the need for cross-subsidization between different zonal railways but is likely to pose various operational issues. Since not all zonal companies would be equally financially viable the government may be required to subsidize them and give grants to bear various social obligations, alongside facing issues related to preserving uniformity in the technical standards and the service-quality levels amongst different zonal companies.

Considering the financial losses and future signs of a diminishing rail share, railway restructuring now seems inescapable. The contemporary trend in both developed and developing nations is towards deregulation and privatization that helps open up new opportunities for all producers to embrace new technologies and for customers to use goods and services that generate maximum satisfaction. The attempt would require a focus on the complete structure of the policy framework rather than the effort to cater to the immediate problem of the paucity of funds. Thus, privatization call for a look beyond the efforts to attract private investments to fund infrastructure and use of private sector skills by management contracts for related activities such as catering.

Any proposal to privatize Indian Railways must carefully take into consideration the effects of privatization on the productive efficiency of the railways along with the government's objectives related to equity-allocation. A focus on productive efficiency should involve cautious attention towards various operational problems; otherwise, it is very much likely that privatization may lead to a host of other problems open for railways rather than providing solutions.

References

Adamson, M. J., Jones, W., & Pratt, R. (1991). Competition issues in privatization: Lessons for the railways' in banister. Transport in Free Market Economy. https://link.springer.com/chapter/https://doi.org/10.1007/978-1-349-11439-9_4.

Alchian, A. A., & Demstetz, H. (1972). Production, information costs and economic organisation. American Economic Review 62(5): 777–795. https://www.jstor.org/stable/1815199?seq=1.

Ananthakrishnan, G. (2020). The Hindu Explains | Why has Indian Railways opened doors for private players? The Hindu. https://www.thehindu.com/business/Industry/the-hindu-explains-why-has-indian-railways-opened-doors-for-private-players/article31991149.ece.

Association of American Railroads (AAR). (2006). Destructive Railroad de-regulation. AAR – Policy and Economics Department. http://www.aar.org/GetFile.asp?File_ID=695.
Ather, K. (2018). Chapter – 1: Introduction and Overview of Indian Railways. In Job Satisfaction Among Railway Employees A Study of Indian Railways Organization Employees with Special Reference to Kota Division of West Central Railway. PhD theses. Bhagwant University. https://shodhganga.inflibnet.ac.in/bitstream/10603/218641/6/15%20chapter%201%20introduction%20and%20overview%20of%20indian%20railways.pdf.
Aurora, B. V. (2020). Privatization of Indian Railways: Why a tragic hero is always at mammoth odds. Outlook. https://www.outlookindia.com/magazine/story/india-news-privatization-of-indian-railways-why-a-tragic-hero-is-always-at-mammoth-odds/302714.
Baumol, W. J. (1982). Contestable markets: An uprising in the theory of industry structure. The American Economic Review. https://www.jstor.org/stable/1808145.
Bera, S. (2020). The plan to privatize parts of the railways. Livemint. https://www.livemint.com/news/india/the-plan-to-privatize-parts-of-the-railways-11583338681991.html.
Carlson, L. H. (2004). The institutional aspects of competitive access in the Western Canadian Rail System. Master Thesis. University of Saskatchewan. http://library.usask.ca/theses/available/etd-01242005-130037/unrestricted/TheThesisJan4b.pdf.
Cowie, J. (2002). Subsidy and productivity in the Privatized British Passenger Railway. Economic Issues Journal Articles. https://thredbo-conference-series.org/downloads/thredbo7_papers/thredbo7-workshopC-Cowie.pdf.
Dalvi, M. Q. (1995). Should Indian Railways be privatized? Economic and Political Weekly. http://www.jstor.com/stable/4402262.
Department of Transport. (1993). Railway finances: Committee Report. Chairman. Sir D. Serpell. https://doi.org/978–0115505904.
European Union. (2004). Energy and transport in figures 2004. European Commission. https://op.europa.eu/en/publication-detail/-/publication/52f721ed-c6b8-11e8-9424-01aa75ed71a1.
Foster, C. (1995). The Economics of Rail Privatization. Centre for the Study of Regulated Industries Discussion Paper.
Government of India. (2019). Report of the Comptroller and Auditor General of India. Union Government (Railways), Railways Finances. https://cag.gov.in/uploads/download_audit_report/2019/Report_No_10_of_2019_Union_Government_(Railways)_Railways_Finances.pdf.
Gunnar, A., & Konstantinos, R. (2013). Rail Liberalisation in Sweden- Policy development in a European Context. *Research in Transportation Business & Management*. https://doi.org/10.1016/j.rtbm.2012.12.004.
Heald, D. (1984). Privatization: Analysing its appeal and limitations. Fiscal Studies. https://www.jstor.org/stable/24434583.
Helm, D., & Thompson, D. (1991). Privatized transport infrastructure and incentive to invest. Journal of Transport Economics and Policy. https://www.jstor.org/stable/20034963.
Hilmola, O. P., & Szekely, B. (2006). Deregulation of Railroads and Future Development Scenarios in Europe – Literature Analysis of Privatization Process Taken Place in US, UK and Sweden. Tutkimusraportti – Research Report 169. Lappeenranta University of Technology. https://doi.org/10.1504/WRITR.2007.013948.
Huang, C. J. K., & Michael, C. (2019). The Privatization of Japan Railways and Japan Post: Why, how, and now. ADBI Working Paper 1039. Tokyo: Asian Development Bank Institute. https://www.adb.org/sites/default/files/publication/539746/adbi-wp1039.pdf.
Indian Railways. (2007). https://www.cs.mcgill.ca/~rwest/wikispeedia/wpcd/wp/i/Indian_Railways.htm#:~:text=Railways%20were%20first%20introduced%20to,distance%20and%20suburban%20rail%20systems.
Indian Railways Fan Club (IRFCA). (2020). IR history: Early days – I. https://www.irfca.org/faq/faq-hist.html.
Indian Railways. (2019). Indian Railways Yearbook 2018–2019. Ministry of Railways, Government of India. https://indianrailways.gov.in/railwayboard/uploads/directorate/stat_econ/Year_Book/Year%20Book%202018-19-English.pdf.

Kay, J. W., & Thompson, D. J. (1986). Privatization: A Policy in Search of a Rationale. *The Economic Journal*. https://doi.org/10.2307/2233423.
Kopicki, R. & Thompson, L. S. (2012). Best Methods of Railway Restructuring and Privatization. CFS Discussion Paper Series 111: 477–3045. http://documents1.worldbank.org/curated/en/761811468761126588/pdf/multi-page.pdf.
Krueger, A. O. & Ito, T. (2004). Governance, Regulation, and Privatization in the Asia-Pacific Region. Bureau of Economic Research. ISBN 0–226–38679–1.
Mahajan, A. S. (2020). Private Trains, public Problems. India Today. https://www.indiatoday.in/india-today-insight/story/private-trains-public-problems-1700598-2020-07-14.
Martin, B. (2002). Railway privatization through Concessions. A report for the International Transport Workers' Federation. Public World. http://www.publicworld.org/files/larailenglish.pdf.
Mattoo, A. (2000). Indian Railways: Agenda for reform. Economic and Political Weekly. https://www.epw.in/journal/2000/10/perspectives/indian-railways-agenda-reform.html.
Ministry of Railways. (2015). Report of Committee for Mobilisation of Resources for Major Railway Projects and Restructuring of Railway Ministry and Railway Board. Indian Railways. https://indianrailways.gov.in/railwayboard/uploads/directorate/HLSRC/FINAL_FILE_Final.pdf.
Mizutani, F. & Nakamura, K. (1997). Privatization of the Japan National Railway: Overview of performance changes. International Journal of Transport Economics. https://www.jstor.org/stable/42747282.
Mizutani, F., & Nakamura, K. (2004). The Japanese Experience with Railway Restructuring. Governance, regulation, and privatization in the Asia-Pacific Region. The University of Chicago Press, Chicago. https://www.nber.org/books-and-chapters/governance-regulation-and-privatization-asia-pacific-region/japanese-experience-railway-restructuring.
Moore, J. (1992). British privatization—Taking capitalism to the people. Harvard Business Review. https://hbr.org/1992/01/british-privatization-taking-capitalism-to-the-people.
Moyer, N., Thompson, E. & Louis, S. (1992). Options for reshaping the Railway. Transport Washington, D.C.: World Bank Group. https://invenio.unidep.org/invenio//record/17166/files/neilm.pdf.
Mufti, I., & Sampal, R. (2019). Why Tejas is '1st private train' — Railways owns it, outside vendors provide food & clean it. The Print. https://theprint.in/india/whats-new-in-first-private-train-tejas-hostesses-refunds-for-delays-no-concessions/301938/.
Nakamura, K. (1996). Privatisation and Beyond- The JR Case. Japan Railway and Transport Review. https://www.ejrcf.or.jp/jrtr/jrtr08/pdf/f04_nak.pdf.
NITI AAYOG, Ministry of Railways. (2020). Draft Documents for discussion on Private Participation in Passenger Trains. https://niti.gov.in/sites/default/files/2020-01/Draft_PIM.pdf.
Pollitt, M. G., & Smith, A. S. J. (2002). The restructuring and privatization of British rail: Was it really that bad? *Fiscal Studies*. https://doi.org/10.1111/j.1475-5890.2002.tb00069.x.
Press Trust of India. (2020). First set of private trains to be introduced by 2023, all 151 such services by 2027: Railways. India Today. https://www.indiatoday.in/india/story/first-set-of-private-trains-to-be-introduced-by-2023-all-151-such-services-by-2027-railways-1702293-2020-07-19.
Preston, J., & Robins, D. (2013). Evaluating the long term impacts of transport policy: The case of passenger rail privatization. *Research in Transport Economics*. https://doi.org/10.1016/j.retrec.2012.05.019.
Private Sector Participation Consult (PSPC). (2004). Wettbewerb im Schienenverkehr-Kaum gewonnen, so zerronnen, Mehr Bahnen (Competition in rail transport - Hardly won, then melted away, More railways). https://repec.cepr.org/repec/cpr/ceprdp/DP5396.pdf.
Rail India Technical and Economic Service (RITES). (1996). Study on the decline in Railway's share in Total Land Traffic Volume. Ministry of Railways, Government if India.
Rennicke, W. J. (2004). Hearing on the Status of the Surface Transportation Board and Railroad Economic Regulation, Discussion Paper, MERCER Management Consulting, USA.
Shaji, S. (2017). Indian Railways on the Fast Track to Total Privatization. NEWS Click. https://www.newsclick.in/indian-railways-fast-track-total-privatization.

Sharma, M. (2020). Private firms to soon run Passenger Trains as Indian Railways moves fast on Privatization. India Today. https://www.indiatoday.in/india/story/private-firms-to-soon-run-passenger-trains-as-indian-railways-moves-fast-on-privatization-1696051-2020-07-02.

Shaw, J., Gibb, R., & Charlton, C. (1998). Competition, regulation, and the privatization of British Rail. Environment and Planning C: Politics and Space 16(6). https://doi.org/10.1068/c160757.

Shrivastva, C. (2020). Rail privatization in India: Possibilities and challenges. http://www.inthelongrun.org/articles/article/railway-privatisation-in-india-opportunities-and-challenges/.

Singh, S. P. (2015). Will privatization help the Railways? Business Standard. https://www.business-standard.com/article/economy-policy/will-privatization-help-the-railways-115041400999_1.html.

Sood, J. (2017). India's first 'private' railway station Habibganj to come up near Bhopal. Livemint. https://www.livemint.com/Politics/6zYTPft0qLmH1veLvzE8VI/Indias-first-private-railway-station-Habibganj-to-come-up.html.

Subir, R. (2019). Railway privatization: India must hasten slowly. Business Line. https://www.icsm.gov.au/education/fundamentals-mapping/types-maps/cadastral-maps-and-plans.

Swede Rail and McKinsey & Company Inc. (1997). Indian Railways: Moving to the fast track.

Thomson Financials. (2006). Thomson ONE Banker - Company Overview. http://banker.thomsonib.com/.

Uni, A. E. (2019). Anand Vihar Rly Station to have privatized parking, ticketing and catering. United News of India. http://www.uniindia.com/anand-vh-railway-st-to-have-privatized-parking-ticketing-catering-from-feb/india/news/1497364.html.

World Bank. (1995). India Transport Sector: Long term issues. Washington, D.C.: World Bank Group. http://documents.worldbank.org/curated/en/522561468771695224/India-Transport-sector-long-term-issues.

World Bank. (2006). Railways database. http://www.worldbank.org/transport/rail/rdb.htm.

World Bank Group. (2017). Railway Reform: Toolkit for Improving Rail Sector Performance. Transport and ICT Global Practice. https://ppiaf.org/ppiaf/sites/ppiaf.org/files/documents/toolkits/railways_toolkit/index.html#:~:text=Railway%20Reform%3A%20Toolkit%20For%20Improving%20Rail%20Sector%20Performance&text=This%20Railway%20Reform%20Toolkit%20aims,and%20execution%20of%20railway%20reforms.

Woxenius, J. (1998). Development of small-scale intermodal transportation in a systems context. Department of Transportation and Logistics, Chalmers University of Technology. https://www.researchgate.net/publication/242704470_Development_of_small-scale_intermodal_freight_transportation_in_a_systems_context.

Wu, C. H., & Zhao, H. P. (1998). The Transport Development Policies and Railways Reform. UIC Seminar, New Delhi. https://www.jstor.org/stable/4408988.

Correction to: Railway Transportation in South Asia

Saptarshi Mitra, Sumana Bandyopadhyay, Stabak Roy, and Tomaz Ponce Dentinho

Correction to:
S. Mitra et al. (eds.), *Railway Transportation in South Asia*, Contemporary South Asian Studies, https://doi.org/10.1007/978-3-030-76878-2

In the original version of the book, the following belated corrections have been incorporated:

The author name "Sumona Bandyopadhyay" has been changed to "Sumana Bandyopadhyay" in the Frontmatter and in the chapter "Introduction: Railway Transportation—Regions, Economy and Development", in which the affiliation "Department of Geography and Disaster Management, Tripura University, Agartala, Tripura, India" of the author "Sumana Bandyopadhyay" has been changed to "Department of Geography, University of Calcutta, Kolkata, India" in the chapter source line, and in the Frontmatter (Contributors). The book has been updated with the changes.

The updated version of the book can be found at
https://doi.org/10.1007/978-3-030-76878-2
https://doi.org/10.1007/978-3-030-76878-2_1

© The Author(s), under exclusive license to Springer Nature Switzerland AG 2021

S. Mitra et al. (eds.), *Railway Transportation in South Asia*, Contemporary South Asian Studies, https://doi.org/10.1007/978-3-030-76878-2_19

The manufacturer's authorised representative in the EU is Springer Nature Customer Service Centre GmbH, Europaplatz 3, 69115 Heidelberg, Germany. If you have any concerns regarding our products, please contact ProductSafety@springernature.com

Printed and bound by CPI Group (UK) Ltd, Croydon, CR0 4YY
30/09/2026
02225508-0013